GRADUATE STUDENT SERIES IN PHYSICS

General Editor: DOUGLAS F. BREWER

Collective Effects in Solids and Liquids

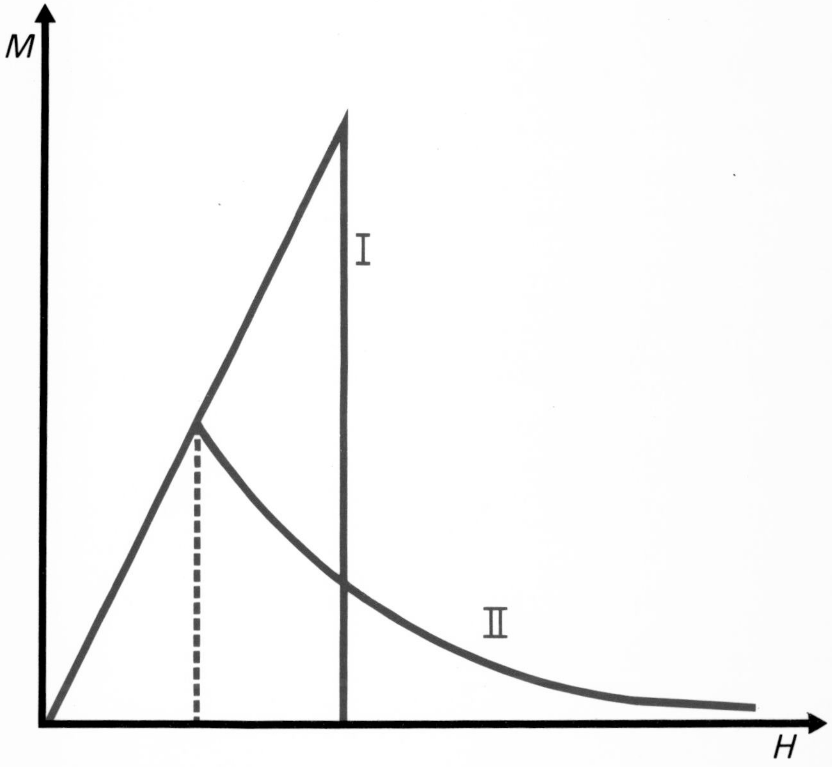

N. H. MARCH &
M. PARRINELLO

COLLECTIVE EFFECTS IN SOLIDS AND LIQUIDS

Graduate Student Series in Physics
Other books in the series

Colour Centres and Imperfections in Insulators and Semiconductors
P D TOWNSEND and J C KELLY

The Superconducting State
A D C GRASSIE

Refrigeration and Thermometry below One Kelvin
D S BETTS

Plasma Physics
E W LAING

Weak Interactions
D BAILIN

Gauge Theories in Particle Physics
I J R AITCHISON and A J G HEY

GRADUATE STUDENT SERIES IN PHYSICS

Series Editor: Professor Douglas F Brewer, M.A., D.Phil.
Professor of Experimental Physics, University of Sussex

COLLECTIVE EFFECTS IN SOLIDS AND LIQUIDS

N H MARCH
Department of Theoretical Chemistry
University of Oxford

M PARRINELLO
International Centre for
Theoretical Physics, Trieste
and
University of Trieste

ADAM HILGER LTD, BRISTOL
Published in association with the University of Sussex Press

British Library Cataloguing in Publication Data

March, N. H.
Collective effects in solids and liquids.—
(Graduate student series in physics, ISSN 0261–7242)
1. Phase transformations (Statistical physics)
I. Title II. Parrinello, M. III. Series
530.4′1 QC176.8.P/

ISBN 0-85274-528-1

Published by Adam Hilger Ltd, Techno House, Redcliffe Way, Bristol BSI 6NX, in association with the University of Sussex Press

The Adam Hilger book-publishing imprint is owned by The Institute of Physics

Phototypeset by Macmillan India Ltd., Bangalore and printed in Great Britain by J W Arrowsmith Ltd, Bristol

PREFACE

This book is intended as an introduction to collective effects in solids and liquids for scientists without specialist knowledge of theory. The background knowledge assumed is first and foremost a working knowledge of quantum mechanics such as is now provided in first degree courses for physicists, materials scientists and physical chemists. Some knowledge of the basic properties of solids and liquids, inevitably, has had to be assumed. For solids, the reader is expected to have as background a knowledge of direct and reciprocal lattices and some acquaintance with the one-electron theory of solids, though the main results needed here are summarised, without proof, in §2.8.2. On the other hand, no knowledge of liquid structure is assumed, but in addition to basic knowledge about liquid properties, some acquaintance with hydrodynamics would help the reader in this area, though it is by no means essential for an overall understanding.

Having in mind graduate students and young research workers who are not specialists in theory, we have not hesitated to supplement derivation of well established results with material reflecting trends in the subject at the present time. We have done this even though at times the account we could give had to be largely descriptive.

The material presented here should be suitable for graduate courses to experimental scientists interested in condensed matter. We have added some problems at the end of each chapter and have given references for further reading, especially on the more recent developments.

Professor G Rickayzen read the whole of the first draft of the manuscript and made many valuable comments for which we are most grateful. Professor L J Sham, Dr A Ferraz and Dr W Andreoni have given comments on single chapters, which we have also benefited from very much. Warm thanks are also due to Mr J Revill of Adam Hilger Ltd who has given us many helpful comments on the presentation. Needless to say, the responsibility for errors and obscurities which remain is solely our own. Finally one of us (NHM) wishes especially to thank Professor S Lundqvist for many valuable discussions on collective phenomena in solids over a long period of time and for keeping him in close touch with the many contributions made by the Gothenburg group in this area.

N H March
M Parrinello

CONTENTS

1

OUTLINE

In this slender volume an account will be given of a variety of collective phenomena which occur in solids and to a lesser extent in liquids. Associated with many collective excitations are quanta called quasi-particles. These have immediate relevance in describing an important class of observations in condensed matter.

The outline of the book is as follows. Chapter 2 is concerned firstly with collective modes of oscillation in the electron assembly formed by the conduction electrons in a metal. These so called plasma oscillations are crucially bound up with the long-range character of the Coulomb interaction between electrons. It is then shown that once account is taken of these organised oscillations in the electron density, which can be directly observed through discrete energy losses suffered by electrons as they penetrate thin metal films, the residual electron–electron interactions are short range. This circumstance allows one to understand the surprising success of the model of independent electrons for other properties of metals. In addition to these oscillations due to the electron–electron Coulomb repulsion, it is convenient also in Chapter 2 to treat the effects of Coulomb attraction between an electron in the conduction band of an insulator and a hole in the valence band, in describing excitations in such materials. The bound electron–hole pair which can often form is strikingly manifested in a number of optical experiments in semiconductors, rare gas crystals and ionic materials. It affords a very useful approach to the discussion of excitations and energy transport in molecular crystals, for example, and certainly will have implications for biological systems, though this latter area is outside the scope of this book. The area of electron–hole droplets, which has led to a whole new field of study in Ge for instance, is also treated and the chapter concludes with a discussion of excitons in metals with particular reference to collective effects involving transient core holes in such experimental situations as soft x-ray emission and photoelectron spectroscopy in metals.

Before tackling the way in which electrons interact with lattice vibrations (Chapter 4), Chapter 3 treats these vibrations as collective modes, following the pioneering work of Debye and of Born and von Karman. The quanta of these lattice waves, called phonons, have dispersion relations which depend on the detailed crystal structure and also on the nature of the force fields in which the atoms vibrate. But the main features follow on general grounds. The chapter on phonons concludes with a brief discussion of the effect of disorder on the atomic dynamics of solids.

Chapter 4 then takes up the problem of an electron in interaction with lattice vibrations. One dramatic case is the polarisation of the ionic lattice around an electron as it moves through a crystal like CsI and we therefore start with that. Again a quasi-particle, the polaron, can be usefully defined as the entity of an electron 'dressed' by the polarisation of the ionic lattice around it. Further quasi-particles, polaritons, are also dealt with and this chapter concludes with a treatment of the effect on lattice structure of an electron gas, the most marked effect of this type being the Peierls transition in a one-dimensional crystal. This tells us that one-dimensional metals cannot exist in nature, being unstable against lattice distortion which leads to an energy gap in the electronic spectrum.

The above discussion leads naturally into the phenomenon of superconductivity (Chapter 5). The electron–lattice interaction results in an effective attraction between electrons and to the formation of a well defined quasi-particle, a bound electron pair or Cooper pair, the two electrons having opposed spins. Josephson junctions are treated briefly and there is a short discussion of type-II superconductors.

Chapter 6 deals then with those solids exhibiting long-range magnetic order. When atoms carry localised spins, at low temperatures one often finds a ferromagnetic array. In a ferromagnet at absolute zero all the localised spins align parallel to one another. To generate the low-lying excited states, we now think of one of these spins having its direction reversed at a particular lattice site. But this misoriented spin is equally likely to be found on any one of the N equivalent sites of the crystal and one must construct travelling waves of spin. These spin waves, which in fact also occur in antiferromagnets, have quanta associated with them called magnons. It turns out that certain features of magnons survive the effects of disorder introduced into magnetic materials. In this same area, with metallic solids one can get cooperative magnetism with positional disorder in the so called spin glasses which are referred to briefly. The discussion concludes with an example of dynamical effects in spin polarisation of conduction electrons by a localised moment, i.e. the Kondo effect.

Having described, albeit briefly, some features of both phonons and magnons in disordered solids, Chapter 7 is devoted to the liquid state, characterised by its short-range order as described by the radial distribution function or the k-space structure factor. Collective mode theory based on independent density fluctuations is outlined for simple liquid metals and used to interpret thermal properties. In one liquid metal, Rb, a well defined collective mode has been observed by neutron scattering and has been reproduced in computer simulation experiments. Collective modes are used to model the dynamical structure factor $S(k, \omega)$, which is essentially the probability that a neutron incident on a liquid will transfer momentum $\hbar k$ and energy $\hbar\omega$, not only in classical liquids but also in the quantal fluid ^{4}He. The relation of the superfluidity of liquid ^{3}He to pairing as in superconductors is also treated.

The final chapter deals with phase transitions. After an introductory discussion of molecular field treatments of magnets and the gas–liquid critical point, transitions from electron liquids to electron crystals are summarised. Structural instabilities are then treated from phonon theory, involving the so called soft modes in which the phonon frequency for a particular wavevector becomes very small. Problems in A15 structures, relating to electron states having some characteristics of one-dimensional systems, are discussed followed by one-dimensional conductors, relating to the electron–phonon interaction and the Peierls transition. The book concludes with a short introduction to solitons; the Josephson junction and vortices in type-II superconductors are briefly referred to. Such non-linear theories having well defined particle properties may eventually play a major role in furthering our understanding of the rich variety of collective behaviour already well established in condensed matter.

2

PLASMONS AND EXCITONS

In this chapter we shall be concerned with collective effects which stem directly from the long range of the Coulomb interaction. The first type of collective behaviour treated is highly relevant in metals and is associated with the collective modes of the conduction electrons (i.e. the 3s electron contributed by each atom in Na metal or the $(3s)^2$ and 3p electrons in the valence shell of Al). Of course, to treat such collective modes, we ought to study the dynamics of the conduction electrons moving in the field of the periodic array of Na^+ ions in body-centred cubic (BCC) Na metal, or of the face-centred cubic (FCC) lattice of trivalent ions in Al metal. Some progress can be made on this problem, but the disadvantage is that one has to do specific quantitative calculations on each individual metal. Fortunately, many of the features of collective effects in the conduction electron bath of simple metals (these are metals with s–p conduction electrons; metals with incomplete d shells are excluded) can be obtained from a simplified model. This consists of smearing out the positive ions uniformly into a constant background, which takes care of overall electrical neutrality—an essential condition. This model, usually referred to as the 'jellium' model of a metal, had its origins in the pioneering studies of Sommerfeld. It is completely characterised by the electron density, n_0 say, or the mean interelectronic spacing r_s defined by

$$n_0 = \frac{1}{\frac{4}{3}\pi r_s^3}. \tag{2.1}$$

For a metal like Al, we therefore take its atomic volume from experiment, find n_0 for the three conduction electrons per atom, and evaluate the average interelectronic distance from equation (2.1) as $r_s \sim 2\,a_0$, a_0 being the Bohr radius $\hbar^2/m_e e^2$. For the simple metals, r_s ranges from this value for Al to about $5.5\,a_0$ for Cs. This then defines the range of metallic densities under normal conditions. By applying high pressures, one can in principle explore a range of $r_s < 2\,a_0$, but in practice the reduction one can achieve in r_s will be quite small.

2.1 Plasma frequency

Now in equilibrium we shall have equal densities of positive and negative charge at any point. But let us suppose that we create an imbalance of charge in some region by displacing electrons locally. Then we have an unneutralised excess of positive charge† in the region from which we have displaced electrons. Clearly electrons will be pulled back into this region by Coulombic attraction, they will overshoot and proceed to oscillate. It is this characteristic oscillation in the plasma which we desire to treat quantitatively. Most important, we need a precise expression for the frequency of this oscillation in the electronic cloud. Clearly, the oscillation frequency depends on the elementary charge e, on the electron mass m_e, since dynamics are involved, and presumably on the only other basic physical variable in the problem, the number of electrons per unit volume n_0. It is clear that we are seeking an equation of motion of the form

$$m_e \ddot{x} = -Kx, \tag{2.2}$$

and we need to understand the physics behind the force constant K in order to get the square of the plasma frequency ω_p from

$$\omega_p^2 = K/m_e. \tag{2.3}$$

Suppose the density of electrons at $\boldsymbol{r}$ at time t is $n(\boldsymbol{r}, t)$, the uniform positive background density being en_0. Then the excess positive charge density is given by $e(n_0 - n)$ and hence if $\mathscr{E}$ is the electric field we have from Maxwell's equations

$$\operatorname{div} \mathscr{E} = 4\pi e(n_0 - n). \tag{2.4}$$

Now we displace the electron gas by $\boldsymbol{x}$ to give a current density $\boldsymbol{j} = n\dot{\boldsymbol{x}}$ and we have the equation of continuity

$$\operatorname{div} \boldsymbol{j} = \operatorname{div}(n\dot{\boldsymbol{x}}) = -\partial n/\partial t. \tag{2.5}$$

If the displacement $\boldsymbol{x}$ is small, equation (2.5) becomes, to first order,

$$n_0 \operatorname{div} \dot{\boldsymbol{x}} = -\partial n/\partial t \tag{2.6}$$

which can be integrated to yield

$$n_0 - n = n_0 \operatorname{div} \boldsymbol{x} \tag{2.7}$$

since $n = n_0$ when $\boldsymbol{x} = 0$. Thus from equations (2.4) and (2.7) we have the result

$$\operatorname{div} \mathscr{E} = 4\pi e n_0 \operatorname{div} \boldsymbol{x} \tag{2.8}$$

and hence we find

$$\mathscr{E} = 4\pi e n_0 \boldsymbol{x}. \tag{2.9}$$

The solution (2.9) obviously satisfies the correct boundary condition $\mathscr{E} = 0$ when $\boldsymbol{x} = 0$. But the Newtonian equation of motion for an electron in an electric field $\mathscr{E}$, namely

$$m_e \ddot{\boldsymbol{x}} = -e\mathscr{E} \tag{2.10}$$

can now be used in equation (2.9) to yield

$$m_e \ddot{\boldsymbol{x}} + 4\pi e^2 n_0 \boldsymbol{x} = 0. \tag{2.11}$$

† We assume that the positive ion background is 'frozen', i.e. it cannot respond to changes in the electronic density distribution.

This has the desired simple harmonic form (2.2) and shows immediately that oscillations in the electron gas can occur, with angular frequency ω_p given by

$$\omega_p = (4\pi n_0 e^2/m_e)^{1/2}. \tag{2.12}$$

If we substitute for n_0 the conduction electron density in a typical metal, it is easily shown from equation (2.12) that $\omega_p \sim 10^{16}\,s^{-1}$. This is a very high frequency compared with a characteristic frequency, say the Debye frequency, of the lattice vibrations discussed in Chapter 3, which is typically $10^{13}\,s^{-1}$.

2.2 Screened interaction between electrons in a metal

It is important to emphasise that such plasma oscillations are a direct consequence of the long-range Coulomb interactions in the plasma. But once these collective motions are recognised, and accounted for, the effective residual interaction between electrons is short ranged. We can see this as follows. Let us suppose that the Coulomb interaction between two electrons separated by a distance r_{ij}, namely e^2/r_{ij}, is screened out over a distance l such that we get the screened Coulomb potential

$$V_{sc} = (e^2/r_{ij}) \exp(-r_{ij}/l). \tag{2.13}$$

Given this 'Yukawa' form, the only remaining question concerns the value of the screening length l. In a degenerate electron gas, we can write l as the product of a characteristic velocity, which is clearly the Fermi velocity v_F, and a characteristic time. We saw above that the plasma oscillations can be viewed as describing electrons rushing in to screen out any electric fields created by charge imbalance. It is clear that the characteristic time must be essentially $2\pi/\omega_p$, the period of these plasma oscillations.

Thus we find the order of magnitude estimate for the screening length

$$l \sim v_F(2\pi/\omega_p), \tag{2.14}$$

the appearance of the plasma frequency directly reflecting the essential role of the Coulomb interaction. If we use the estimate $\omega_p \sim 10^{16}\,s^{-1}$ given above, then we need only an appropriate value of the Fermi velocity v_F to estimate the screening length l. Using free-electron theory (cf. equation (A2.1.2) in Appendix 2.1), v_F can be estimated from the electron density n_0 in equation (2.1), and a typical value is $v_F \sim 10^{-2}c$, where c is the velocity of light. In atomic units† $v_F \sim 1$ au and then l is found to be of the order of 1 Å. This shows that there is substantial screening out of the Coulomb repulsion, and we have here a manifestation of the elementary fact that long-range electric fields cannot exist in a conducting medium.

† Atomic units are defined such that $\hbar = m_e = e = 1$, and then the unit of length is evidently the Bohr radius $a_0 = \hbar^2/m_e e^2$, the unit of energy is $e^2/a_0 = 27$ eV, while $c = 137$.

There is an alternative, and often convenient, way of representing such screening. This is to work in Fourier transform and write the Fourier component $V(k)$ of the screened potential $V_{sc}(r) = (e^2/r) \exp(-qr)$, with $q = l^{-1}$, as

$$V(k) = \frac{4\pi e^2}{k^2 \epsilon(k)} \tag{2.15}$$

where $\epsilon(k)$ is the k-dependent dielectric constant. By straightforward Fourier transform of the screened Coulomb potential, one finds

$$V(k) = \frac{4\pi e^2}{k^2 + q^2} \tag{2.16}$$

and hence by comparing equations (2.15) and (2.16) one obtains

$$\epsilon(k) = \frac{k^2 + q^2}{k^2}. \tag{2.17}$$

Of course the physical argument leading to equation (2.14) is not immediately conclusive, and therefore a quantitative treatment yielding $l = q^{-1}$ in equation (2.13) is set out in Appendix 2.1 (cf. equations (A2.1.6) and (A2.1.9)). It is clear already, however, from the above qualitative argument that the two factors which prevent the Coulomb interaction playing a crucial role in simple metals are: (a) total degeneracy of the high-density electron gas; and (b) screening.

2.3 Single-particle and collective excitations

Having included the major dynamical effect of Coulomb interactions through the plasma oscillations, we may now, as mentioned in Chapter 1, describe the remaining excitations usefully in an independent-particle model. In the simplest free-electron version of this model, the wavefunction can be expressed as a Slater determinant of plane waves (see §7.10). In this antisymmetric independent-particle wavefunction the only correlations between electrons that one takes into account are those that follow from the Pauli principle: namely the Fermi hole that an electron with upward spin digs around itself by effectively excluding other electrons with the same spin direction from its own vicinity. The ground-state wavefunction is then readily constructed by filling all states in momentum space for which $\hbar|\boldsymbol{k}| \leqslant \hbar k_F$, k_F being the Fermi wavenumber, related to the velocity v_F in equation (2.14) by $\hbar k_F = m_e v_F$. The simplest possible excited states that one can construct within this model can be obtained by taking a single electron from an occupied state with momentum $\hbar|\boldsymbol{k}| \leqslant \hbar k_F$ and exciting it into an unoccupied state of momentum $\hbar|\boldsymbol{k}'| > \hbar k_F$. In doing so the total momentum of the system, which in the ground state is zero, changes by an amount $\hbar \boldsymbol{q} = \hbar(\boldsymbol{k}' - \boldsymbol{k})$ while the energy is increased by

$$E = \frac{\hbar^2}{2m_e}(\boldsymbol{k}+\boldsymbol{q})^2 - \frac{\hbar^2}{2m_e}\boldsymbol{k}^2. \tag{2.18}$$

Taking into account the restrictions $k \leqslant k_F$ and $|\boldsymbol{k}'| = |\boldsymbol{k}+\boldsymbol{q}| > k_F$, one therefore finds that for a given value of the total momentum $\hbar q$, not all excitation energies are permitted. In fact, these are bounded between a maximum value

$$E_{max} = \frac{\hbar^2}{2m_e}(q^2 + 2k_F q) \tag{2.19}$$

and a minimum value

$$E_{min} = \begin{cases} 0 & 0 \leqslant q \leqslant 2k_F \\ \dfrac{\hbar^2}{2m_e}(q^2 - 2k_F q) & q > 2k_F. \end{cases} \tag{2.20}$$

These restrictions are illustrated in figure 2.1 where the allowed region is bounded by E_{max} of equation (2.19) and E_{min} of equation (2.20). Also related to figure 2.1 is the plasmon dispersion curve $\hbar\omega_p(q)$, which for small wavevectors can be shown to have the form (see Appendix 2.2 for a brief discussion)

$$\omega_p^2(q) = \omega_p^2 + \alpha q^2 + \ldots . \tag{2.21}$$

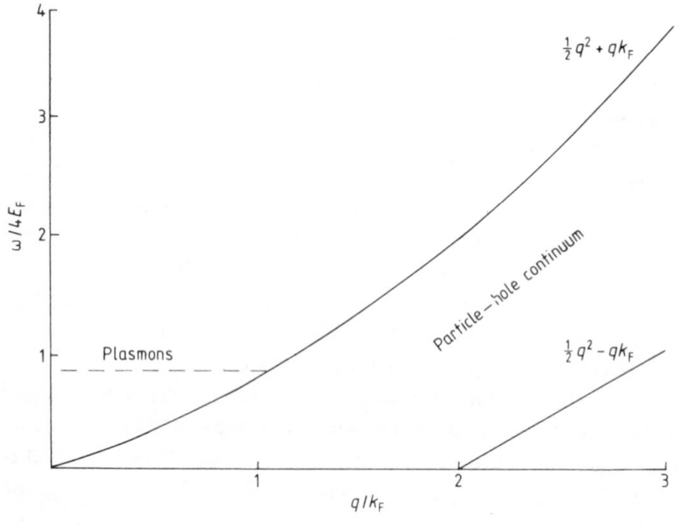

Figure 2.1 Excitation spectrum of uniform interacting electron gas. Pair excitations (electron–hole) occur between the lines labelled $\frac{1}{2}q^2 + qk_F$ and $\frac{1}{2}q^2 - qk_F$. The plasma mode intersecting the particle–hole continuum is indicated schematically.

This equation evidently generalises to small, non-zero q the result obtained in Section 2.1. However, while the value of the plasma frequency ω_p is given precisely from theory by equation (2.12), the coefficient α of the q^2 term is known only approximately. A possible approach to calculating α is briefly sketched in Appendix 2.2. Here we want first to note that α has the dimensions of a velocity squared and therefore we expect it to be of the order of v_F^2, where $v_F = \hbar k_F/m_e$ is as usual the Fermi velocity. Elementary theory predicts, in fact, that

$$\alpha = \tfrac{3}{5} v_F^2 \tag{2.22}$$

which is in only fair agreement with the experiments that have been carried out in simple metals (cf. §2.4.3).

At larger wavenumbers q, equation (2.21) eventually ceases to be valid, higher powers of q being required in order to properly describe the dispersion relation $\omega_p(q)$. However, more dramatic effects can be expected when a critical wavenumber, q_c say, is reached. This is given by

$$\hbar\omega(q_c) = \frac{\hbar^2}{2m_e}(q_c^2 + 2k_F q_c), \tag{2.23}$$

which defines the point at which the plasmon dispersion curve first touches the upper edge of the region in which single-particle excitations are allowed (see figure 2.1). Therefore, for $q > q_c$, the plasmon can decay into single-particle excitations. This implies that once it has been created a plasmon can live for a limited time, τ say. If this time is smaller than the characteristic oscillation time of the plasmon ($\sim 2\pi/\omega_p$), then obviously the plasmon cannot exist as a coherent motion of all the electrons in the charge cloud and it will no longer be an observable entity.

In spite of this, in many cases plasmon excitations have been observed for $q > q_c$. However, due to the strong interaction between plasmon and single-particle excitations, it is found (see §2.4.3) that both the dispersion and the lifetime of the plasmon change drastically when q_c is reached.

For $q < q_c$, the simple theory developed above would predict that the plasmon is undamped (i.e. has infinite lifetime). But the experiments show that this is not so (see §2.4.3), and therefore sources of damping different from the single-pair excitations must be invoked in order to explain the observations. In a perfect crystal, one such source is the interaction of the electrons with the periodic lattice potential, which has been neglected by working with the smeared ion 'jellium' model. Another source of damping could be the simultaneous excitation of several electron–hole pairs. These so called multiparticle excitations are no longer confined to lie in the strip of the (ω, q) plane defined by equations (2.19) and (2.20) and can provide an effective source of damping in the range $0 < q < q_c$. However, this effect is often small compared to the electron–lattice interaction.

2.4 Experimental observations of plasmons

The existence of plasmons has important consequences for the optical properties of metals. This can be understood as follows. Consider the jellium model, and apply a weak, time-varying external field. Under these circumstances, equation (2.11) has to be modified, to read, with $\boldsymbol{D}$ denoting the electric displacement,

$$m_e\ddot{\boldsymbol{x}} + 4\pi e^2 n_0 \boldsymbol{x} = -e\boldsymbol{D} \tag{2.24}$$

in order to account for the new applied external force. Equation (2.10) remains unchanged and if we Fourier analyse equations (2.24) and (2.10) we obtain

$$(-m_e\omega^2 + 4\pi e^2 n_0)\boldsymbol{x}(\omega) = -e\boldsymbol{D}(\omega) \tag{2.25}$$

and

$$m_e\omega^2 \boldsymbol{x}(\omega) = e\mathscr{E}(\omega) \tag{2.26}$$

where $\boldsymbol{x}(\omega)$ and $\mathscr{E}(\omega)$ are the Fourier transforms of $\boldsymbol{x}(t)$ and $\mathscr{E}(t)$ respectively. Combining equations (2.25) and (2.26) and introducing ω_p from equation (2.12), one obtains

$$\mathscr{E}(\omega)\left(1 - \frac{\omega_p^2}{\omega^2}\right) = \boldsymbol{D}(\omega). \tag{2.27}$$

This will be related to the optical properties immediately below.

2.4.1 *Frequency-dependent dielectric function $\epsilon(\omega)$*

Recalling the definition of the dielectric function through

$$\boldsymbol{D}(\omega) = \epsilon(\omega)\mathscr{E}(\omega) \tag{2.28}$$

one obtains from equation (2.27) the frequency-dependent dielectric function $\epsilon(\omega)$ as

$$\epsilon(\omega) = 1 - \frac{\omega_p^2}{\omega^2} \tag{2.29}$$

which evidently vanishes at the plasma frequency ω_p.

The connection with the optical properties is easily made by recalling that the optical constants n and K are related to the real and imaginary parts of the dielectric constant

$$\epsilon(\omega) = \epsilon_1(\omega) + \mathrm{i}\epsilon_2(\omega) \tag{2.30}$$

by the relation

$$[n(\omega) + \mathrm{i}K(\omega)]^2 = \epsilon_1(\omega) + \mathrm{i}\epsilon_2(\omega) \tag{2.31}$$

In the present case, $\epsilon_2 = 0$ and therefore $K(\omega)$ vanishes for $\omega > \omega_p$, which means that the system (the jellium model) is perfectly transparent at high frequencies. A straightforward application of these results to real metals is not possible, since there the interaction with the lattice renders equation (2.29) inapplicable. However, at sufficiently high energies, these interactions contribute negligibly and equation (2.29) is approximately valid. One can therefore measure ω_p by seeking the points at which $\epsilon(\omega) = \epsilon_1(\omega) + i\epsilon_2(\omega)$ is approximately zero. From the above considerations, it follows also that alkali metals, for which the jellium model is a good approximation, and which have a plasma frequency of the order of $10^{16}\,s^{-1}$, are expected to be transparent to ultraviolet radiation. This is indeed a well known experimental result (see the plot of $K(\omega)$ for a simple metal in figure 2.2).

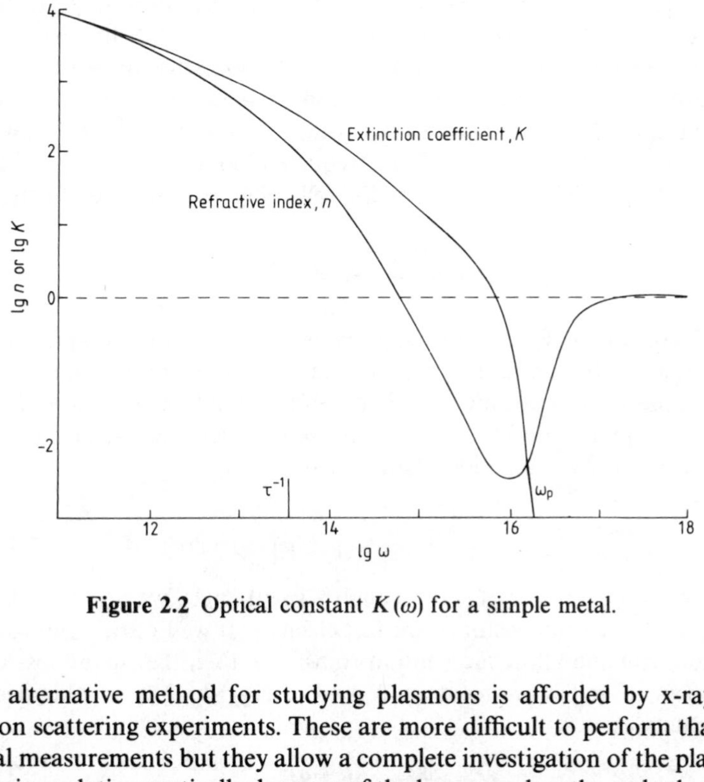

Figure 2.2 Optical constant $K(\omega)$ for a simple metal.

An alternative method for studying plasmons is afforded by x-ray and electron scattering experiments. These are more difficult to perform than the optical measurements but they allow a complete investigation of the plasmon dispersion relation; optically, because of the long wavelengths, only the $\boldsymbol{q} = 0$ region is sampled. The basic principles of the method are similar to those described in Section 3.6 for determining the dispersion of lattice vibrational waves. However, the different nature of the system calls for the use of different probes. For instance, because the plasmon energies are so much higher than the phonon energies, to be discussed in Chapter 3, it is feasible to perform inelastic x-ray scattering. Indeed, experiments of this kind have been successfully carried out. However, for various reasons, most of the experiments on plasmons have been carried out by fast-electron scattering experiments, the interpretation of which will be discussed immediately below. In practice, in these experiments, the energy loss suffered by the impinging electrons when traversing a thin metallic film is measured as a function of the scattering angle. Later we shall describe in some detail the results of one experiment to determine the plasmon dispersion relation.

2.4.2 *Energy losses of fast electrons*

It will be shown here that a fast electron traversing an electron gas will suffer characteristic energy losses which are directly related to the plasma oscillations. The electron must be fast enough to allow lowest-order perturbation theory to be used and then it turns out that the rate of loss of energy to the metal is related directly to a generalisation of the dielectric function discussed previously. There we discussed the k-dependent dielectric function $\epsilon(k)$ in § 2.2 and the frequency-dependent dielectric function $\epsilon(\omega)$ in § 2.4.1.The results of the present section require the full frequency- and wavevector-dependent dielectric function $\epsilon(\boldsymbol{k}, \omega)$ defined by the following generalisation of equation (2.28):

$$\boldsymbol{D}(\boldsymbol{k}, \omega) = \epsilon(\boldsymbol{k}, \omega)\, \mathscr{E}(\boldsymbol{k}, \omega). \tag{2.32}$$

We shall content ourselves in the main text with an outline of the approach and a statement of the central results; the details are given in Appendix 2.3.

If we denote the electronic coordinates of the metal electrons by $\boldsymbol{r}_i$, $i = 1$ to N, then the interaction of the fast electron at $\boldsymbol{R}$ with the metal is evidently represented by the interaction Hamiltonian

$$H_{\text{int}} = \sum_i \frac{e^2}{|\boldsymbol{r}_i - \boldsymbol{R}|}. \tag{2.33}$$

If the electron is only scattered through a small angle, we can write $\boldsymbol{R} = \boldsymbol{v}t$, where $\boldsymbol{v}$ is the velocity vector of the fast electron. If we Fourier transform (cf Appendix 2.3) and allow momentum transfer $\boldsymbol{q}$, then the energy loss of the electron with initial momentum $\boldsymbol{k}$ and energy $k^2/2m_e$ is obtained approximately as

$$\begin{aligned}\omega &= \frac{k^2}{2m_e} - \frac{(\boldsymbol{k}-\boldsymbol{q})^2}{2m_e} \\ &= \frac{\boldsymbol{k}\cdot\boldsymbol{q}}{m_e} - \frac{q^2}{2m_e} \simeq \boldsymbol{v}\cdot\boldsymbol{q}.\end{aligned} \tag{2.34}$$

In Appendix 2.3, classical electrodynamics is used to calculate the probability that the electron transfers momentum $\boldsymbol{q}$ and energy $\omega = \boldsymbol{q}\cdot\boldsymbol{v}$ to the metal. The desired result is that, apart from factors given in equation (A2.3.7), the energy loss δE is determined by

$$\delta E \propto \mathrm{Im}\left(\frac{1}{\epsilon(\boldsymbol{q}, \omega)}\right). \tag{2.35}$$

Therefore theoretical determination of the energy loss requires the above knowledge of the frequency- and wavevector-dependent dielectric function $\epsilon(\boldsymbol{q}, \omega)$. Clearly there will be a peak in the power dissipation whenever collective modes are excited. This happens when $\epsilon(\boldsymbol{q}, \omega) = 0$. This condition gives the plasmon excitations, as can be seen from equation (2.32). Thus, this equation can be obeyed for $\boldsymbol{D}(\boldsymbol{q}, \omega) = 0$ (no external charges) only if either $\epsilon(\boldsymbol{q}, \omega)$ or $\mathscr{E}(\boldsymbol{q}, \omega)$ is zero. The latter condition is trivial while in the former case one can have $\mathscr{E}(\boldsymbol{q}, \omega) \neq 0$ even in the absence of external charges. This condition is indeed realised when a plasma oscillation is set up in the system.

2.4.3 Experimental demonstration of anisotropy of plasmon dispersion

It is now well established from a variety of experiments that the position of the plasmon energy $\hbar\omega_p$ is displaced to higher energy values with increasing wavevector $\boldsymbol{q}$ in a manner approximately proportional to q^2 (see e.g. Kloos 1973), in general accord with the theoretical form of equation (2.21). Agreement with theory for the coefficient α of the q^2 term is not quantitative however. Thus, for the alkali metals Li, Na and K, α (in suitable units) has been measured as 0.22, 0.22 and 0.12 respectively, whereas using the same units equation (2.22) would give 0.36, 0.31 and 0.28 respectively.

Furthermore, experiments on single crystals of Al (Urner-Wille and Raether 1976) have shown that the dispersion coefficient α is a function of crystal direction. This indicates the presence of effects arising from the lattice potential in determining the plasmon dispersion and these effects will have to be incorporated in the theory to get quantitative agreement with experiment.

Probably the most accurate measurements available, at the time of writing, on the anisotropy of the plasmon dispersion are those of Stiebling and Raether (1978) on Si. Although the four valence electrons in Si are bound (energy gap ~ 1 eV) the plasmon energy $\hbar\omega_p(q = 0) \sim 17$ eV is large enough to treat these electrons as nearly free. In this sense, from the present point of view of collective oscillations, the valence electrons of Si behave rather like those of Al. In particular, in Si, a large loss peak due to these plasma oscillations is observed in the loss spectrum. Therefore Stiebling and Raether (1978) have studied the dispersion of plasmons in Si at 16.7 eV with 50 keV electrons in the directions $\langle 100 \rangle$ and $\langle 111 \rangle$ up to the critical wavenumber $q_c \simeq 1.1$ Å^{-1}. Some marked anisotropy is thereby revealed. Along the $\langle 100 \rangle$ direction, the q dependence could be measured up to large q vectors of around 2.4 Å^{-1}. At $q \gtrsim q_c$, they found that the slope of the dispersion curve, plotted as a quadratic function of q, decreases strongly. Such an effect had also been seen earlier for Al and Be.

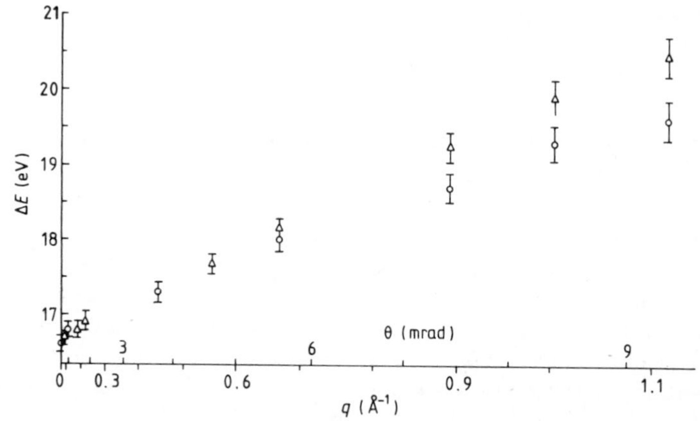

Figure 2.3 Plasmon dispersion relation for Si, as determined by Stiebling and Raether (1978) by electron scattering. Shows anisotropy, upper points corresponding to ⟨100⟩ direction and lower ones to ⟨111⟩ direction. (From Stiebling J and Raether H 1978 *Phys. Rev. Lett.* **40** 1293.)

To illustrate the observed anisotropy, figure 2.3 shows the results of Stiebling and Raether (1978) in Si, in which the measured dispersion curve $\hbar\omega_p(q)$ up to q_c is plotted on a quadratic scale of the wavenumber q. It can be seen that above $q \sim 0.6\,\text{Å}^{-1}$ the losses in the different directions are well separated and two values of α can be extracted. In suitable units these are $\alpha_{\langle 100 \rangle} = 0.41 \pm 0.01$ and $\alpha_{\langle 111 \rangle} = 0.32 \pm 0.02$. For the nearly-free-electron gas in the same units α is 0.4, if a correction for the interaction between electrons in the gas is included. The highest α value in Si occurs in the ⟨100⟩ direction, whereas in Al it occurs in the ⟨110⟩ direction.

For larger q the observations can be summarised as follows. There exists a region around q_c, visible by the change of the slope in the dependence of $\hbar\omega_p(q)$ as a function of q^2, which can be interpreted as the transition from collective excitations to single-particle excitations. This interpretation is supported by measurements of the linewidth. The electron energy loss experiments on Al demonstrated that for $q > q_c$ the width of the plasma loss line increases strongly (Zacharias 1974, 1975, Gibbons *et al* 1976) indicating the decay of the plasmon into an electron–hole pair (single-particle excitation) which, as we have seen, becomes possible for $q \geqslant q_c$. The way the plasma lifetime is seen experimentally will be clear from figure 2.4, when we recognise that the widths W shown there are related to the lifetime by the uncertainty principle $\Delta W \Delta\tau \sim 1$.

To summarise, the predictions of the jellium model are well borne out by the experiments on plasmons at a qualitative and often semiquantitative level. But the anisotropy revealed by the experiments described above arises from the influence of the ionic lattice on the electron gas and cannot be accounted for in the jellium model in which the ions are smeared out to yield a uniform

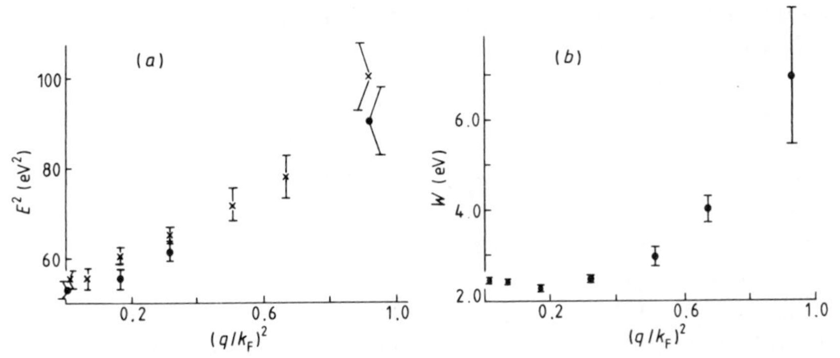

Figure 2.4 (a) Dispersion of lithium plasmon; (b) widths (W) of plasmon as determined experimentally for metallic lithium. (From Gibbons *et al* 1976.)

neutralising background of positive charge. In a number of cases, the fact that we have granular ions in a lattice can be taken into account in a perturbative manner assuming a weak electron–ion interaction. Even in the results shown in figure 2.3 for Si, where the electron–ion interaction is larger than in the simple metals, lattice effects, though of consequence, are not dominant. Progress in the theory of plasmons propagating in periodic lattices has been made (see e.g. March and Tosi 1972; this method has been used by Girlanda, Parrinello and Tosatti (1976) for a variety of metals and semiconductors). For the interested reader a brief summary of a semiclassical version of this theory, due to Bloch (1933, 1934), is contained in Appendix 2.2.

2.5 Non-degenerate plasma: Debye length

Up to now, our discussion has been about the plasmon in a degenerate Fermi gas. We shall not deal with the non-degenerate plasma in the same detail. However, it is of considerable interest and two examples which may be cited are:

(a) the screening of a charged impurity (say P) in a semiconductor (say Si) (cf Dingle 1955, Mansfield 1956);
(b) the electrical conductivity of a hydrogen plasma, around $T = 2 \times 10^4$ K.

In case (a) the system is non-degenerate because the carrier density is very low, while in (b) the non-degeneracy is because of the very high temperature.

The order of magnitude argument given above, namely that the screening (Debye) length l_D is the product of a characteristic velocity and a characteristic time, gives

$$l_D \sim v_{th}/\omega_p \tag{2.36}$$

in the non-degenerate electron assembly under discussion, where

$v_{th} = (3k_BT/m)^{1/2}$. Thus, in the screened Coulomb potential form (e^2/r) $\exp(-q_Dr)$ we can write

$$q_D^2 = l_D^{-2} = \frac{4\pi n_0 e^2}{k_BT} \tag{2.37}$$

where n_0 is the free-electron density and k_B is the Boltzmann constant.

Though we should have to go through the full theory to verify the constant, because the above argument is order of magnitude, the value of l_D as given above is in fact correct.

This screening length l_D is in a suitable form to use for obtaining the screened Coulomb potential round a proton in a non-degenerate hydrogen plasma. For a P impurity in Si, with dielectric constant ϵ, the charged impurity potential $e^2/\epsilon r$ goes into $(e^2/\epsilon r)\exp(-q_Dr)$ with q_D^2 now given by

$$q_D^2 = \frac{4\pi n_0 e^2}{\epsilon k_BT}. \tag{2.38}$$

As an example, let us return to the case of P, a pentavalent impurity, in Si. A silicon crystal has 5×10^{22} Si atoms/cm^3. Suppose now that one impurity phosphorus atom is introduced for every 10^5 Si atoms, resulting in an electron density available for screening of 5×10^{17} cm^{-3}. Assuming the electronic system to be non-degenerate at room temperature, the Debye length comes out to be around 60 Å, the dielectric constant of Si being 12.

This length can easily be employed, in conjunction with lowest-order perturbation theory (Born approximation) to estimate the ionised impurity scattering in a lightly doped semiconductor. Hence the impurity contribution to the electrical resistivity can be calculated (cf Jones and March 1973).

The same method can also be employed to calculate the electrical conductivity of a hydrogen plasma†. Measurements have been carried out under controlled conditions, and at temperatures around 20 000 K it seems a fair assumption that the short-range order of the protons is not of prime importance. Then the quantum degeneracy of the electrons can also be neglected, and therefore the scattering of the plasma electrons off the screened protons can proceed in the same way as that described above for impurities in semiconductors.

2.6 Surface plasmons; inhomogeneous electron gas

The previous discussion has been concerned with plasmons in bulk metals, even though, as we have seen in the previous section, it is often convenient to

† Metallic hydrogen is believed to exist in the centre of Jupiter, and the transport processes in a hydrogen plasma, including electrical conductivity, therefore have some astrophysical interest (cf Brown and March 1972).

study them experimentally by firing fast electrons into thin metal films and observing their energy losses. In this section, we turn to the phenomenon of surface plasmons. Such collective modes, the nature of which is to be described below, occur at the interface between a metal and a vacuum. Since the result for the frequency of the surface collective mode will prove to be dependent on the geometry of the system, we shall treat the case of a planar metal surface below. The modifications required if one works in spherical geometry are referred to in problem 2.2.

2.6.1 Classical treatment for long wavelength

The derivation below will be based on Maxwell's equations. Retardation effects, due to the finite value of the velocity of electromagnetic wave propagation, will be neglected for simplicity of presentation. As already remarked, we shall consider specifically a planar interface between metal (M) and vacuum (V). Referring to figure 2.5, we wish to solve Maxwell's equations in these two media, the vacuum having a dielectric constant of unity while the metal is characterised by the frequency-dependent form $\epsilon(\omega)$. Naturally, we shall eventually need to specify the precise form of $\epsilon(\omega)$ for the model employed. But for the present we wish to study the way in which the surface plasmon frequency can be calculated from a given dielectric function $\epsilon(\omega)$.

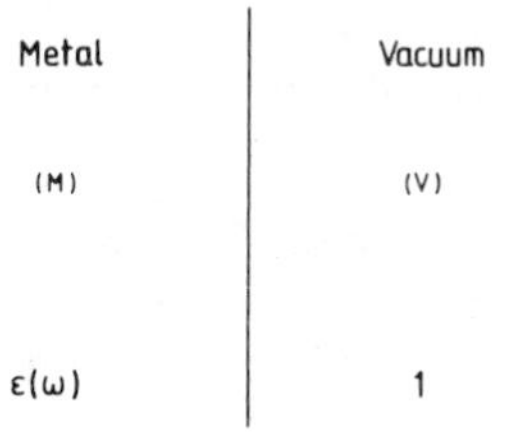

Figure 2.5 Indicates planar interface between metal (M) and vacuum (V). Dielectric constants of the two media are shown.

In the two media, the electrostatic potential ϕ can be determined from Laplace's equation:

$$\nabla^2 \phi = 0. \tag{2.39}$$

The surface will be taken in the (x, y) plane and vectors in the surface will be denoted by $\boldsymbol{X}$. In Fourier transform space the analogue of the vector $\boldsymbol{X}$ will be denoted by $\boldsymbol{Q}$. Since we are seeking now a collective mode localised in the surface, it is clear that the appropriate solution of equation (2.39) will have the form

$$\phi = \text{constant} \cdot \exp(\mathrm{i}\boldsymbol{Q} \cdot \boldsymbol{X}) \exp(-Q|z|) \tag{2.40}$$

as is readily verified from Laplace's equation (equation (2.39)). In writing the solution (2.40), we have naturally adopted a running wave form in the (x, y) plane, but in the z-direction perpendicular to the planar surface we have a very localised disturbance for non-zero $\boldsymbol{Q}$.

This is the point at which the boundary conditions at the surface between M and V must be invoked. Thus, one must satisfy the requirements that:

(a) the normal component of the electric displacement must be continuous across the interface;
(b) the component of electric field parallel to the interface, $E_{\parallel}$ say, must be continuous.

Condition (b) is automatically satisfied by assuming the form (2.40) so that we must now impose (a) on the solution. To find $\boldsymbol{D}$, one simply calculates the electric field $\mathscr{E} = -\text{grad}\,\phi$, which has then to be multiplied by the appropriate dielectric function for M and V respectively. Almost immediately one finds

$$Q\epsilon(\omega)\phi = -Q\phi \tag{2.41}$$

and we see that the collective mode can exist provided

$$\epsilon(\omega) + 1 = 0 \tag{2.42}$$

since equation (2.41) then permits a solution with finite amplitude of ϕ. Equation (2.42) is the central result for the surface plasmon.

2.6.2 *Model of dielectric function $\epsilon(\omega)$ in terms of bulk plasmon*

To calculate the frequency of the surface plasmon, as remarked already, we must make a model of $\epsilon(\omega)$. The simplest one is to base it on the bulk plasma frequency and express $\epsilon(\omega)$ through equation (2.29). Substituting this model $\epsilon(\omega)$ into the surface plasmon condition (2.42) we find immediately

$$\omega_s = \omega_p/\sqrt{2} \tag{2.43}$$

which is clearly the frequency of the surface plasmon. The factor $1/\sqrt{2}$ by which the bulk plasma frequency is reduced is characteristic of planar geometry; for a sphere the factor is $1/\sqrt{3}$.

2.6.3 *Comparison with experiment*

Inserting into equation (2.12) the number of valence electrons per atom (to calculate n_0) together with the values of e and m_e, for simple metals one finds theoretical values of ω_s shown in the second row of table 2.1. There is seen to be really excellent agreement between experiment and the simple classical theory given above.

We should comment however that, in reality, the surface plasmon shows dispersion, whereas obviously the above argument yields ω_s independent of

Table 2.1 Surface plasmon energy $\hbar\omega_s$ for planar surface (in eV).

	Mg	Al	In	Ga
Experiment	7.1	10.6	8.7	10.3
Theory (equation (2.43))	7.7	11.2	8.9	10.3

wavevector Q. This will be discussed briefly in § 2.6.6 and also in Appendix 2.2 where the dispersion of the surface plasmon is shown to be proportional to Q.

2.6.4 Form of charge density in surface plasmon mode

In connection with the above deviations from the classical theory, we note that from the potential derived above we can calculate the charge density disturbance associated with the plasmon mode. Indeed, the result is obvious. Since we have solved Laplace's equation (not Poisson's equation), it is clear that the charge disturbance must be completely localised in the surface, that is it has the form of a sharp spike or a delta function. If we now assign a finite width to this spike, then we do get deviations from the classical theory proportional to Q (cf Inglesfield and Wikborg 1973).

Secondly, let us comment on the way the experimental values for the surface plasmon frequencies given in the first row of table 2.1 were obtained. Again, one uses inelastic electron scattering. The transmission experiment is very simple as regards the theoretical interpretation. But surfaces are irregular in practice, because it is necessary to use evaporated films.

The most satisfactory approach is to use a reflection method, at grazing incidence. It then turns out that the probability of creating a surface plasmon is inversely proportional to the (small) angle of incidence and this allows the intensity of the surface plasmon peaks to be enhanced relative to the bulk plasmon. At about 1°, surface plasmons are dominant, whereas if the angle of incidence is increased to 5° say, then the bulk contribution begins to make its presence felt.

For a more quantitative study of the relation between the probability of creating a surface plasmon in reflection and the angle of incidence, reference should be made to the work of Muscat and Newns (1976).

2.6.5 Relation to surface energy of metals

This discussion of plasmons in metals has a limited relevance to the problem of the surface energy of metals. This is a complex problem, which we know to involve both single-electron excitations and collective effects due to the Coulomb interaction. However, a physically appealing picture (though we now know it to be an oversimplification) of the variation of the surface energy from one metal to another was given by Schmidt and Lucas (1972). Their argument was based on the change in the zero-point energy of the plasma oscillations when the free surface is formed. In terms of the radius of a sphere containing one valence electron, i.e. r_s in equation (2.1), they estimate that in going from bulk modes with no free surface to modes involving surface plasmons the zero-point energy change is proportional to $r_s^{-5/2}$, which they then identify as the variation of surface energy with r_s. In obtaining this result though, one has to invoke a cut-off wavevector for collective surface oscillations and the quantitative details of the theory are sensitive to the choice of the cut-off (cf Barton 1979). Nevertheless the agreement between theory and experiment is (somewhat fortuitously) good, as figure 2.6 shows. The interested reader will find further details in the review by Brown and March (1976).

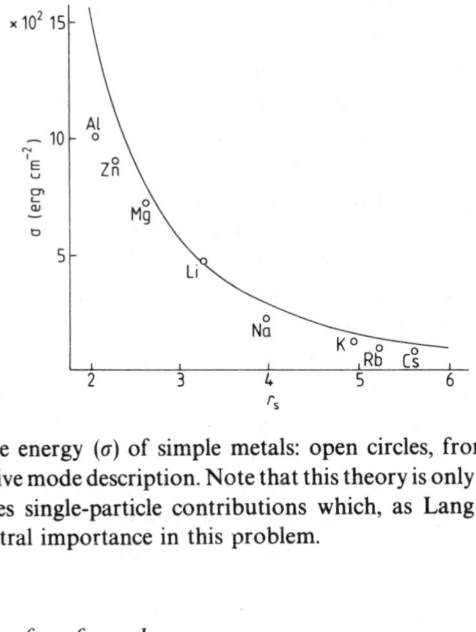

Figure 2.6 Surface energy (σ) of simple metals: open circles, from experiment; full curve, from collective mode description. Note that this theory is only part of the story. A full theory requires single-particle contributions which, as Lang and Kohn (1970) discuss, are of central importance in this problem.

2.6.6 Dispersion of surface plasmon

As we have seen, the limiting long-wavelength value (2.43) of ω_s for planar surfaces has been verified experimentally, but there are not many experimental data bearing directly on the dispersion of the surface plasmon. However, one datum is given by the electron energy loss experiment on Mg films carried out by Kunz (1966). This provides a determination of the real part of the surface plasmon frequency, say $\mathrm{Re}[\omega_s(Q)]$, where Q is as usual the momentum parallel to the surface. It is found from the experimental data for $\mathrm{Re}(\omega_s)$ for these films that there is first a decrease and then an increase with increasing Q.

On the other hand, theoretical calculations for a semi-infinite electron gas (Ritchie 1963, Ritchie and Marusak 1966) yield the result that Re (ω_s) increases linearly with increasing Q (see Appendix 2.2).

There are at least two factors which must be taken into account when comparing experiments on the surface plasmon dispersion relation with the theory for a semi-infinite electron gas:

(a) the dispersion relation for a thin film is more complex than that for a free surface (Raether 1967, Economou 1969) since interference occurs between the two surfaces of the film;
(b) the detailed variation of the electron density through the metal surface is involved in calculating the surface plasmon dispersion.

For a discussion of the imaginary part of the surface plasmon frequency $\mathrm{Im}[\omega_s(Q)]$, reference may be made to the review by Brown and March (1976) and to the other references given there.

2.7 Density oscillations round a test charge in a bulk metal and at metal surfaces

The screened Coulomb potential derived previously shows that, in a degenerate electron gas such as in a simple metal, there is a characteristic screening length of the order of 1 Å. However, we must qualify that result by considering the detailed wave nature of the Fermi electrons.

To do so, let us consider a one-dimensional box with free particles moving between $x = 0$ and $x = l$ and having wavefunctions

$$\psi_n(x) = \left(\frac{2}{l}\right)^{1/2} \sin\left(\frac{n\pi x}{l}\right) \qquad n = 1, 2, 3, \ldots. \tag{2.44}$$

The electron density $n(x)$ associated with filling the lowest N levels singly is given by

$$n(x) = \sum_{n=1}^{N} \left(\frac{2}{l}\right) \sin^2\left(\frac{n\pi x}{l}\right) = \frac{N+\frac{1}{2}}{l} - \frac{\sin[2\pi(N+\frac{1}{2})x/l]}{2l\sin(\pi x/l)}. \tag{2.45}$$

The summation can be completed exactly as indicated, but as l gets very large we can in fact replace the summation by an integration. Introducing the Fermi wavenumber k_F through the usual phase space requirement that (cf Appendix 2.1)

$$\text{area of occupied phase space} = 2k_F l, \tag{2.46}$$

k-space being occupied from $-k_F$ to k_F, the number of cells of size h is therefore $2k_F l/h$, which must equal the number of particles for singly occupied levels. After integrating from $-k_F$ to k_F, we obtain the result

$$n(x) = n_0 - n_0 \frac{\sin(2k_F x)}{2k_F x} \tag{2.47}$$

where $n_0 = N/l$, showing that long-range oscillations of wavelength π/k_F are induced in the one-dimensional Fermi gas by the 'perturbation' due to the wall of the box at $x = 0$, the other end of the box having gone to $+\infty$.

This is a general type of result for a perturbation in a degenerate Fermi gas, and for a test charge in a three-dimensional system we obtain the asymptotic behaviour

$$n(r) - n_0 \sim \text{constant}\cdot\frac{\cos(2k_F r)}{r^3}; \tag{2.48}$$

a derivation of this result is given in Appendix 2.1.

This result for the charge displaced by an impurity, or test charge, would obviously be relevant in calculating the interaction energy between a pair of charged defects in a metal (Corless and March 1961) or indeed in obtaining the pair force between ions in a solid or liquid metal (see e.g. Jones and March 1973). The oscillations in equation (2.48) are known as Friedel oscillations.

If the perturbation is strong, these oscillations are modified to include a phase factor, i.e. $n(r) - n_0 \sim [\cos(2k_F r + \phi)]/r^3$. If one is dealing with a charge in a surface, then for a direction parallel to a planar metal surface the decay with distance is r^{-5} times an oscillatory factor, but we shall not develop that further here (see e.g. Flores *et al* 1977, Lau and Kohn 1978, Einstein 1978). Fermi surface topology is also known to influence the range of the oscillatory behaviour in metals with and without surfaces (see e.g. Flores *et al* 1979).

Having discussed collective excitations in simple metals at some length, even though we pointed out the relevance for plasmon studies in a covalent semiconductor like Si, we now turn to the problem of describing excited states in insulators, starting from the pioneering work of Frenkel.

2.8 Frenkel excitons in insulating crystals

In §2.3 we discussed the excited states that can be obtained in a metal by promoting an electron to a higher energy state, thereby leaving behind a hole. Because of screening (cf equation (2.13)), the effective Coulombic interaction between the negatively charged electron and the positively charged hole is weak and in most cases the electron and the hole can be viewed as separate entities moving independently of one another (see, however, §2.13).

A different situation can arise in insulating or semiconducting solids. There, excitations of this kind are obtained by promoting in one atom (or in a covalently bonded crystal like Ge or Si, in one bond) one electron to a higher energy state. In contrast to the metallic solid, here the other electrons are not free to move, and therefore they can only partially screen the attraction between electron and hole. Thus the formation of a bound electron–hole state is possible, and these two particles are linked together in their motion through the crystal. This can occur because the electron–hole pair originally created on one atom can jump to another atom. The amplitude probability for such an event to occur is of the form

$$V_{ij} = \langle \psi_{R_i}(r_e, r_h) W \psi_{R_j}(r_e, r_h) \rangle \qquad i \neq j \tag{2.49}$$

where $\psi_{R_i}(r_e, r_h)$ is the wavefunction that describes the electron–hole pair on atom i, and W is the potential created by one electron–hole pair. If this last quantity has the right symmetry, then W has the form of the potential due to a dipole and therefore it decreases as the cube of the distance $|R_i - R_j|$ between the atoms. It follows that if the electron–hole wavefunctions $\psi(r_e, r_h)$ are very well localised on each atom, V_{ij} is significantly different from zero only if $|R_i - R_j|$ is small. This implies in many cases that only jumps between nearest neighbours are of importance. This type of running wave of tightly bound electron–hole pairs is termed a Frenkel exciton.

It is appropriate to add here that Frenkel-like excitons are typical of large energy gap insulators, where there is weak screening of the electron–hole interaction. For example, the lowest exciton state in the solid rare gases is well described in this way. This has been shown in the specific calculations of Knox (1963).

2.8.1 Dipole–dipole interaction and energy of Frenkel excitons

Following Heller and Marcus (1951), let us make the above discussion quantitative. These workers follow the pioneering approach of Frenkel, but emphasise the feature discussed above: namely the electromagnetic coupling between atoms. For simplicity, they consider a cubic crystal of identical one-electron atoms. As a starting point a determinantal wavefunction of all the atoms is assumed, which ignores spin effects and neglects overlap of the ground-state wavefunctions on different atoms.

For convenience, it will be assumed that the ground-state atomic functions are S-like and that the corresponding excited state wavefunctions have P-character. One can write the ground-state wavefunction as the Slater determinant Ψ_0, for a crystal with N atoms:

$$\Psi_0 = \frac{1}{(N!)^{1/2}} \begin{vmatrix} \psi_1^0(r_1) & \cdots & \psi_1^0(r_N) \\ \vdots & & \vdots \\ \psi_N^0(r_1) & \cdots & \psi_N^0(r_N) \end{vmatrix} \tag{2.50}$$

while a typical zeroth-order excited-state wavefunction will be

$$\Psi_J^n = \frac{1}{(N!)^{1/2}} \begin{vmatrix} \psi_1^0(r_1) & \cdots & \psi_1^0(r_N) \\ \vdots & & \vdots \\ \psi_J^n(r_1) & & \psi_J^n(r_N) \\ \vdots & & \vdots \\ \psi_N^0(r_1) & \cdots & \psi_N^0(r_N) \end{vmatrix} \tag{2.51}$$

The notation is such that the superscript n refers to the nth excited atomic state of energy ϵ_n, the subscript labelling the crystal site involved. The energy corresponding to the ground-state wavefunction Ψ_0 is then found to be

$$E^0 = N\epsilon_0 - \tfrac{1}{2}e^2 \sum_{I,J}{}' \int \psi_I^0(\boldsymbol{r}_1)\psi_J^0(\boldsymbol{r}_2)\frac{1}{r_{12}}\psi_I^0(\boldsymbol{r}_2)\psi_J^0(\boldsymbol{r}_1)\,\mathrm{d}\boldsymbol{r}_1\,\mathrm{d}\boldsymbol{r}_2, \qquad (2.52)$$

where the prime denotes the exclusion of the case $I = J$ from the sum. Consistent with the assumption that no overlapping of the ground-state wavefunction occurs, the total Coulomb interaction of neighbouring atoms has been neglected. The Hamiltonian employed is

$$H = \sum_I H_I + \sum_{I,J}{}' H_{IJ} \qquad (2.53)$$

which is a sum of one-centre terms H_I given by

$$H_I = -\frac{e^2}{r_{iI}} - \frac{\hbar^2}{2m_\mathrm{e}}\nabla_i^2 \qquad (2.54)$$

and a part involving, among others, Coulomb repulsion terms e^2/r_{ij}, namely

$$\sum_{I,J}{}' H_{IJ} = \sum_{I,J}{}' \frac{e^2}{R_{IJ}} - \sum_{j,I}\frac{e^2}{r_{jI}} + \sum_{i,j}{}'\frac{e^2}{r_{ij}}. \qquad (2.55)$$

Here ∇_i^2 refers to the ith electron, which is taken to be associated with the Ith atom, the lower case letters referring to electrons and the capitals to sites.

The energy of the excited-state wavefunction Ψ_J^n can now be calculated as

$$\begin{aligned} E_{JJ}^n = {} & N\epsilon_0 + \epsilon_n - \epsilon_0 \\ & - e^2 \sum_{L \neq 0} \int \psi_{L+J}^0(\boldsymbol{r}_2)\psi_{L+J}^0(\boldsymbol{r}_1)\frac{1}{r_{12}}\psi_J^n(\boldsymbol{r}_1)\psi_J^n(\boldsymbol{r}_2)\,\mathrm{d}\boldsymbol{r}_1\,\mathrm{d}\boldsymbol{r}_2 \\ & - \frac{e^2}{2}\sum_{\substack{I \neq M \\ I \neq J \\ J \neq M}}{}'' \int \psi_I^0(\boldsymbol{r}_1)\psi_M^0(\boldsymbol{r}_2)\frac{1}{r_{12}}\psi_I^0(\boldsymbol{r}_2)\psi_M^0(\boldsymbol{r}_1)\,\mathrm{d}\boldsymbol{r}_1\,\mathrm{d}\boldsymbol{r}_2. \end{aligned} \qquad (2.56)$$

As already remarked, in this equation the energies of the isolated atom in its ground and excited states are respectively ϵ_0 and ϵ_n, while N is the total number of atoms (equal to the number of electrons). The excited-state wavefunctions do not diagonalise the Hamiltonian and are, of course, degenerate. The off-diagonal matrix elements are typified by E_{JL} given by

$$E_{JL} = \int \ldots \int \Psi_J^n H \Psi_L^n \,\mathrm{d}\boldsymbol{r}_1 \ldots \mathrm{d}\boldsymbol{r}_N. \qquad (2.57)$$

These reduce to

$$E_{JL} = e^2 \int \psi_J^n(\boldsymbol{r}_1)\psi_L^0(\boldsymbol{r}_2)\frac{1}{r_{12}}\psi_J^0(\boldsymbol{r}_1)\psi_L^n(\boldsymbol{r}_2)\,\mathrm{d}\boldsymbol{r}_1\,\mathrm{d}\boldsymbol{r}_2$$

Coulomb term

$$-e^2 \int \psi_J^n(\boldsymbol{r}_1)\psi_L^0(\boldsymbol{r}_2)\frac{1}{r_{12}}\psi_J^n(\boldsymbol{r}_2)\psi_L^0(\boldsymbol{r}_1)\,\mathrm{d}\boldsymbol{r}_1\,\mathrm{d}\boldsymbol{r}_2. \tag{2.58}$$

exchange term

To obtain a correct zeroth-order wavefunction from the degenerate set Ψ_J^n, the following linear combination is now formed (cf Bloch's theorem in equation (2.68)):

$$\psi_{\boldsymbol{k}}^n = \frac{1}{N^{1/2}}\sum_J \exp(2\pi\mathrm{i}\boldsymbol{k}\cdot\boldsymbol{r}_J)\Psi_J^n. \tag{2.59}$$

By making use of the relation (see e.g. Ziman 1976)

$$\sum_J \exp[2\pi\mathrm{i}(\boldsymbol{k}-\boldsymbol{k}')\cdot\boldsymbol{r}_J] = \delta_{\boldsymbol{k}-\boldsymbol{k}',0} \tag{2.60}$$

it can be shown that the off-diagonal matrix elements of two wavefunctions with quantum numbers $\boldsymbol{k}$ and $\boldsymbol{k}'$ do indeed vanish. The new diagonal energy is then found to be

$$E_{\boldsymbol{k}}^n = E_{JJ}^n + \sum_L{}' E_{JL}\exp[2\pi\mathrm{i}\boldsymbol{k}\cdot(\boldsymbol{r}_J-\boldsymbol{r}_L)], \tag{2.61}$$

the prime as usual meaning that $J \neq L$, while the sum in equation (2.61) is to be taken over all lattice sites.

The exchange terms in equation (2.58) will in general depend exponentially on the lattice parameter. The Coulomb contribution to E_{JL} can be seen to be the interaction between two charge clouds, each with density equal to the product of the excited- and ground-state wavefunctions of an isolated atom. This term can conveniently be expanded in the distance R between atoms. Writing

$$\begin{aligned}\boldsymbol{\mu}_1 &= e\int \psi_J^n(\boldsymbol{r}_1)(\boldsymbol{r}_1-\boldsymbol{r}_J)\psi_J^0(\boldsymbol{r}_1)\mathrm{d}\boldsymbol{r}_1\\ \boldsymbol{\mu}_2 &= e\int \psi_L^n(\boldsymbol{r}_2)(\boldsymbol{r}_2-\boldsymbol{r}_L)\psi_L^0(\boldsymbol{r}_2)\mathrm{d}\boldsymbol{r}_2\end{aligned} \tag{2.62}$$

one finds readily, since $\boldsymbol{\mu}_1$ and $\boldsymbol{\mu}_2$ have the same direction and magnitude for all atoms, that

$$\sum_L{}' E_{JL}\exp[2\pi\mathrm{i}\boldsymbol{k}\cdot(\boldsymbol{r}_J-\boldsymbol{r}_L)]$$

$$\simeq -|\boldsymbol{\mu}|^2\sum_L{}'\left(\frac{3\cos^2(\boldsymbol{\mu},\boldsymbol{r}_J-\boldsymbol{r}_L)-1}{|\boldsymbol{r}_L-\boldsymbol{r}_J|^3}\right)\exp[2\pi\mathrm{i}\boldsymbol{k}\cdot(\boldsymbol{r}_J-\boldsymbol{r}_L)]. \tag{2.63}$$

This confirms what was said above, i.e. that the dipole–dipole term is the leading contribution. The sum, when carried out for $|\boldsymbol{k}| \simeq 0$ (i.e. $|\boldsymbol{k}|R \ll 1$) is the interaction energy of a lattice of static dipoles, with dipole moment of magnitude $|\boldsymbol{\mu}|$. If we neglect outer surface (depolarisation) terms and replace the summation by an integral m, both approximations being valid when $|\boldsymbol{k}| \simeq 0$, then one finds the result given by Heller and Marcus (1951):

$$E_{\boldsymbol{k}}^n = E_{JJ}^n + (8\pi/3)P_2(\cos(\boldsymbol{\mu},\boldsymbol{k}))\,n_0|\boldsymbol{\mu}|^2[j_0(\rho)-j_2(\rho)], \tag{2.64}$$

where P_2 is the Legendre polynomial of order two, n_0 is the density of atoms while j_0 and j_2 are spherical Bessel functions of argument $\rho = 2\pi|\boldsymbol{k}|R_0$, where R_0 is determined by $(4\pi/3)R_0^3 = 1/n_0$.

The accuracy of the approximation of the integral to the sum for particular directions relative to the cubic axes in a face-centred cubic crystal, which was used in arriving at equation (2.64), has been examined numerically by Heller and Marcus (1951) and details may be found in their paper.

It should be added here that the Heller–Marcus interaction term is responsible for the energy splitting of the longitudinal and transverse excitons (a general fact, no matter how localised the excitons are). This is an interesting point, because such splitting is observable and is a measure of the localisation of the excitons in the sense that the more localised the excitons, the larger is the longitudinal–transverse splitting, just as also the larger is the oscillator strength (see immediately below).

2.8.2 Effective mass and energy transport properties

One important result of this development of the Frenkel exciton picture is now seen by calculating the reciprocal effective mass of the exciton near (but not at) the bottom of the $E(\boldsymbol{k})$ curve. One then obtains (for a more general result for the effective mass tensor, see equation (2.72))

$$\left.\frac{1}{m_{\mathrm{eff}}}\right|_{\boldsymbol{k}\simeq 0} = \left(\frac{1}{\hbar^2}\frac{\partial^2 E}{\partial k^2}\right)_{|\boldsymbol{k}|\simeq 0}$$

$$= \frac{8\pi}{3}P_2(\cos(\boldsymbol{\mu},\boldsymbol{k}))n_0\frac{|\boldsymbol{\mu}|^2}{\hbar^2}\left(-\frac{4\pi^2 R_0^2}{5}\right). \qquad (2.65)$$

It is useful to introduce a dimensionless quantity, well known in atomic theory and characteristic of an isolated atom, namely

$$f_{n0} = \frac{(\epsilon_n - \epsilon_0)m_{\mathrm{e}}|\boldsymbol{\mu}|^2}{3\hbar^2 e^2}. \qquad (2.66)$$

In the jargon of atomic physics this is known as the oscillator strength and in terms of it one can write the approximate form for the effective mass of the exciton m_{ex} in terms of the electronic mass m_{e} as

$$m_{\mathrm{ex}} \sim \frac{m_{\mathrm{e}}}{f_{n0}}\frac{R_0}{a} \qquad (2.67)$$

where f_{n0}, as emphasised, is a characteristic of the corresponding $0 \to n$ isolated atomic transition, a is the equivalent Bohr radius of the internal exciton orbit in the crystal, while R_0 as defined above is merely given in terms of the atomic density n_0 by $R_0 = (\frac{3}{4}\pi n_0)^{1/3}$.

The important conclusion from this argument is that overlapping of electronic wavefunctions on adjacent atoms or molecules in crystals is not necessary for effective transfer of excitation energy by the exciton mechanism.

This is due to the fact that the principal interaction is electromagnetic, falling off with distance between atoms as R^{-3}.

The above discussion has qualitative value for energy transfer by excitons, and indeed the ability of Frenkel excitons to move through a crystal with a definite energy and momentum is of considerable importance in describing certain biological and chemical processes.

As one example, we take a molecular crystal whose constituent molecules may be photochemically reactive and where light absorption may be the first step leading to chemical change. For instance, anthracene is photoexcited by reaction with oxygen included in the lattice. Energy transport by Frenkel excitons from the absorption site to the reaction site is one important step in describing such changes (cf Craig 1974).

In summary, the above treatment focuses on the way in which, for atomic excitation waves (Frenkel excitons), the structure of the exciton band is determined by the dipole–dipole terms of the electron–hole interaction.

On the other hand for excitons of large radius (non-localised excitons), to which we now turn, Wannier's work shows the importance of the structure of the valence and conduction bands in determining the characteristics of such excitons. This is therefore a convenient point to summarise briefly the essential results of band theory that will be drawn on. These are:

(a) The quantum numbers describing the one-electron wavefunctions are the components of the wavevector $\boldsymbol{k}$. (We have already seen the usefulness of the $\boldsymbol{k}$-space, or momentum $\boldsymbol{p}(=\hbar\boldsymbol{k})$-space description for free-particle systems.)

(b) The one-electron wavefunctions $\psi_{\boldsymbol{k}}(\boldsymbol{r})$ have the form determined by Bloch's theorem, namely

$$\psi_{\boldsymbol{k}}(\boldsymbol{r}) = \exp(\mathrm{i}\boldsymbol{k}\cdot\boldsymbol{r})\, u_{\boldsymbol{k}}(\boldsymbol{r}) \tag{2.68}$$

where $u_{\boldsymbol{k}}(\boldsymbol{r})$ is periodic with the period of the crystal lattice. Forming the electron density $\psi_{\boldsymbol{k}}^{*}\psi_{\boldsymbol{k}}$ from equation (2.68), it is seen to be periodic, expressing the equivalence of the same point in any unit cell of the crystal. When the periodic potential is reduced to zero, equation (2.68) reduces to the plane wave $\exp(\mathrm{i}\boldsymbol{k}\cdot\boldsymbol{r})$, as it obviously must do.

(c) Each energy band is described by a dispersion relation $E_n(\boldsymbol{k})$, n labelling the band, where $E_n(\boldsymbol{k})$ is periodic in the reciprocal lattice. It is therefore only necessary to describe $E_n(\boldsymbol{k})$ within one unit cell of the reciprocal lattice; this is the first Brillouin zone.

(d) Each band has density of states such that $N(E)\mathrm{d}E$ is the number of energy levels lying in the range E to $E+\mathrm{d}E$, where

$$N(E) \propto \int \delta(E-E(\boldsymbol{k}))\mathrm{d}\boldsymbol{k}. \tag{2.69}$$

An equivalent, and often useful, form is, for a crystal of volume V,

$$N(E) = \frac{2V}{(2\pi)^3}\int \frac{\mathrm{d}S}{|\nabla_{\boldsymbol{k}}E(\boldsymbol{k})|} \tag{2.70}$$

where the integral dS is over a constant-energy surface. Equation (2.70) is obtained using the fact that the density of states in $\boldsymbol{k}$-space per unit volume of crystal is $1/(2\pi)^3$ (cf the phase-space argument for free electrons in Appendix 2.1).

(e) The group velocity (cf the elementary formula $\mathrm{d}\omega/\mathrm{d}k$ in any wave motion) of a Bloch electron is $\nabla_k E(\boldsymbol{k})$, which appears in equation (2.70). In typical insulators and semiconductors, one is interested in band extrema and in particular the lowest states of the conduction bands and the highest valence band states. The band extrema occur at points, $\boldsymbol{k}_{n,\alpha}$ say, where the group velocity $\nabla_k E_n(\boldsymbol{k})$ vanishes. Often, the lowest conduction band has just one minimum at the centre of the zone $\boldsymbol{k}=0$; this is the case for most alkali halides, for II–VI and III–V compounds. In other important cases, however, the conduction band has several equivalent extrema away from $\boldsymbol{k}=0$; one is interested in establishing such degeneracy as well as in the form of the energy surface near the extrema.

(f) The energy surface near the extremum of a non-degenerate band can be obtained by expansion as

$$F_n(\boldsymbol{k}) \simeq E_n(\boldsymbol{k}_n)+\tfrac{1}{2}\sum_{i,j}(\boldsymbol{k}-\boldsymbol{k}_n)_i \left.\frac{\mathrm{d}^2E_n(\boldsymbol{k})}{\mathrm{d}k_i\mathrm{d}k_j}\right|_{\boldsymbol{k}=\boldsymbol{k}_n}(\boldsymbol{k}-\boldsymbol{k}_n)_j+\ldots. \tag{2.71}$$

If the expansion is cut off at this point, one has the effective mass approximation, with effective mass tensor

$$\left(\frac{1}{m_{\mathrm{eff}}}\right)_{ij}=\frac{1}{\hbar^2}\left.\frac{\mathrm{d}^2E_n(\boldsymbol{k})}{\mathrm{d}k_i\mathrm{d}k_j}\right|_{\boldsymbol{k}=\boldsymbol{k}_n} \tag{2.72}$$

which is the desired generalisation of the elementary result already employed in equation (2.65).

(g) In some semiconductors, one can usefully parametrise the effect of the lattice by regarding electrons as moving freely, with modified effective mass. In particular, in formulating the theory of Wannier excitons below, it will be assumed that the effect of the lattice on electrons (and holes) can be represented in such a manner, plus knowledge of the dielectric constant. It should be remarked that through these parameters it is possible to subsume some of the many-body effects into the theory.

2.9 Wannier excitons

As we saw in the previous section, the exciton is the quasi-particle of the running waves set up as an atomic excitation, created on a given site, and proceeds to propagate through the crystal. The implication in the previous discussion of Frenkel excitons was that they were highly localised.

Complementary to the discussion of Frenkel was that of Wannier, who showed that a different description could be given in which a hole in the valence band resulting from the excitation of an electron across the gap in the insulator could be bound to this electron. In this case, the state is described by the total wavevector $\boldsymbol{k}$.

However, Wannier's ideas were formulated with particular reference to weakly bound electron–hole pairs and therefore the case in which the electron and hole are separated by distances large compared with the lattice constant is now referred to as the Wannier exciton.

Figure 2.7 illustrates schematically the case of the large-radius (Wannier) exciton. The problem is obviously simpler than in the case of the Frenkel exciton, since the electron is seeing the hole plus an average lattice potential, whereas when the exciton radius becomes of the order of the lattice spacing, the detail of the lattice potential becomes important, in addition to the effect of the hole. Such excitons in which the electron–hole pair has a radius greater than the lattice parameter occur frequently in solids. Therefore, we now turn to deal with a quantitative treatment of the Wannier exciton.

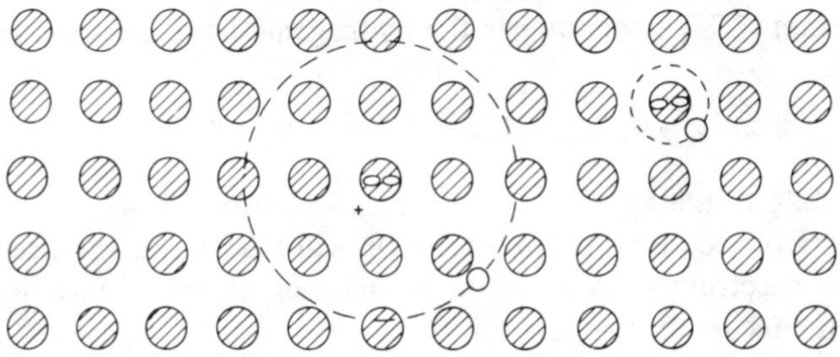

Figure 2.7 Shows large-radius (Wannier) exciton, where electron sees hole plus an average lattice potential. In contrast, in the small (Frenkel) exciton case, the detail of the lattice potential is important.

2.9.1 Effective mass equation for large-radius exciton

Though the detailed behaviour of the lattice potential is not important in this case, we must evidently attribute effective masses m_e^* and m_h^* to the electron and hole respectively, which reflect the structure of the conduction and valence bands in the crystal under consideration, as we briefly summarised previously. Wannier's idea can then be formulated in a useful approximate manner by neglecting particles other than the excited electron and the hole, except through these effective masses m_e^* and m_h^* and the effective interaction between electron and hole. This latter is taken as $-e^2/\epsilon r$, where ϵ is the dielectric constant.

The Schrödinger equation is therefore written for the two-particle problem of electron and hole

$$\left(-\frac{\hbar^2}{2m_e^*}\nabla_e^2-\frac{\hbar^2}{2m_h^*}\nabla_h^2-\frac{e^2}{\epsilon r_{eh}}\right)\psi = E\psi \tag{2.73}$$

where r_{eh} is evidently the electron–hole separation. But this is exactly the problem of the hydrogen atom, in which the mass of the proton is taken as finite. As shown in many introductory works on quantum mechanics (see e.g. Pauling and Wilson 1935, p. 113), the essential point is that we can now separate out the centre of mass motion.

We shall deal explicitly here with the case when the momentum $\boldsymbol{K}$ of the centre of mass is zero. This is of considerable interest already, because optical absorption from the ground state can create excitons near $\boldsymbol{K} = 0$. In this case, we have to solve the Schrödinger equation

$$\left(\frac{p^2}{2\mu} - \frac{e^2}{\epsilon r}\right)\psi(r) = E\psi(r) \tag{2.74}$$

where μ is evidently the reduced mass given by

$$\frac{1}{\mu} = \frac{1}{m_e^*} + \frac{1}{m_h^*}. \tag{2.75}$$

This equation (2.74) evidently has the hydrogenic level spectrum given by

$$E_n = -\frac{\mu e^4}{2\hbar^2 \epsilon^2 n^2} \qquad n = 1, 2, 3 \ldots \tag{2.76}$$

corresponding to the appropriate hydrogenic wavefunctions.

It is clear from equation (2.76) that for optical absorption, the direct exciton absorption spectrum is expected to be a series of lines below the optical absorption edge of the crystal.

Experimental results for Cu_2O are shown in figure 2.8, the principal quantum numbers associated with the hydrogenic spectrum being as discussed there. The $n = 1$ case is missing (i.e. has very weak intensity) since the band–band transition is forbidden. This is discussed by Gross (1962); it is outside the scope of this discussion to go into further detail.

We want to comment now on the dielectric constant to be used in the Schrödinger equation (2.74). The situation depends again on the electron–hole separation. If the electron and hole are close together, their internal kinetic energy is high. The other valence electrons, which can be assigned an effective resonant frequency E_g/h, E_g being the magnitude of the energy gap, cannot follow the internal motion of the pair and the dielectric constant can be put equal to unity. But as the exciton radius gets large, this situation no longer obtains and eventually even the ions can respond to the motion of the pair. Then $\epsilon \to \epsilon_\infty$, the high-frequency dielectric constant. The way in which the transition from $-e^2/r$ to $-e^2/\epsilon_\infty r$ takes place has been studied by Haken and Schottky (1958) and by Englert (1959).

2.9.2 Optical absorption as modified by electron–hole interaction

One way in which electron–hole interaction affects observable properties of

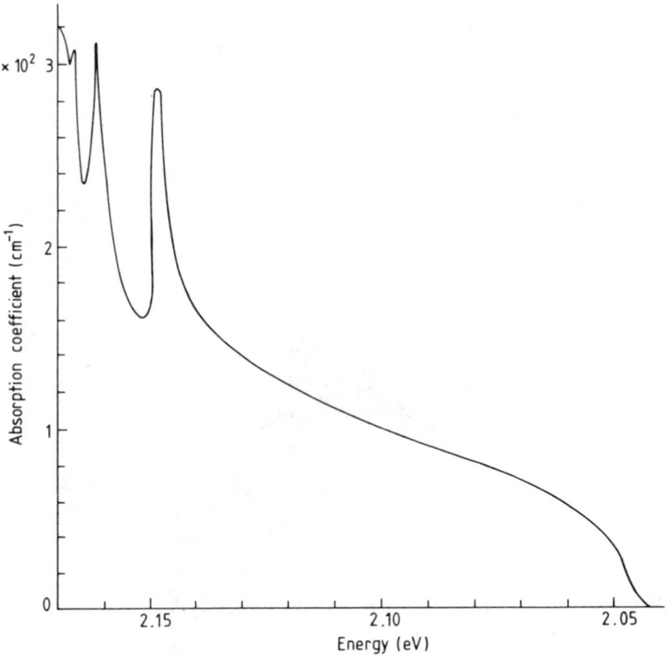

Figure 2.8 Experimentally observed exciton spectrum in Cu_2O. Principal quantum numbers n associated with 'hydrogenic' spectrum are 2 for peak at lowest energy and 3 for next peak.

insulating solids is in the optical absorption as already mentioned. We shall give a fairly detailed account of this below, both for continuous absorption and for transitions between discrete exciton states. The discussion follows that of McLean (1960) (see also Elliott and Gibson 1974).

It should be noted first of all that except at very low temperatures, exciton lines are not observed in materials of high dielectric constant and low effective mass. This is because the exciton binding energy given by equation (2.76) is very much less than the thermal energy $k_B T$ and the exciton levels cannot be distinguished from the conduction band.

Let us therefore briefly consider the absorption edge of an insulating solid in which the energy bands are assumed to be parabolic and the minimum conduction band energy occurs at the same $\boldsymbol{k}$ value as the maximum valence band energy (see figure 2.9). At first, electron–hole interaction will be neglected. Wavevector conservation allows only near-vertical transitions to take place between states separated in energy by

$$\hbar\omega = E_g + \frac{\hbar^2 k^2}{2m_h^*} + \frac{\hbar^2 k^2}{2m_e^*} \tag{2.77}$$

where m^* as usual denotes the appropriate effective mass of electrons (e) and holes (h), while E_g is the energy gap shown in figure 2.9. If the valence band is

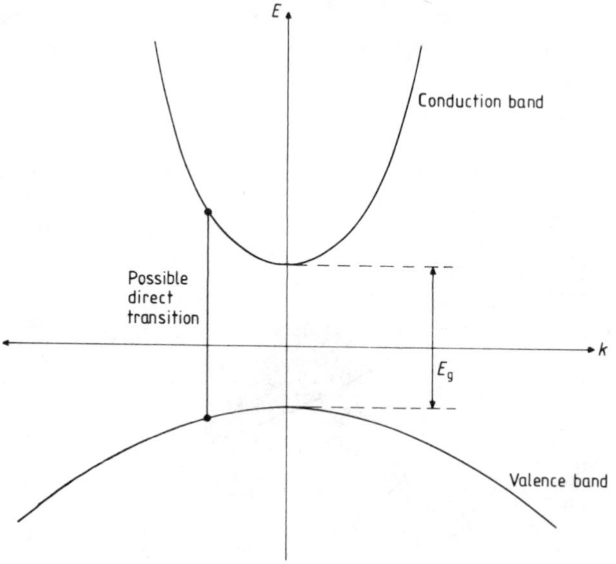

Figure 2.9 Insulator with parabolic bands and band edges at same point ($\boldsymbol{k} = 0$) in $\boldsymbol{k}$-space.

full, and the conduction band states are empty, and if the matrix element of the transition is assumed independent of $\boldsymbol{k}$, the absorption will simply be proportional to the density of states at $\boldsymbol{k}$. For parabolic bands, this varies as $E^{1/2}\,dE$, if E is the energy measured from the band edge. As a consequence, the absorption will be proportional to

$$(\hbar\omega - E_g)^{1/2} \qquad \text{for} \quad \hbar\omega > E_g \tag{2.78}$$

and zero otherwise. When one considers experimental results on the direct gap semiconductor InSb, in which the exciton binding energy is of the order of 10^{-3} eV and which shows therefore no exciton lines even at moderately low temperatures, the result (2.78) is found to hold over a substantial frequency range.

The above result is modified if the matrix element for the transition varies with $\boldsymbol{k}$. In particular, if conditions are such that the transition is forbidden near points of high symmetry like $\boldsymbol{k} = 0$, then the matrix element may increase linearly with $\boldsymbol{k}$ due to second-order effects. If so, the absorption varies as

$$(\hbar\omega - E_g)^{3/2} \qquad \text{(forbidden transitions).} \tag{2.79}$$

As discussed by McLean (1960), it is now quantitatively interesting to study the effect of electron–hole interaction on the above results (2.78) and (2.79). It turns out that in writing these equations, equal weight has been assigned to all the possible states of the electron–hole pair formed in the absorption process, regardless of the relative motion of the particles in these states. This is correct, but in reality, when the pair is formed the electron and hole must be at the same position in the crystal and so they will be preferentially formed into states in which they are most likely to be together. Thus a final state in which the wavevector of relative motion is $\boldsymbol{k}$ will be weighted by the probability, $|\phi(0)|^2$ say, that the electron and hole can be found at the same point in space in that state. If $\boldsymbol{K}$ is the total wavevector of the pair, $\boldsymbol{K} = 0$ in the wavefunction ϕ since electron–hole pairs can be formed only with a zero total wavevector in direct transitions. When the electron–hole interaction is neglected, $\phi(\boldsymbol{r})$ describes a plane wave so that all final states have equal weight. However, when the electron–hole interaction is taken into account, one can solve the appropriate Schrödinger equation (2.73) for $\phi(\boldsymbol{r})$ (see problem 2.5) to obtain (cf Elliott 1957 for the detailed theory) for the unbound solutions

$$|\phi(0)|^2 = \frac{\pi\alpha \exp(\pi\alpha)}{\sinh(\pi\alpha)} \tag{2.80}$$

where (compare equation (2.76) with $n = 1$)

$$\alpha = \left(\frac{2\mu E_\mathrm{b}}{\hbar^2 k^2}\right)^{1/2} \qquad E_\mathrm{b} = \frac{\mu e^4}{2\hbar^2 \epsilon^2}. \tag{2.81}$$

Such a weighting factor $|\phi(0)|^2$ must be introduced into the formula (2.78) for allowed transitions. The main features of the resulting formula, due to Elliott, when this weighting is properly accounted for are:

(a) When $\hbar\omega - E_\mathrm{g} \gg \pi^2 E_\mathrm{b}$, the energy put into the relative motion of the electron and the hole is much greater than the Coulomb binding energy, so that the effects of electron–hole interaction are negligible and the form (2.78) for allowed transitions is regained.

(b) When $\hbar\omega - E_\mathrm{g} \to 0$, the absorption coefficient now tends to a non-zero value, in contrast to equation (2.78), the absorption increasing as the strength of the electron–hole interaction increases.

As already referred to, over and above this continuous absorption due to allowed transitions, absorption can also be produced by transitions into bound exciton states. Although each exciton state has a band of energies associated with it, absorption by direct transitions can excite only one discrete state in each band, namely the state with zero wavevector. We shall therefore have a series of discrete absorption lines at energies E_b/n^2, $n = 1, 2, 3, \ldots$, below the continuous absorption threshold, i.e. below the energy E_g normally regarded as the threshold for direct transitions. Absorption into state n will be proportional again to $|\phi(0)|^2$ which is clearly non-zero only for s-states of the exciton and then it can be shown to take the value $(\mu e^2/\hbar^2\epsilon)^3/\pi n^3$ (see problem 2.5). The absorption into the exciton states will therefore fall in magnitude as n^{-3} and will eventually merge into the continuous absorption discussed in some detail previously.

To illustrate some of the above points, in figure 2.10 we show an example of GaSe, which is a direct gap semiconductor. The absorption is shown at two temperatures: at the higher temperature $E_b \sim k_B T$, but at the lower temperature two Wannier-type excitons can be seen. As was anticipated in the above discussion, one expects a smooth curve following some power law, with the exciton peaks at the lower end, in general accord with the experimental results.

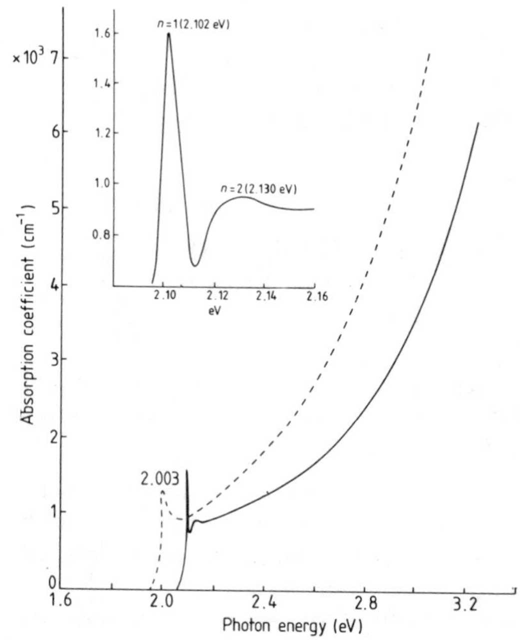

Figure 2.10 Optical absorption in direct gap semiconductor GaSe: broken curve, room temperature; full curve, 77 K (redrawn from Elliott and Gibson (1974)).

In summarising this discussion of excitons, let us re-emphasise first some common features of the limiting cases of Frenkel (localised) and Wannier (large-radius) excitons. These are:

(a) the energy states form quasi-continuous bands, described by dispersion relations $E(\boldsymbol{k})$, where $\boldsymbol{k}$ is the wavevector;
(b) related to (a), the exciton is distributed throughout the entire crystal, when its stationary states are considered;
(c) properties of the excitons depend in an important way on the crystal structure.

These common characteristics of exciton states are central in determining the properties of excitons in interaction with light, with the vibrations of the atomic or molecular lattice (see Chapter 3), etc.

2.9.3 *Intermediately bound excitons*

We must mention that more recent work has been concerned with intermediately bound excitons. Here the direct electron–hole interaction is still large (as in Frenkel excitons, the screening being weak) but the dispersion of the electron (and/or hole) bands is important, as in Wannier excitons. As for the latter point (cf §2.9.1 in the case of Wannier excitons), both the electron and the hole wavefunctions are considered as being localised in $\boldsymbol{k}$-space and so, only a particular point (say $\boldsymbol{k}_0$) in the Brillouin zone is considered. Therefore the band dispersion which is taken into account is limited to the neighbourhood of $\boldsymbol{k}_0$ (through the appropriate effective mass). Now the opposite limit of narrow bands would correspond to Frenkel excitons. The intermediately bound exciton lies between the above two limits.

In this connection, we should cite the work of Andreoni *et al* (1975). There, the appropriate integral equation for the exciton is formulated and solved in detail for the lowest exciton state. These workers calculate Wannier functions and explicitly take into account the dispersion of the conduction band. In this work, which is subsequent to earlier studies of Takeuti (1957; see also Altarelli and Bassani 1971), the relative importance of the different contributions to the binding energy is discussed. These approaches can be used to give a quantitatively correct description of excitons in rare gas solids. Intermediately bound exciton theory is also appropriate for the alkali halides. Problems of the hole relaxation, and of the appropriate screening to use, remain in this area. We note that Egri (1979) has presented a fairly simple model by which excitons of arbitrary binding and localisation can be studied and the interested reader is referred to his paper also.

2.9.4 *Indirect excitons*

Previously we have explicitly dealt with excitons of total momentum $\boldsymbol{k} = 0$. It is however possible to create an exciton of non-zero wavevector $\boldsymbol{k}$ by binding a hole at $\boldsymbol{k} = 0$ to an electron excited in a minimum of the conduction band at a point $\boldsymbol{k}_{\mathrm{m}}$ in the Brillouin zone (see figure 2.11). The corresponding bound state is referred to as an indirect exciton.

The lifetime of such an indirect exciton is usually much longer than that of a normal exciton. In fact, the indirect exciton cannot decay by direct emission of a photon because such a transition would not conserve momentum, the momentum $\hbar\boldsymbol{k}$ of a photon of the required energy being very much less than $\hbar\boldsymbol{k}_{\mathrm{m}}$. Thus the radiative recombination of an indirect exciton is only possible if, somehow, a momentum $-\hbar\boldsymbol{k}_{\mathrm{m}}$ is simultaneously carried away. This is possible via the vibrating lattice, the modes and the quasi-particles associated with these vibrations being fully dealt with in the next chapter. It turns out that the

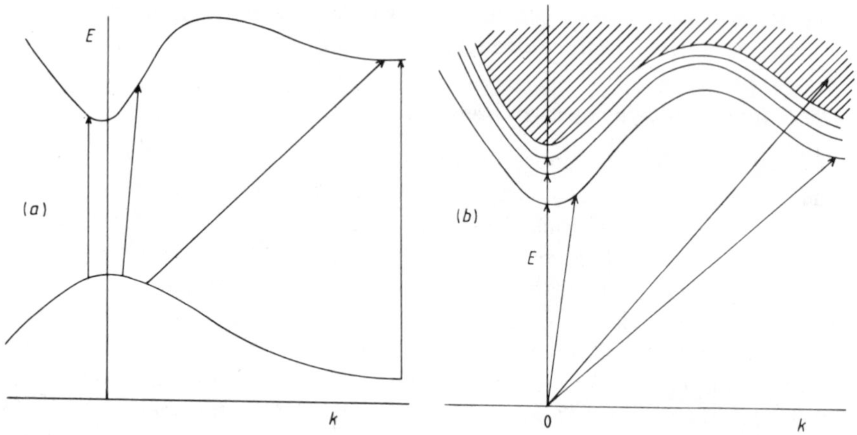

Figure 2.11 (a) Energy-band plot of one-particle states: vertical lines, direct transitions; diagonal lines, indirect transitions. (b) Exciton bands, ground state is at origin: vertical lines on axis, direct transitions; diagonal lines, indirect.

probability of such a process in which, via the lattice, the required momentum is carried away is relatively low.

This suggests that it may be feasible to observe states in which many excitons are created simultaneously, and this turns out to be the case. Such a situation is usually achieved by illuminating the crystal with laser radiation of appropriate wavelength and intensity in such a manner that a steady-state situation is reached where in unit time the number of excitons created is equal to the number that decay by radiative recombination. This has led to a good deal of interesting work and indeed many different possibilities have been explored. We will restrict ourselves to two regimes where the theory can be worked out.

2.10 Low-density regime: molecular state

In this regime, two excitons can interact and are expected to form a bound state of molecular character, rather like an H_2 molecule. However, the following differences need stressing:

(a) The ratio of electron mass to proton mass is a factor of about 2000 smaller than that between electrons and holes in a semiconductor, where $m_e^*/m_h^* \sim 1$. This fact obviously increases the kinetic energy of the bi-excitonic molecule, thereby lowering the dissociation energy.
(b) The underlying complexities of the band structure, the symmetries of which affect those of the bi-excitonic ground state and modify the energy level fine structure, are an intrinsic feature of the electron–hole problem.

2.11 High-density regime: electron–hole liquid

More exciting are the problems opened up in the high-density regime. Here, the excitonic wavefunctions can overlap so strongly that the system is expected to undergo a phase transition. Though we shall discuss this topic generally in Chapter 8, it is important to note here for this specific problem that this can be a transition either to an ordered state (excitonic crystal) or to a liquid state with, naturally, only short-range order. However, the light masses of the particles involved make the conditions for the crystal to form very difficult to satisfy. Some unexpected observations of a giant diamagnetic effect (cf the Meissner effect in Chapter 5) in CuCl have tentatively been interpreted as resulting from the formation of an excitonic crystal (see Abrikosov 1978).

On the other hand, overwhelming evidence exists for the occurrence of a metallic electron–hole liquid. Such a situation can obtain when large numbers of electrons and holes are created by optical pumping in a semiconductor at low temperature. In this latter situation, essentially no free carriers would normally be present because of the existence of the energy gap separating the occupied valence bands from the unoccupied conduction bands. By illuminating with light of appropriate frequency, electrons can be excited across this energy gap into the conduction band, leaving behind holes in the valence band.

These electrons and holes rapidly thermalise, so that their energies correspond to those of the edges of the conduction and valence bands. But in the semiconductors Ge and Si, these edges are at different wavevectors so that the recombination rate is slow. Thus a high density of thermalised electrons and holes can be established at low temperatures. To a first approximation, in which all band structure complications are, for the moment, put aside, one can consider that the fluid of electrons and holes is composed of two spatially interpenetrating Fermi liquids, with particles of masses m_{e}^* and m_{h}^*, interacting through the Coulomb potential $e^2/\epsilon r$, with ϵ the static dielectric constant. As already remarked, compelling experimental evidence exists to demonstrate that the electrons and holes condense to form a high-density metallic liquid. Here we have then a type of metal in which both electrons and holes are itinerant, and we must now discuss how to set up the theory of such a metallic state.

2.11.1 Hamiltonian for an electron–hole liquid

We require the Hamiltonian to describe this interacting assembly of electrons and holes, with effective masses determined by local curvature of the energy bands close to the band edges (cf §2.8.2). Since the particles are moving in Ge or Si say, the dielectric constant must enter the Hamiltonian as mentioned above, and we have

$$H = -\sum_{\mathrm{e}}\frac{\hbar^2\nabla_{\mathrm{e}}^2}{2m_{\mathrm{e}}^*} - \sum_{\mathrm{h}}\frac{\hbar^2\nabla_{\mathrm{h}}^2}{2m_{\mathrm{h}}^*} - \sum\frac{e^2}{\epsilon|\boldsymbol{r}_{\mathrm{e}}-\boldsymbol{r}_{\mathrm{h}}|} + \sum_{i<j}\frac{e^2}{\epsilon r_{ij}}. \tag{2.82}$$

The first two terms are obviously the kinetic energies of electrons and holes respectively, the third term represents the Coulombic attraction between electrons and holes, while the final term represents the Coulomb repulsion between like particles separated by distance r_{ij}.

Of course, as we have emphasised above, a single electron and hole will bind to form a simple hydrogenic exciton with energy (and radius) given by the usual Rydberg form (2.76). In Ge, for example, the binding energy of the exciton is 4.15 meV, while the exciton radius is 133 Å, while in Si the energy is 14.7 me V and the radius is 44 Å. We must emphasise (cf § 2.8.2) that it is the smallness of the energy scale, and the correspondingly large exciton radii, which makes the effective mass description used in the Hamiltonian (2.82) for the kinetic and potential energies quantitatively useful.

Clearly, the formation of such a metallic electron–hole liquid will be favourable whenever the energy of the metallic state is less than that of the phase consisting of a gas of excitons. As we have seen, the latter energy is of the order of $-\mu e^4/2\hbar^2\epsilon^2$ per electron–hole pair, μ being the reduced mass, which must now be compared with the energy of the metallic liquid.

The kinetic energy (KE) per particle for a free Fermi gas is simply (cf Appendix 2.1)

$$\text{KE per particle} = \tfrac{3}{5} E_{\text{F}} \tag{2.83}$$

while E_{F} is related to the number of particles n per unit volume by

$$E_{\text{F}} = \frac{\hbar^2 (3\pi^2)^{2/3}}{2m} n^{2/3}. \tag{2.84}$$

Hence for the idealised system under consideration, the kinetic energy per electron–hole pair is obtained by adding the electron and the hole contributions to yield (putting $3/\pi \sim 1$)

$$\begin{aligned} \text{KE per electron–hole pair} &\simeq \frac{\hbar^2\pi^2}{m_{\text{e}}^*} n^{2/3} + \frac{\hbar^2\pi^2}{m_{\text{h}}^*} n^{2/3} \\ &= 2\left(\frac{\hbar^2\pi^2}{2\mu}\right) n^{2/3} \end{aligned} \tag{2.85}$$

where n is now the number of electron–hole pairs per unit volume.

The potential energy V can be estimated simply to be of the order $-e^2/\epsilon r$, where r is some average electron–hole distance. Taking $r \sim n^{-1/3}$ we find

$$V \simeq \frac{-e^2}{\epsilon} n^{1/3}. \tag{2.86}$$

Putting kinetic and potential energies together, we obtain the following estimate of the total energy E:

$$E \simeq 2\left(\frac{\hbar^2\pi^2}{2\mu}\right) n^{2/3} - \frac{e^2}{\epsilon} n^{1/3}. \tag{2.87}$$

Equation (2.87) now has to be minimised with respect to n in order to find the equilibrium density. The corresponding value for E is found to be

$$E \simeq -0.1\,\mu e^4/2\hbar^2\epsilon^2 \tag{2.88}$$

which is higher than the excitonic binding energy.

On the basis of this argument alone, one would therefore have to conclude that the electron–hole liquid is not energetically favourable. However, different calculations have shown that, if the band structure anisotropies are properly taken into account, in cases such as Ge or InSb the positive kinetic energy contribution in equation (2.87) is greatly reduced and the metallic liquid state is energetically favoured.

This has been amply confirmed experimentally and the situation found is that there exists a critical temperature T_c (of the order of a few degrees kelvin) below which a gas of excitons undergoes a phase transition and condenses into tiny metallic droplets in equilibrium with a vapour of excitons. This can be likened to water droplets in equilibrium with water vapour on a foggy day.

To show the kind of agreement that can be obtained between theory and experiment, we give results taken from Haken and Nikitine (1975) in table 2.2. In view of the complexities referred to above, the outcome is seen to be rather satisfactory. For further details on this intriguing area the reader is referred to reviews of the theory by Rice (1977, 1979), experiment by Hensel *et al* (1977) and to other references given in these articles.

Table 2.2 Comparison between ground-state calculations and experiment for properties of electron–hole droplets in Ge and Si. The source of the experimental values is given by Rice (1977) and Hensel *et al* (1977).

Property	Ge	Si	Reference
Energy (meV)			
theory	1.7	5.7	Brinkman *et al* (1972)
	2.5	6.3	Combescot and Nozières (1972)
	1.8	7.3	Vashista, Das and Singwi (1974)
experiment	1.5–2.0	8.2	
Density (cm^{-3})			
theory	1.8×10^{17}	3.4×10^{18}	Brinkman *et al* (1972)
	2.0×10^{17}	3.1×10^{18}	Combescot and Nozières (1972)
	2.2×10^{17}	3.2×10^{18}	Vashista, Das and Singwi (1974)
experiment	2.4×10^{17}	3.7×10^{18}	

2.12 Coupling between plasmons and excitons

Though the concept of plasma oscillation is most easily understood when applied to metals, the idea of a collective motion of the electronic charges is also applicable to insulating materials. In this case, however, one expects the plasmon to be substantially broadened because the strong interaction of the electron with ions tends to prevent free charge oscillation.

In spite of this, features that can be associated with the existence of plasmons have been identified even in the spectra of rare gas crystals. As has been pointed out by Giaquinta *et al* (1976), the simultaneous presence in these systems of plasmons and excitons brings in the possibility of interaction between these two modes. This interaction is due to the fact that the plasma oscillation has associated with it a macroscopic electric field $\mathscr{E}$ (cf §2.1) while the exciton (in rare gas crystals actually an intermediate exciton) can be approximated roughly as an electric dipole. Neglecting damping effects, this problem can be treated as that of two coupled oscillators. If the unperturbed frequencies are ω_P and ω_{ex} and the coupling constant is λ, it can be shown that the resulting frequencies are given by

$$\omega_{\pm}^2 = \tfrac{1}{2}\{(\omega_P^2 + \omega_{ex}^2) \pm [(\omega_P^2 - \omega_{ex}^2)^2 + 4\lambda^2]^{1/2}\}. \qquad (2.89)$$

Estimates can be made for the unperturbed frequencies ω_P and ω_{ex} of plasmon and exciton modes, and for the coupling constant λ at $\boldsymbol{q} = 0$. Given the crudeness of the model, the resulting agreement with optical experiments carried out on the rare gas crystals is rather good. This example well illustrates the usefulness of the collective mode approach, as a rather complicated problem has been reduced to the manageable and familiar form of two coupled oscillators.

2.13 Excitons in metals

We have argued previously that usually, in a metal, once the plasmon has been accounted for, the residual Coulomb interaction is screened and of short range. There are circumstances, however, in which this residual interaction must be incorporated.

A striking example is provided by the so called edge singularities. These occur when an x-ray is emitted or absorbed by a metal. Let us consider the emission process to be definite. This occurs when an electron is ejected from one of the inner atomic shells. The emission process is due to an electron falling from the conduction band to the empty core level.

In the absence of any interaction, the lowest frequency of the x-rays emitted in this process would be equal to the energy difference between the bottom of the conduction band and the core level, divided by Planck's constant. However, an empty core level can be considered as a positive hole and therefore its positive charge will strongly affect the state of the conduction electrons. In order to understand the processes involved, it will be helpful to follow the effect of the electron–positive hole interaction as the density of conduction electrons is continuously increased from zero to its actual value in the metal in question.

If the number of conduction electrons is very small, then the situation is similar to that occurring in semiconductors where a bound excitonic state is formed (see § 2.9 especially). The only qualitative difference is the fact that the effective mass of the core hole can be taken as infinite in the present discussion. In fact, once a core hole is created on one atomic site, it has a very small probability of hopping to a neighbouring atom, since core wavefunctions centred on different atomic sites have a vanishingly small overlap. But as the conduction electron density is increased, we must expect that screening discussed above will progressively reduce the effective electron–hole interaction. This leads to a smaller binding energy and a shorter lifetime of the excitonic complex, which in turn will be reflected in a broader exciton emission line. This would suggest that excitonic lines could not be observed in metals. However, there is another effect which counteracts this broadening of the excitonic line, as was first pointed out by Mahan (1967). Due to the very small mobility of the core hole, the net effect of the creation of a hole can be likened to the sudden switching on of a localised perturbation in the electron gas. The perturbation can be represented by a potential which is the difference in atomic potential before and after the hole is created.

The switching on of this localised perturbation causes a sudden change in the electronic wavefunction from $\Psi(\boldsymbol{r}_1, \ldots, \boldsymbol{r}_N)$ to say $\Psi'(\boldsymbol{r}_1, \ldots, \boldsymbol{r}_N)$ as a result of the conduction electrons rushing in to screen out the field of the positive hole. This transient phenomenon has a singular character as a result of the Pauli exclusion principle. In fact, in a Fermi system, it is possible to create electron–hole pairs of very low energy and large momentum by transferring one electron from just below to just above the Fermi surface. Since the hole with its infinite effective mass can absorb any momentum, it can excite large numbers of electron–hole pairs at zero cost in energy. This can lead to a singular behaviour of the x-ray emission/absorption as was first calculated quantitatively by Nozières and de Dominicis (1969). The details of their calculation are too advanced to present here, but their prediction for the above singular behaviour is that the form of the absorption spectrum in the neighbourhood of the absorption edge is (Mahan 1974)

$$A(\omega) \simeq f(\omega) \frac{1}{(\omega - \omega_{\mathrm{E}})^{\alpha_l}}, \tag{2.90}$$

where $f(\omega)$ is a smooth function of ω, the frequency at the edge is ω_{E}, while α_l is†

† Subscript l refers to the orbital angular momentum of the conduction electron which is either excited or de-excited in absorption or emission.

determined from the changes that the hole causes in the wavefunctions of the conduction electrons at the Fermi energy. The transition from the excitonic regime (cf §2.8) to the situation described by Nozières and de Dominicis is illustrated in figure 2.12.

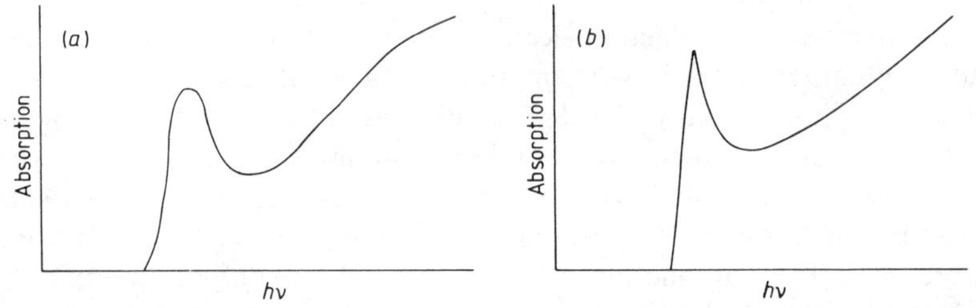

Figure 2.12 (a) Exciton absorption line with ineffective carrier screening; (b) line as broadened in a normal metal.

It is convenient to take two contrasting examples to illustrate the above considerations: the L_2 emission from Na and the K emission from Li. They show very different behaviour which can best be seen by considering the explicit expression for α_l in terms of the phase shifts δ_l of the conduction electrons at the Fermi energy due to scattering off the hole potential. From conventional scattering theory, this merely means that if we expand a plane wave $\exp(\mathrm{i}\boldsymbol{k}\cdot\boldsymbol{r})$ in spherical waves, by Bauer's expansion

$$\exp(\mathrm{i}\boldsymbol{k}\cdot\boldsymbol{r}) = \sum_l (2l+1)\mathrm{i}^l j_l(kr) P_l(\cos\theta), \tag{2.91}$$

where j_l and P_l are spherical Bessel functions and Legendre polynomials respectively, then the asymptotic form of the radial wavefunction for a given l is changed from $(kr)^{-1}\sin(kr - \frac{1}{2}l\pi)$ to $(kr)^{-1}\sin(kr - \frac{1}{2}l\pi + \delta_l)$ by the spherical scattering potential. To ensure that the hole is properly screened, these phase shifts δ_l at the Fermi level must satisfy the Friedel sum rule set out in problem 2.7. In terms of these phase shifts at the Fermi level it can be shown that†

$$\alpha_l = \frac{2\delta_l}{\pi} - \sum_{l'} 2(2l'+1)\frac{\delta_{l'}^2}{\pi^2}. \tag{2.92}$$

Now let us take the case of the L_2 line from Na. Here one is observing the transition to a 2p core state, and therefore because of the dipole selection rule only s-like electrons in the Na conduction band will be involved in the transition. Thus it follows from equation (2.92) that we have $l = 0$. In the case of simple metals like Na, δ_0 is expected to be the largest phase shift and therefore from equation (2.92) we have that $\alpha_0 > 0$ which, from equation (2.90), gives a spike at the edge $\omega \sim \omega_E$. However, the K emission from Li is due to transitions from p-like conduction band states into the empty 1s core

† The summation in equation (2.92) is often referred to as the Anderson (1967) singularity index. It is observable directly in x-ray photoemission (see below).

level. Hence we must take $l = 1$ in equation (2.92) and, since δ_1 is small, α_1 can be negative, leading through equation (2.90) to the disappearance of the spike and a diminution of the spectrum near the threshold, as is indeed observed in the experimental soft x-ray spectrum from Li metal (figure 2.13). Both predictions are confirmed by the experiments which allow one to extract $\alpha_0 = 0.37$ for the L_2 emission from Na and $\alpha_1 = -0.1$ for the K emission from Li†.

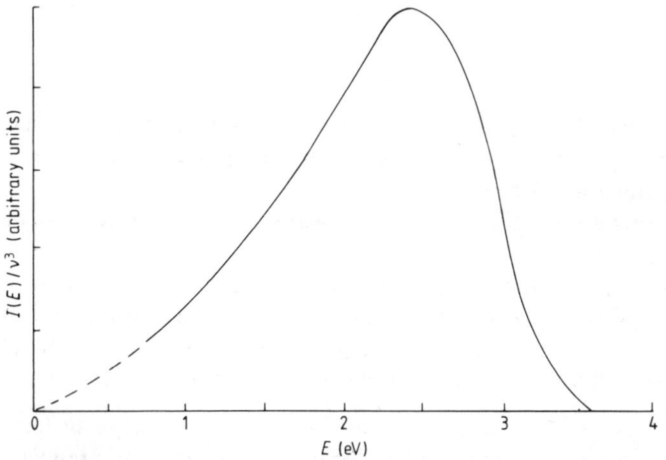

Figure 2.13 Observed soft x-ray emission intensity for metallic Li.

It is of interest to add here that an effect similar in character to the Mahan–Nozières–de Dominicis phenomenon has been predicted to occur also in photoemission (Doniach and Šunjić 1970), where it leads to asymmetric

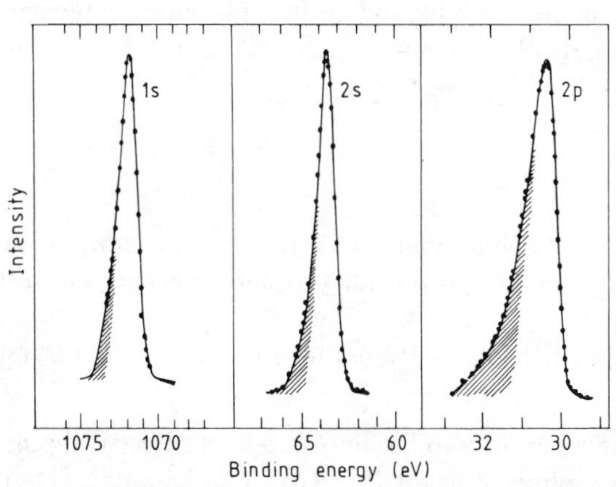

Figure 2.14 X-ray photoemission lineshapes for Na, for 1s, 2s and 2p core states. Note especially the asymmetry of lines shown by shading (From Citrin P H 1973 *Phys. Rev.* B **8** 5545.)

† See, however, Citrin *et al* (1979) who propose $\alpha \doteqdot 0$ for Li.

lineshapes. An example of this type of effect is shown in figure 2.14 which is taken from the work of Citrin (1973).

Finally in this connection we wish to point out that considerable impetus to the study of this area of collective behaviour has been afforded by the advent of synchrotron radiation which provides a very convenient source of x-ray radiation.

Problems

2.1 Estimate the frequency of plasma oscillations in bulk Al, given that the radius of a sphere containing one valence electron is $r_s = 2\,a_0$, a_0 being the first Bohr radius for hydrogen (0.529 Å).

How long does it take for a charged disturbance, introduced into bulk Al, to be screened out?

If one creates a hole in the K shell of an Al atom in Al metal, how long must the hole survive for its field to be shielded out by the metal valence electrons? Will such screening actually occur in soft x-ray emission from Al? Which electrons in the Al valence band will contribute to such emission?

2.2 In connection with the discussion of surface energy in terms of the change in zero-point energy of collective oscillations, you are given:

(a) The corresponding surface energy is

$$\frac{\hbar\omega_p k_c^2}{8\pi}\left(\frac{\omega_s}{\omega_p}-\frac{1}{2}\right)$$

where ω_p and ω_s are the bulk and surface plasmon frequencies respectively, while k_c is the cut-off wavenumber referred to in the discussion in §2.6.5.

(b) The cut-off wavenumber is

$$k_c = (\omega_p/\sqrt{2}v_F) \qquad \text{(P2.1)}$$

where v_F is the Fermi velocity.

Use (a) and (b) to show that the surface energy is proportional to $r_s^{-5/2}$. Calculate the constant of proportionality and compare your result with figure 2.6.

Can you explain (b), apart from a factor of the order of unity, in physical terms?

2.3 Assuming an exciton to be simulated by an electric dipole of moment μ calculate the coupling constant λ appearing in equation (2.89).

2.4 Find the density of states $N(E)$ near the maximum and the minimum energies of a band whose dispersion relation $E(\boldsymbol{k})$ is given by

$$E(\boldsymbol{k}) = -A^2[\cos(k_x a)+\cos(k_y a)+\cos(k_z a)]. \qquad \text{(P2.2)}$$

2.5 From the exact solutions of the hydrogenic problem in equation (2.74) demonstrate that for the bound solution with s symmetry and principle quantum number n the square of the amplitude of the normalised wavefunction at the origin is given by

$$|\phi(0)|^2 = \frac{(\mu e^2/\hbar^2 \epsilon)^3}{\pi n^3} \tag{P2.3}$$

while for the unbound states equation (2.80) holds. (This latter part requires continuum solutions of the hydrogenic equation such as are given by Landau and Lifshitz (1962a).)

2.6 Given that the band structure is ellipsoidal rather than spherical, estimate how the kinetic energy of the electron–hole liquid discussed in §2.11 would be affected.

[*Hint.* Adopt a form for the energy band structure of the conduction electrons which is given by

$$E(\boldsymbol{k}) = \frac{\hbar^2}{2m}[k_x^2 + \beta(k_y^2 + k_z^2)], \tag{P2.4}$$

β being a measure of the anisotropy of the band. Assuming the validity of the free-electron approximation, calculate the total kinetic energy. Focus on the way the kinetic energy varies as a function of β.]

2.7 Suppose the scattering of conduction electrons at the Fermi energy $\frac{1}{2}k_{\mathrm{F}}^2$ off a core hole is described by a change in the asymptotic form of the radial wavefunction for orbital angular momentum l, from the free-electron form $(k_{\mathrm{F}}r)^{-1}\sin(k_{\mathrm{F}}r - \frac{1}{2}l\pi)$ to $(k_{\mathrm{F}}r)^{-1}\sin(k_{\mathrm{F}}r + \delta_l - \frac{1}{2}l\pi)$ where δ_l is evidently the phase shift of this partial wave of angular momentum l. Then Nozières and de Dominicis (1969) have shown that the singularity index α_l in equation (2.90) is related to the phase shifts by equation (2.92). Now the requirement that the hole is perfectly screened by the metallic electrons relates the phase shifts at the Fermi level by the so called Friedel sum rule

$$\sum_{j=0}^{\infty} 2(2j+1)(\delta_j/\pi) = 1 \tag{P2.5}$$

(*Note.* If a charge z had been shielded, rather than a charge unity, the right-hand side would be z instead of 1).

Assume that the phase shifts are all zero except for the s(α_0) and p(α_1) phase shifts. Plot the threshold exponents α_0, α_1 and $\alpha_{l>1}$ as functions of the p-wave phase shift (cf March 1973, Dow 1973, Citrin, Wertheim and Baer 1975).

2.8 Separate out the centre-of-mass motion in equation (2.73) by writing $\psi(\boldsymbol{R},\boldsymbol{r}) = \exp(\mathrm{i}\boldsymbol{K}\cdot\boldsymbol{R})f(\boldsymbol{r})$ where $\boldsymbol{R} = \frac{1}{2}(\boldsymbol{r}_{\mathrm{e}} + \boldsymbol{r}_{\mathrm{h}})$ and $\boldsymbol{r} = \boldsymbol{r}_{\mathrm{e}} - \boldsymbol{r}_{\mathrm{h}}$ represent centre-of-mass of pair and electron–hole separation. Hence show that for each value of $\boldsymbol{K}$ there are allowed energies

$$E_n(\boldsymbol{K}) = \frac{-\mu e^4}{2\hbar^2\epsilon^2 n^2} + \frac{\hbar^2 K^2}{2(m_{\mathrm{e}}^* + m_{\mathrm{h}}^*)} \tag{P2.6}$$

3

PHONONS IN SOLIDS

In Chapter 2 we treated the organised oscillations that can occur in the conduction electron charge cloud in a metal. There the positive ion distribution was modelled according to Sommerfeld, by smearing out the positive charge into a uniform neutralising background. Of course, in real solids, the discrete ions vibrate about equilibrium positions and the aim in this chapter is to discuss the collective modes of these vibrations. We need in no way restrict the discussion to metals; even though the force laws operating between ions in metals (see problem 3.7) and atoms in insulators are very different in nature, the collective modes of vibration of a crystal lattice still have general features which are common to metals, semiconductors and insulators. Naturally, fully quantitative work must incorporate the force laws specific to a chosen material (say the d band metal Fe or the covalently bonded semiconductor Si).

While discussing, for the first time, the theory of the lattice specific heat of crystals many years ago, Einstein adopted as a starting point a model in which every atom on a monatomic crystal lattice vibrated quite independently of the other atoms. It is then a simple matter to calculate the internal energy of the crystal. Thus, the mean energy of a linear harmonic oscillator of frequency ν say, in an assembly of independent oscillators in thermal equilibrium at temperature T is given by Planck's formula (omitting zero-point energy):

$$\text{average energy per oscillator} = \frac{h\nu}{\exp(h\nu/k_B T) - 1}. \tag{3.1}$$

In the high-temperature limit when $k_B T \gg h\nu$, it is clear that $\exp(h\nu/k_B T) - 1 \sim h\nu/k_B T$ and we regain the equipartition of energy result that $E = k_B T$, half of this energy being kinetic and half potential, i.e. $\frac{1}{2}k_B T$ energy/degree of freedom†.

Now, for the internal energy E of the crystal of N atoms (i.e. $3N$ linear oscillators) at temperature T, we have

$$E = \frac{3Nh\nu}{\exp(h\nu/k_B T) - 1}. \tag{3.2}$$

Obviously from the first law of thermodynamics

$$dQ = dE + p\,dV, \tag{3.3}$$

† If we define the number of degrees of freedom as the number of squared terms in the expression for the energy, then we have two degrees of freedom for a harmonic oscillator, one from the kinetic term and the other from the potential energy.

where dQ is the heat absorbed by the system, dE is the increase in internal energy, and $p\,dV$ is the work done by the system against an external pressure p. But

$$c_V = \left(\frac{dQ}{dT}\right)_V = \left(\frac{\partial E}{\partial T}\right)_V \tag{3.4}$$

and hence we can calculate the specific heat at constant volume (c_V) by differentiating equation (3.2) with respect to T. The result is correct at high temperature (since $E \rightarrow 3Nk_BT$ we regain the Dulong and Petit law that $c_V \rightarrow 3Nk_B$) but at low temperatures the behaviour is in marked disagreement with experiment giving an exponential decrease instead of a T^3 form. Thus, the Einstein model of independent oscillators cannot be correct for a vibrating crystal lattice at low temperatures.

3.1 Debye model

Debye proposed therefore an alternative model in which he argued that, because of the coupling together of the atoms in the crystal, a better picture would be to liken the crystal to an elastic continuum, and to study the possible modes of vibration of such a medium.

It is a straightforward matter to calculate in such a medium the number of modes $g(\nu)\,d\nu$ (often referred to as the density of states) having frequencies in the range between ν and $\nu + d\nu$, the result being (Ashcroft and Mermin 1976)

$$g(\nu)d\nu = A\nu^2 d\nu. \tag{3.5}$$

Now in a crystal structure with N atoms, the total number of frequencies must be $3N$ and hence

$$\int_{\substack{\text{all allowed}\\ \text{frequencies}}} g(\nu)d\nu = 3N. \tag{3.6}$$

In equation (3.6) we have written the integral over all allowed frequencies, for it is clear that the integral of $g(\nu)$ in equation (3.5) would diverge to infinity if we allowed the integral in (3.6) to extend to infinitely high frequencies.

Debye assumed therefore that equation (3.5) should be used only out to a maximum frequency, now called the Debye frequency, ν_D. Thus, in Debye's model, modes of allowed vibrations exist in the frequency range 0 to ν_D, there being, by assumption, no modes of higher frequency.

Inserting equation (3.5) into (3.6) we obtain immediately

$$A\nu_D^3/3 = 3N \quad \text{or} \quad A = 9N/\nu_D^3. \tag{3.7}$$

At this stage, it is an elementary matter to calculate the internal energy E from the Debye model using equations (3.1), (3.5) and (3.7) as

$$E = \frac{9Nh}{\nu_D^3}\int_0^{\nu_D} \frac{\nu^3 \mathrm{d}\nu}{\exp(h\nu/k_B T) - 1}. \tag{3.8}$$

This integral can be evaluated from tables, but we will focus immediately on the low temperature specific heat which can be derived analytically. At low temperatures $e^{h\nu/k_B T} - 1 \sim e^{h\nu/k_B T}$ and we find

$$E \sim \frac{9Nh}{\nu_D^3}\int_0^{\nu_D} \nu^3 \exp\left(-\frac{h\nu}{k_B T}\right)\mathrm{d}\nu. \tag{3.9}$$

Putting $x = h\nu/k_B T$, we see that

$$E \sim \frac{9Nh}{\nu_D^3}\int_0^{x_D}\left(\frac{k_B T}{h}\right)^4 x^3 \exp(-x)\,\mathrm{d}x$$

$$\sim T^4 \qquad \text{for small } T. \tag{3.10}$$

Hence $\mathrm{d}E/\mathrm{d}T = c_V \propto T^3$; this is the famous Debye T^3 law, which is in agreement with experiment at low temperatures†.

This is a truly major change from the Einstein model of independent atomic motions: the observed T^3 law, confirming the low-temperature behaviour of the Debye model, is a direct manifestation of the collective character of atomic vibrations of a crystal lattice.

If we use equation (3.8) for all temperatures however, we do not get complete agreement with experiment. The reason is that (3.5) is only valid for frequencies which correspond to wavelengths that are long compared with the lattice spacing. Or put another way, the theory so far presented neglects the detailed atomic structure of the crystal.

3.2 Dispersion of lattice waves

Before going on to discuss dispersion, let us dwell for a moment on the problem of two coupled oscillators.

Suppose we have two identical oscillators with equal natural frequencies ν. Now we couple them together and discuss the new vibrational modes. We find two new frequencies, as shown schematically in figure 3.1. If we now bring up a third oscillator, then we get three new coupled modes. As we bring up more and more oscillators, the frequencies become denser and we eventually find a band of allowed frequencies. We turn now to discuss this quantitatively for a linear chain of spacing a, with equal masses m, and with Hooke's law forces between near-neighbour atoms. In this simple case the general atom l (see figure 3.2)

† The reservation must be made that this is the low-temperature contribution from the lattice vibrations. In metals the electronic specific heat is proportional to T and at sufficiently low temperatures will clearly dominate the lattice contribution.

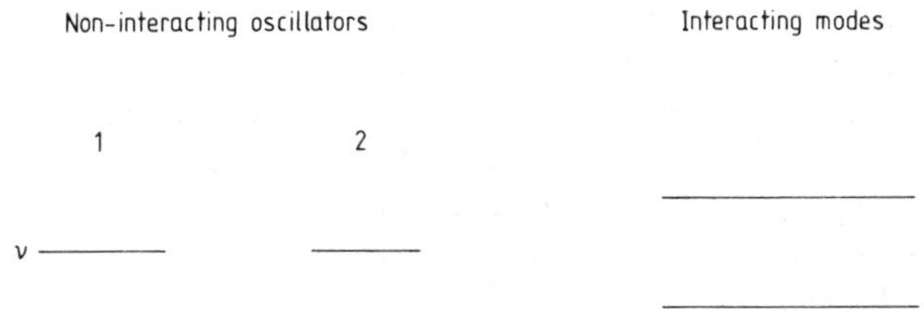

Figure 3.1

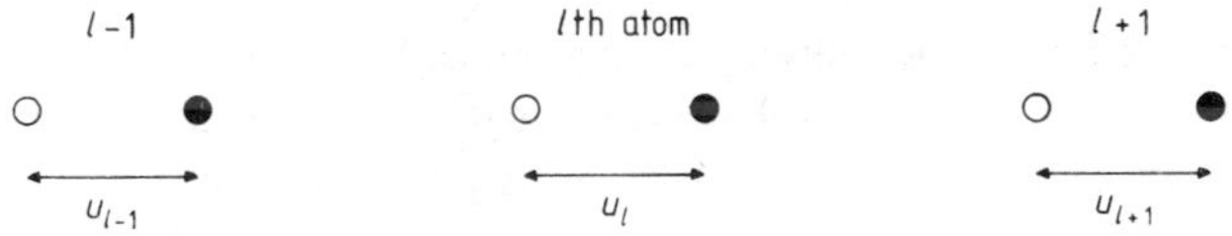

Figure 3.2 Illustrating atomic displacements in the vibrational modes of a one-dimensional monatomic lattice. ○, equilibrium positions; ●, displaced positions; u_l, displacement of lth atom from equilibrium.

will feel only two forces. The one which is due to the atom $l-1$ on its left, can be written as

$$\gamma[u_{l-1}(t)-u_l(t)] \tag{3.11}$$

where γ is the elastic constant and $u_{l-1}(t)$ and $u_l(t)$ are the displacements at time t of the $(l-1)$th and lth atoms respectively from their equilibrium positions. In the same notation the force which is due to the atom $l+1$ on the right is written as

$$-\gamma[u_l(t)-u_{l+1}(t)]. \tag{3.12}$$

The resultant of the forces on atom l is clearly the sum of (3.11) and (3.12) and the equation of motion is

$$m\ddot{u}_l(t) = -\gamma[2u_l(t)-u_{l+1}(t)-u_{l-1}(t)]. \tag{3.13}$$

We want to find a solution of equation (3.13) which has the form of a travelling wave and therefore we put

$$u_{l,k}(t) = u(0)\exp[\mathrm{i}(kla-\omega t)]. \tag{3.14}$$

Inserting (3.14) into (3.13) we find that (3.13) is satisfied only if the (angular) frequency ω and the wavenumber k are connected through the equation

$$m\omega^2 = 2\gamma[1-\cos(ka)]. \tag{3.16}$$

Equation (3.16) obviously gives the required dispersion relation which can be put in the compact form

$$\omega(k) = 2\left(\frac{\gamma}{m}\right)^{1/2}\left|\sin\left(\frac{ka}{2}\right)\right|. \tag{3.17}$$

Equation (3.17) has been derived under very restricted assumptions: linear chain and near-neighbour interactions; nevertheless it illustrates some general features of lattice wave dispersion which are worth examining. In the limit of small wavenumbers, $ka \ll 1$, the wavelength $\lambda = 2\pi/k$ is so much longer than the interatomic spacing a that the details of the lattice are unimportant and the continuum elasticity theory becomes a good approximation. In fact for $ka \ll 1$ we can approximate $\sin(ka/2)$ by its argument and equation (3.17) becomes:

$$\omega(k) \simeq a\left(\frac{\gamma}{m}\right)^{1/2} k, \tag{3.18}$$

$a(\gamma/m)^{1/2}$ being the sound velocity in our simplified model (cf Appendix 3.1, where $\omega = ck$, with c the sound velocity, is discussed in the Debye model).

The other important property of $\omega(k)$ is its periodicity. From (3.17) we have in fact

$$\omega(k) = \omega\left(k + \frac{2\pi n}{a}\right) \qquad n = 0, \pm 1, \ldots . \tag{3.19}$$

This means that only one interval of extent $2\pi/a$ is sufficient to describe equation (3.17). The choice of this interval is irrelevant in principle. In practice however, the most convenient choice is the interval

$$-\pi/a \leqslant k \leqslant \pi/a, \tag{3.20}$$

defining the Brillouin zone of the lattice (cf §2.8.2). Another important feature in our problem is that any interval of extent $2\pi/a$, and in particular (3.20), contains all the possible different vibrations of the system. In fact these are given by $u_{l,k}(t)$ in (3.14), which is also a periodic function of k through the factor $\exp(ikla)$:

$$u_{l,k+2\pi n/a} = u_{l,k} \tag{3.21}$$

as can be seen by using the property (3.19) and the fact that $\exp(2\pi i l n) = 1$, l and n being integers. Equations (3.19) and (3.21) are consequences of the lattice periodicity (cf §2.8.2 for analogous results for electron states) and as such will remain valid in more complicated situations in which the lattice is no longer a linear chain and the interatomic forces have a range greater than the first neighbour distance.

3.3 Optical phonons

Let us now consider a slightly more complicated situation in which we have two atoms, A and B, per unit cell, as depicted in figure 3.3. Provided that we know the masses and the force constants, we can calculate the dispersion relation along lines similar to those we followed in the monatomic case.

In the interests of simplicity we shall assume that the AB linear chain has a single lattice distance d, that the atoms interact through the same force

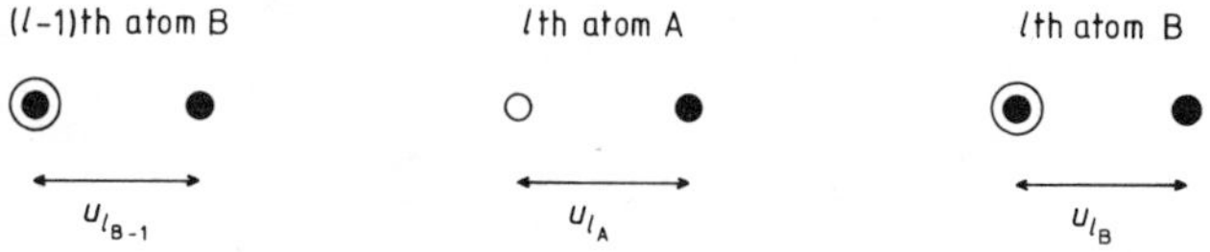

Figure 3.3 Dynamics of diatomic lattice, with atoms A and B. Equilibrium positions: ○, type A atoms; ⊙, type B atoms. Displacements from equilibrium positions (•) are labelled in accord with equations (3.22).

constant, and that therefore the sole difference from the monatomic case is that A and B have masses m_A and m_B.

In terms of displacements from the equilibrium positions (see figure 3.3) we have the equations of motion

$$\left.\begin{aligned} m_A \ddot{u}_{l_A} &= -\gamma(2u_{l_A} - u_{l_B} - u_{l_{B-1}}) \\ \text{and} \quad m_B \ddot{u}_{l_B} &= -\gamma(2u_{l_B} - u_{l_{A+1}} - u_{l_A}). \end{aligned}\right\} \tag{3.22}$$

These equations admit wave-like solutions of the form (see problem 3.8)

$$\left.\begin{aligned} u_{l_A} &= u_A(0)\exp[\mathrm{i}(kl_A d - \omega t)] \\ \text{and} \quad u_{l_B} &= u_B(0)\exp[\mathrm{i}(kl_B d - \omega t)] \end{aligned}\right\} \tag{3.23}$$

only if the following conditions are satisfied:

$$\left.\begin{aligned} (-m_A\omega^2 + 2\gamma)u_A(0) + 2\gamma\cos(kd)u_B(0) &= 0 \\ \text{and} \quad 2\gamma\cos(kd)u_A(0) + (-m_B\omega^2 + 2\gamma)u_B(0) &= 0. \end{aligned}\right\} \tag{3.24}$$

These homogeneous equations (3.24) can be solved for $u_A(0)$ and $u_B(0) \neq 0$ only if the determinant of the coefficients vanishes. The eigenfrequencies $\omega(k)$ are then obtained as

$$\omega_\pm^2(k) = \gamma\left(\frac{1}{m_A} + \frac{1}{m_B}\right) \pm \gamma\left[\left(\frac{1}{m_A} + \frac{1}{m_B}\right)^2 - \frac{4\sin^2(kd)}{m_A m_B}\right]^{1/2}. \tag{3.25}$$

We note that the periodicity of these two roots is reduced, being now π/d instead of $2\pi/d$ as in the monatomic case. This arises from the fact that in the diatomic chain the length of a unit cell is doubled relative to the previous monatomic lattice. The Brillouin zone (BZ) is therefore defined by $-\pi/2d \leqslant k \leqslant \pi/2d$ and in figure 3.4 we indicate schematically the two curves $\omega_+(k)$ and $\omega_-(k)$ in this zone.

It can be seen that the lower curve is not qualitatively different from the dispersion relation found for the monatomic case. This lower branch is called acoustic, since it reduces to ordinary sound waves in the long-wavelength limit.

The novel feature of the diatomic chain is, of course, the upper branch

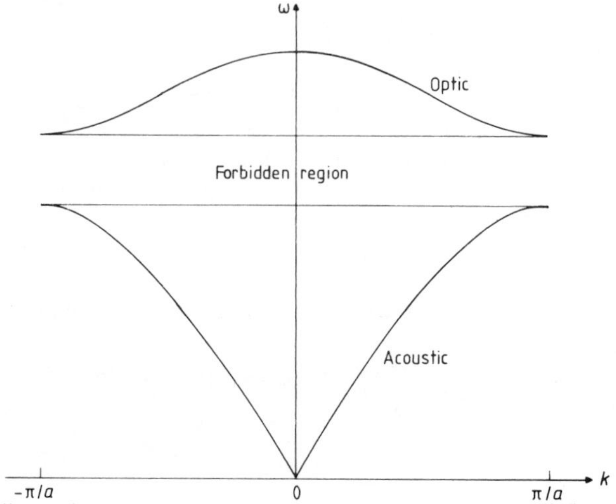

Figure 3.4 Shows two bands of allowed vibrational frequencies in linear diatomic case. Note that a is the size of the unit cell, i.e. $a = 2d$.

(called optic), the corresponding angular frequencies always being non-zero. In order to understand more about the physical distinctions between the two solutions, let us go back to equation (3.24) and solve for the amplitude ratio $u_{l_A}(0)/u_{l_B}(0)$. This ratio gives the phase relationship between the atomic vibrations inside the same unit cell and is different in the two branches depicted in figure 3.4. In particular, for the long-wavelength limit $k \to 0$, $(u_{l_A}(0)/u_{l_B}(0))_- = 1$ for the acoustic branch, while $(u_{l_A}(0)/u_{l_B}(0))_+ = -m_B/m_A$ for the optic branch. This means that whereas in a long-wavelength acoustic vibration the atoms inside the unit cell vibrate in phase, they have opposed phases in a long-wavelength optic mode of vibration. This circumstance is important when A and B are oppositely charged, as in the example of the alkali halide crystals, because in such a case their relative motion creates an electric field which is easily detected in optical experiments. We shall return to phonons in the NaCl structure later in this chapter.

3.4 Dynamical matrix: three-dimensional case

Though we have so far discussed only very idealised cases, it turns out that the main conclusions remain qualitatively valid when we turn to real crystals.

Of course, atomic force laws in crystals are complicated, but nevertheless, provided the temperature is low enough for the displacement $\boldsymbol{u}_i$ of the atoms from their equilibrium positions $\boldsymbol{R}_i$ to be small, the potential energy V can be expanded in powers of the displacements.

Below, we shall assume V to be built from a pairwise potential $\Phi_{ij}(\boldsymbol{r})$, which is often a very useful approximation in crystals. Secondly, we treat only the case of those monatomic crystals having one atom per unit cell. Then we have

$$V = \tfrac{1}{2} \sum_{i \neq j} \Phi(x_i - x_j) \qquad x_i = R_i + u_i. \tag{3.26}$$

Expanding in the displacements $\boldsymbol{u}$, one obtains the desired result as ($\boldsymbol{u} = u^1, u^2, u^3$)

$$V = \sum_{i \neq j} \tfrac{1}{2} \Phi(\boldsymbol{R}_i - \boldsymbol{R}_j) + \text{terms linear in } u$$

$$+ \tfrac{1}{2} \sum_{i \neq j} \sum_{\alpha, \beta = 1}^{3} \nabla^{\alpha}_{\boldsymbol{R}_i} \nabla^{\beta}_{\boldsymbol{R}_j} \Phi(\boldsymbol{R}_i - \boldsymbol{R}_j) u_i^{\alpha} u_j^{\beta}. \tag{3.27}$$

Only the last term of this expansion is important for the present purposes since:

(a) the first term, say V_0, on the right-hand side is an uninteresting constant, i.e. it is independent of the displacements;
(b) the second term vanishes, due to the equilibrium condition $\nabla_{\boldsymbol{R}_i} V = 0$.

This approximation, based solely on the last term of equation (3.27), is called harmonic because it reduces the lattice dynamical problem to that of a set of coupled harmonic oscillators.

The procedure followed in the one-dimensional cases treated above can now be generalised as follows. On account of the equilibrium condition, equation (3.27) can be rewritten as:

$$V = V_0 + \tfrac{1}{2} \sum_{i \neq j} \sum_{\alpha, \beta = 1}^{3} V_{ij}^{\alpha\beta} u_i^{\alpha} u_j^{\beta} \tag{3.28}$$

where

$$V_{ij}^{\alpha\beta} = \nabla^{\alpha}_{\boldsymbol{R}_i} \nabla^{\beta}_{\boldsymbol{R}_j} \Phi(\boldsymbol{R}_i - \boldsymbol{R}_j). \tag{3.29}$$

For pairwise potentials, $V_{ij}^{\alpha\beta}$ depends only on the relative distance $\boldsymbol{R}_i - \boldsymbol{R}_j$. The equations of motion for the displacements u_i^{α} then read

$$m\ddot{u}_i^{\alpha} = - \sum_{j, \beta} V^{\alpha\beta}(\boldsymbol{R}_i - \boldsymbol{R}_j) u_j^{\beta}. \tag{3.30}$$

As for the one-dimensional chain, we seek solutions in the form of running waves:

$$u_i^{\alpha} = u^{\alpha}(0) \exp[\mathrm{i}(\boldsymbol{k} \cdot \boldsymbol{r}_i - \omega t)] \tag{3.31}$$

and substitution into the equation of motion yields

$$m\omega^2 u^{\alpha}(0) = \sum_{j \neq i, \beta} V^{\alpha\beta}(\boldsymbol{R}_j - \boldsymbol{R}_i) \exp[-\mathrm{i}\boldsymbol{k} \cdot (\boldsymbol{R}_i - \boldsymbol{R}_j)] u^{\beta}(0). \tag{3.32}$$

The sum on the right-hand side of this equation is clearly independent of i and therefore equation (3.32) reduces to a 3×3 homogeneous system of linear equations. Defining the dynamical matrix **D** by

$$D_{\alpha\beta}(\boldsymbol{k}) = \frac{1}{m} \sum_{\boldsymbol{l}}' V^{\alpha\beta}(\boldsymbol{l})(\exp(\mathrm{i}\boldsymbol{k}\cdot\boldsymbol{l}) \tag{3.33}$$

where the sum runs over all the direct lattice vectors $\boldsymbol{l}$ different from zero, the condition for the existence of non-trivial solutions then reads

$$\det|\omega^2 \delta_{\alpha\beta} - D_{\alpha\beta}(\boldsymbol{k})| = 0 \tag{3.34}$$

where $\delta_{\alpha\beta}$ is the Kronecker delta (i.e. 1 if $\alpha = \beta$, zero otherwise). This is a cubic equation in ω^2 from which three different roots can be found. The corresponding solutions for $u^{\alpha}(0)$ will then determine the displacement directions and thence the polarisation of the solutions.

3.4.1 Polarisation of modes

In an elastic continuum (cf. Appendix 3.1), three different types of acoustic modes can be propagated. The essential difference between these three acoustic modes is in their polarisation. Thus, if we treat an isotropic continuum, one mode is longitudinally polarised, the displacement vector of each atom being along the direction of wave propagation. The other two modes have the same velocity and have transverse polarisation, the atoms moving in the planes normal to the wavevector. It is generally the case that the longitudinal mode has a higher velocity than the transverse modes (see figure 3.7) (the dispersion curves in the alloy $Sm_{1-x}Y_xS$ represent an interesting exception as discussed by Guntherodt *et al* 1978).

But real crystals are not isotropic in their macroscopic elastic properties. The velocity of a wave of a given type will depend on the direction of propagation. The transverse modes are not, in general, degenerate, except in certain special symmetry directions (see figure 3.7). The formal classification into transverse and longitudinal modes does not apply rigorously, i.e. the polarisation vector ($\boldsymbol{e}_{\boldsymbol{q}}$ say) describing the displacement of atoms in a given mode need not be strictly along or normal to $\boldsymbol{q}$.

These results can be generalised to crystals with p atoms per unit cell. It will be convenient to mention the results of this generalisation in the course of a brief summary of the main properties of phonons in periodic lattices. Thus we have the following results:

(i) With p atoms per unit cell there are in general $3p$ phonon branches. Three of these are acoustic in character, while the remaining $3(p-1)$ branches are optic modes. The factor p comes from the p different modes of vibration inside the unit cell while the factor 3 arises from the three possible different polarisations of the atomic vibrations relative to the wavevector $\boldsymbol{q}$.

(ii) As already remarked, acoustic modes reduce to the usual anisotropic continuum elastic waves in the long-wavelength limit. Such elastic waves have a frequency $\omega(\boldsymbol{q})$ which is linear in the wavevector $\boldsymbol{q}$.

(iii) Optic modes primarily involve the relative motion of atoms within each cell. Optic branches have frequencies which are, very roughly, constant over the zone and which usually, but not invariably, decrease with increasing wavevector $\boldsymbol{q}$. At long wavelengths, the longitudinal and transverse optic modes of ionic crystals are split by the electric field associated with the longitudinal modes (for an experimental example on KBr, see the following section). The longitudinal and transverse optic modes are degenerate in cubic valence crystals or cubic rare-gas crystals.

(iv) Optic mode frequencies and acoustic mode frequencies may overlap, resulting in a quasi-continuous spectrum. However, there may be a gap between the highest acoustic and lowest optic mode frequencies. This is most likely when the masses of the atoms in the unit cell differ widely. Thus, in problem 3.6, it can be seen that for a linear chain with nearest-neighbour interactions only, the gap in the frequency spectrum, say ω_{gap}, is related to the maximum frequency $\omega_{\max}$ by

$$\omega_{\text{gap}} = \omega_{\max} \left| \frac{\mu - 1}{\mu + 1} \right| \tag{3.35}$$

where μ is the ratio of the two masses. As we show from experiment in the next section, a gap exists for KBr, but turns out not to be there in KCl although both have closely similar atomic masses.

3.5 Thermodynamic properties

Once one has solved the harmonic lattice dynamical problem for the $3p$ dispersion curves, it is straightforward, at least in principle, to calculate the thermodynamic properties of such a harmonic crystal.

For such a crystal, each mode of vibration corresponds to an exact eigenfrequency and therefore it makes an independent contribution to the total internal energy E of the crystal. Thus E is simply a sum over all the different modes, weighted by the proper statistical factor, namely (cf equation 3.8)

$$E = \sum_{s=1}^{3p} \sum_{\boldsymbol{k}} \frac{\hbar\omega^{(s)}(\boldsymbol{k})}{\exp[\hbar\omega^{(s)}(\boldsymbol{k})/k_B T] - 1} \tag{3.36}$$

where the summation over $\boldsymbol{k}$ is restricted to the (first) BZ (Brillouin zone) while the index s runs over all the different branches of the dispersion curves.

This equation is apparently a good deal more complicated than the one we wrote down earlier for the Debye model. However, in the low-temperature limit the exponential factor will, essentially, remove the contributions from all the optical branches from the sum, and one is left simply with the low-lying acoustic modes. These are well approximated by the continuum limit $\omega \propto k$. Then we regain the Debye approximation, which is thereby demonstrated to be correct in the low-temperature limit. If the temperature is raised sufficiently, the Debye approximation is no longer valid and proper account must be taken of the microscopic behaviour of the crystal.

To see this, let us write the internal energy E as (cf equation (3.8))

$$E = \int \mathrm{d}\omega g(\omega) \frac{\hbar\omega}{\exp(\hbar\omega/k_{\mathrm{B}}T) - 1}. \tag{3.37}$$

In equation (3.37) we have introduced the frequency spectrum or phonon density of states function, which is important to discuss now in some detail.

3.5.1 Phonon density of states

More precisely, the phonon density of states $g(\omega)$, defined such that there are $g(\omega)\mathrm{d}\omega$ modes in the frequency range between ω and $\omega + \mathrm{d}\omega$, can be written in terms of the dispersion relation $\omega_s(\boldsymbol{q})$ for the sth branch as

$$g(\omega) = \sum_{\boldsymbol{q},s} \delta(\omega - \omega_s(\boldsymbol{q})). \tag{3.38}$$

For small values of ω, only the low-lying acoustic modes can contribute to the sum (3.38). As we have emphasised, these have the form $\omega \propto q$ and therefore the Debye approximation (cf equation (3.5)) is recovered. However at higher frequencies, the detailed form of the dispersion curves $\omega_s(\boldsymbol{q})$ obviously enters the calculation of $g(\omega)$ and deviations from equation (3.5) must be expected.

Just as for electron states (cf equation (2.70)), it is convenient for some purposes to replace the summation over $\boldsymbol{q}$ in equation (3.38) by an integral over the constant-frequency surface ω. Let us in fact construct two such constant-frequency surfaces in $\boldsymbol{q}$-space, labelled by slightly differing frequencies ω and $\omega + \mathrm{d}\omega$. The spacing between them is then $\mathrm{d}\omega/|\nabla_{\boldsymbol{q}}\omega|$, where $\nabla_{\boldsymbol{q}}\omega$ is the gradient of ω in $\boldsymbol{q}$-space, this spacing of course varying from point to point. The volume of $\boldsymbol{q}$-space enclosed between the two surfaces is

$$\mathrm{d}\omega \int \frac{\mathrm{d}S}{|\nabla_{\boldsymbol{q}}\omega|} \equiv \mathrm{d}\omega \int \frac{\mathrm{d}S}{v_{\mathrm{g}}}, \tag{3.39}$$

the integration being over the constant-frequency surface. In writing equation (3.39) we have utilised the fact (cf. the discussion of electronic density of states in §2.8.2) that $\nabla_{\boldsymbol{q}}\omega$ is the group velocity and have written its magnitude $|\nabla_{\boldsymbol{q}}\omega|$ as v_{g} in equation (3.39).

In the summary of electron state properties in Chapter 2 we saw that, for a particular branch of the spectrum, the density of allowed values of $\boldsymbol{q}$ in the zone is $V/(2\pi)^3$, where V is the volume of the crystal. Thus we finally obtain for the number of modes between ω and $\omega + \mathrm{d}\omega$

$$g(\omega)\mathrm{d}\omega = \mathrm{d}\omega \frac{V}{(2\pi)^3} \int \frac{\mathrm{d}S}{|\nabla_{\boldsymbol{q}}\omega|}. \tag{3.40}$$

This equation shows that, since the area of the constant-frequency surface and the group velocity are smooth functions of ω, the function $g(\omega)$ will generally be a smoothly varying function, except that special attention must be given to values of the frequency ω where the surface intersects critical points, which by definition are those for which the gradient $\nabla_{\boldsymbol{q}}\omega$ is zero. This occurs, for example, at the zone boundary, from the fact that $\omega_s(\boldsymbol{q})$ is periodic in $\boldsymbol{q}$-space. This is equivalent to thc statement that the group velocity is zero at the zone boundary. These singularities, the critical points, were discussed in detail by Van Hove (1953, see also Phillips 1956) and some typical shapes of $g(\omega)$ are displayed in figure 3.5. It is worth stressing that their number and character depend crucially on the dimensionality of the system. In particular, in one-dimensional systems, Van Hove singularities lead to divergencies in $g(\omega)$, as shown in figure 3.16 (see also problem 3.1).

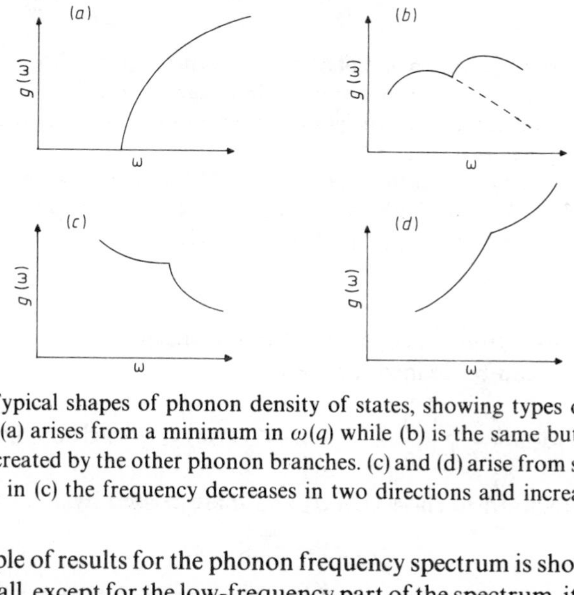

Figure 3.5 Typical shapes of phonon density of states, showing types of Van Hove singularities: (a) arises from a minimum in $\omega(q)$ while (b) is the same but occurs on a background created by the other phonon branches. (c) and (d) arise from saddlepoints. For example, in (c) the frequency decreases in two directions and increases in one.

An example of results for the phonon frequency spectrum is shown in figure 3.6. First of all, except for the low-frequency part of the spectrum, it can be seen that the Debye theory, with $g(\omega) \propto \omega^2$ out to the Debye frequency, is at best a gross approximation. Secondly, the points of non-analyticity, or Van Hove singularities, are shown in the full curve in figure 3.6. These can be significant in the interpretation of optical absorption, both in perfect and in defect crystals (see e.g. Stoneham 1975). In §3.9 we shall see that when disorder is introduced into the lattice, the Van Hove singularities, which are fundamentally a consequence of perfect periodicity, are smeared out.

As an explicit example of the use of the dynamical matrix, the phonon dispersion relations $\omega_s(\boldsymbol{q})$ along the three symmetry directions in an FCC lattice with central short-range forces are worked out fully in Appendix 3.2.

Sometimes it proves useful to define a density of states in ω^2, analogous to

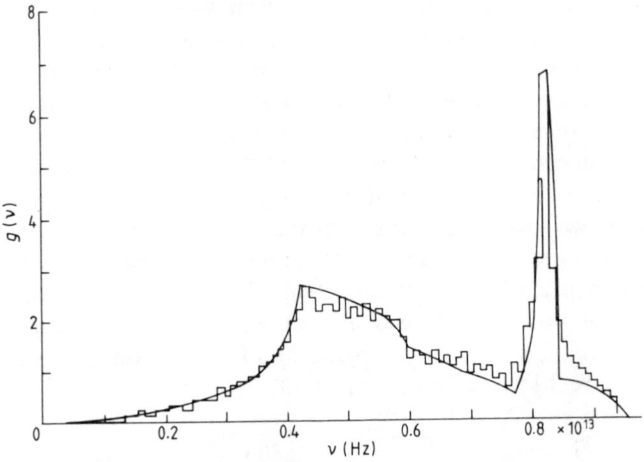

Figure 3.6 Short-range force model of Al frequency spectrum $g(\nu)$. The histogram, due to Walker (1956), is directly calculated from the dynamical matrix. Full curve, due to Phillips (1956), is obtained by interpolating between (Van Hove) singularities predicted on topological grounds.

Note that the actual frequency spectrum in Al metal is different, because the interionic forces in this metal are long range (cf the Friedel oscillations in equation (2.48)). (From Phillips J C 1956 *Phys. Rev.* **104** 1263.)

$g(\omega)$ which is the density of states in ω. For p atoms per unit cell, the density in ω^2, say $g_2(\omega^2)$ can be defined through

$$g_2(\omega^2)\mathrm{d}(\omega^2) = \sum_{\boldsymbol{q}} \frac{1}{3Np} \qquad \omega^2 \leqslant \omega^2(\boldsymbol{q}) \leqslant \omega^2 + \mathrm{d}(\omega^2) \text{ in } \sum_{\boldsymbol{q}}. \tag{3.41}$$

It is straightforward to show that g_2 and g are related by

$$g_2(\omega^2) = g(\omega)/2\omega. \tag{3.42}$$

In band gaps and at frequencies outside the phonon spectrum, it follows that both g and g_2 are identically zero.

It should be clear from the above discussion that to calculate the thermodynamical properties associated with lattice vibrations over a wide temperature range from equation (3.37), and in particular the lattice specific heat, full knowledge of the phonon frequency spectrum $g(\omega)$ is required. The question has been raised as to whether an inversion of measured specific heat data could be performed to yield $g(\omega)$, but the basic difficulties facing such a programme have been underlined by Chambers (1961). Fortunately, other methods than those based on thermodynamic data are available to study lattice vibrational modes experimentally, as will now be discussed in some detail.

3.6 Phonon dispersion curves from scattering experiments

It is, of course, important to test the theory of collective modes of vibration in crystals by confrontation with experiment. While it is not our purpose here to go into the detailed theory of the way in which scattering experiments can be used to extract phonon dispersion curves, we will sketch below the principles involved.

Consider first the case of x-ray scattering. Suppose that $\boldsymbol{k}_0$ and $\boldsymbol{k}$ are the wavevectors of the incident and the scattered x-ray beams respectively. Then, in the case of the zero-order or Bragg scattering one can write

$$\boldsymbol{k} - \boldsymbol{k}_0 = \boldsymbol{G} \tag{3.43}$$

where $\boldsymbol{G}$ is a reciprocal lattice vector. In the presence of the lattice vibrations, one has additionally a first-order scattering process in which

$$\boldsymbol{k} - \boldsymbol{k}_0 = \boldsymbol{G} - \boldsymbol{q} \tag{3.44}$$

where $\boldsymbol{q}$ is now the phonon wavevector. In addition to this equation one must also write a condition for energy conservation, namely

$$\hbar c(\boldsymbol{k} - \boldsymbol{k}_0) = \pm \hbar\omega_s(\boldsymbol{q}) \tag{3.45}$$

where the plus sign corresponds to the disappearance of a $(\boldsymbol{q}, s)$ phonon with an increase in the x-ray energy, while the minus sign corresponds to the opposite situation.

The severe disadvantage of the x-ray method is that because $\hbar c\boldsymbol{k}_0$ is very large compared to $\hbar\omega_s(\boldsymbol{q})$, the change in the x-ray energy is relatively small. There is obvious difficulty then in accurately determining $\omega_s(\boldsymbol{q})$ by this method.

For the scattering of neutrons, similar conditions must be satisfied. For first-order coherent inelastic scattering the same equation (3.44) holds. But the energy conservation condition reads

$$\frac{\hbar^2}{2M_n}(k^2 - k_0^2) = \pm \hbar\omega_s(\boldsymbol{q}), \tag{3.46}$$

M_n being the mass of the neutron. If one makes use of thermal neutrons with wavelengths of a few ångstroms, the relative change in neutron energy is quite considerable. Evidently then, accurate measurements of phonon dispersion relations are entirely possible with neutrons, and indeed neutron scattering experiments are the main source of the known phonon dispersion curves in crystals.

In discussing the scattering of thermal neutrons by the vibrating crystal lattice, a full account of which is found in the book by Marshall and Lovesey (1971), we have a dynamical structure factor (cf the discussion in Chapter 7 for liquids), which represents the probability that a neutron incident on the crystal will transfer energy $\hbar\omega$ and momentum $\hbar\boldsymbol{k}$ to it, and is denoted by $S(\boldsymbol{k}, \omega)$.

The result of such a calculation, which is summarised in Appendix 3.3, contains three parts: the first is coherent elastic scattering, containing the Bragg peaks; the second and third describe scattering in which phonon emission and absorption occur respectively (cf form (7.7) for liquids). The entire function $S(\boldsymbol{k},\omega)$ involving the sum of these three terms is multiplied by a factor $\exp(-2W)$ known as the Debye–Waller factor. Under simplified circumstances, this can be evaluated to yield, in the low-temperature limit,

$$2W = \frac{3}{2}\left(\frac{\hbar^2 k^2}{2M}\right)\left(\frac{1}{k_B \Theta_D}\right)\left[1 + \frac{2\pi^2}{3}\left(\frac{T}{\Theta_D}\right)^2 + \ldots\right] \tag{3.47}$$

where Θ_D is the Debye temperature (cf Appendix 3.3). The magnitude of the Debye–Waller factor depends on the ratio of two energies. If we were to replace the atomic mass M by the neutron mass M_n, the factor $\hbar^2 k^2/2M_n$ would be the energy of a neutron with momentum $\hbar\boldsymbol{k}$. This has to be reduced by M_n/M and compared to $k_B\Theta_D$. This $\boldsymbol{k}$ dependence has the effect of reducing the intensity of coherent processes involving large wavevectors compared to those with small $\boldsymbol{k}$. We also note that the term $(T/\Theta_D)^2$ results in all coherent processes decreasing in intensity with increasing temperature.

Finally, in this connection, it can be shown that if we return to the full formula (A3.3.4), then in the Debye approximation, the high-temperature limit of the Debye–Waller factor is given by (see Appendix A3.3)

$$W = 3\left(\frac{\hbar^2 k^2}{2M}\right)\left(\frac{1}{k_B \Theta_D}\right)\left(\frac{T}{\Theta_D}\right). \tag{3.48}$$

3.6.1 Interplanar force constants in metals

The first example we shall deal with briefly is the measurement of phonon dispersion relations in FCC Pb by Brockhouse *et al* (1962). An illustration of their results is given in figure 3.7 for the $[\zeta 00]$, $[1\zeta 0]$, $[\zeta\zeta 0]$ and $[\zeta\zeta\zeta]$ directions.

They analysed their data in terms of interplanar force constants as suggested by Foreman and Lomer (1957) whose method we shall summarise briefly below. These workers pointed out that when the wavevector $\boldsymbol{q}$ lies along one of the symmetry directions, for example $\langle 100\rangle$, $\langle 110\rangle$ and $\langle 111\rangle$ directions for an FCC crystal, then the equation determining the phonon frequencies (cf. equation (3.34)) factorises and may be solved directly (see also the short-range force example of Appendix 3.2). The interpretation of this follows since lattice waves propagating along a symmetry direction correspond to the vibrations of complete planes of atoms. Thus the problem is like the linear chain treated in §3.2, where each particle there now represents a plane of atoms, and each plane is connected to its neighbours by harmonic forces.

Moreover, it is not difficult to show that as a result of the symmetry of the atoms in the $\{100\}$, $\{110\}$ and $\{111\}$ planes of an FCC crystal, the longitudinal and two transverse modes of relative displacement are independent of one

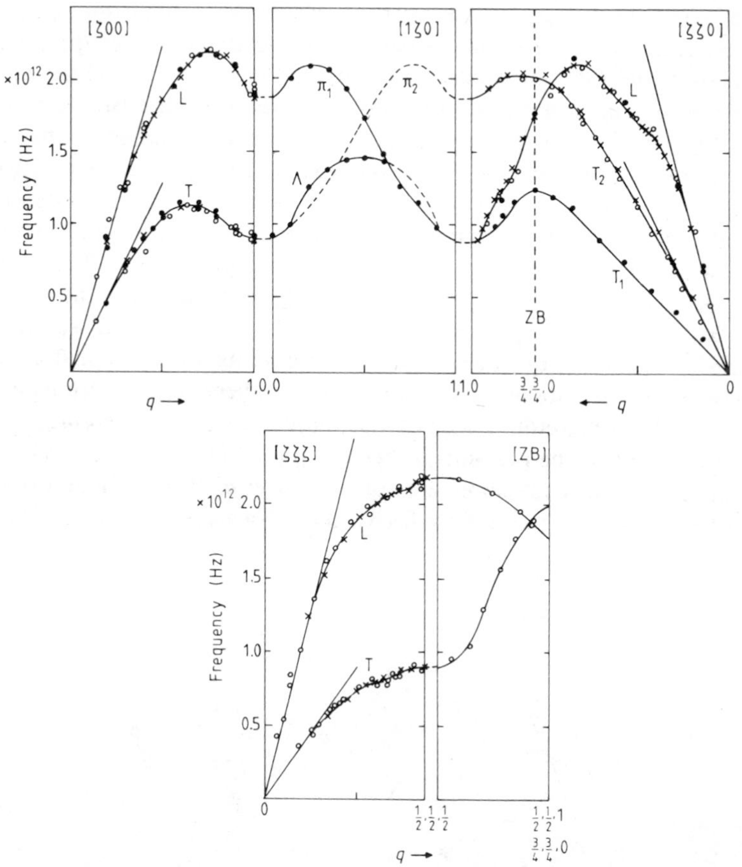

Figure 3.7 Measured phonon dispersion relations for Pb by neutron spectroscopy along various symmetry directions. Of most interest for the present purpose are the experimental points denoted by crosses and open and full circles. $\zeta = aq/2\pi$. (From Brockhouse B N *et al* 1962 *Phys. Rev.* **128** 1099.)

another for each type of plane, so that each longitudinal and transverse mode of vibration may be treated separately as a linear chain problem.

If we refer to the analysis of the one-dimensional chain in §3.2, and merely relax the requirement there of nearest-neighbour interactions, then for such a linear chain having only one degree of freedom the phonon frequencies are determined by

$$M\omega^2(\boldsymbol{q}) = 2\sum_{n=0}^{\infty} P(n)\cos(nqa) \tag{3.49}$$

where a is the 'particle' spacing, $P(n)$ is the harmonic force constant per atom acting between nth neighbour planes and M is the mass of the atoms. Thus the force constants between the planes of atoms normal to the $\langle 100\rangle$, $\langle 110\rangle$ and

⟨111⟩ symmetry directions of an FCC crystal are given simply by the coefficients in the Fourier series expansion of the square of the appropriate dispersion curve. Of course such an expansion reflects the periodicity of the dispersion relation in $\boldsymbol{q}$-space, with the period being the first Brillouin zone.

Brockhouse *et al* (1962) noted that when they attempted to fit their measured dispersion curves on Pb with such a Fourier series, they needed to go out to a large number of neighbours, clearly demonstrating the existence of long-range forces in Pb. This is connected with the oscillations in the potential around an ion in an electron gas (cf equation (2.48) and problem 3.7). Indeed, if one considers Na, where the free-electron gas approximation to the conduction electrons is more appropriate than for Pb, then we shall see in the following chapter that the interplanar force constants for planes that are not first and second neighbours can be understood in general terms from such a picture. We shall therefore conclude this discussion by commenting on the Fourier series fit of the phonons in Na along the [111] direction, from the neutron spectroscopy measurements of Woods *et al* (1962). These workers found that, though the range of the forces in BCC Na was significantly less than

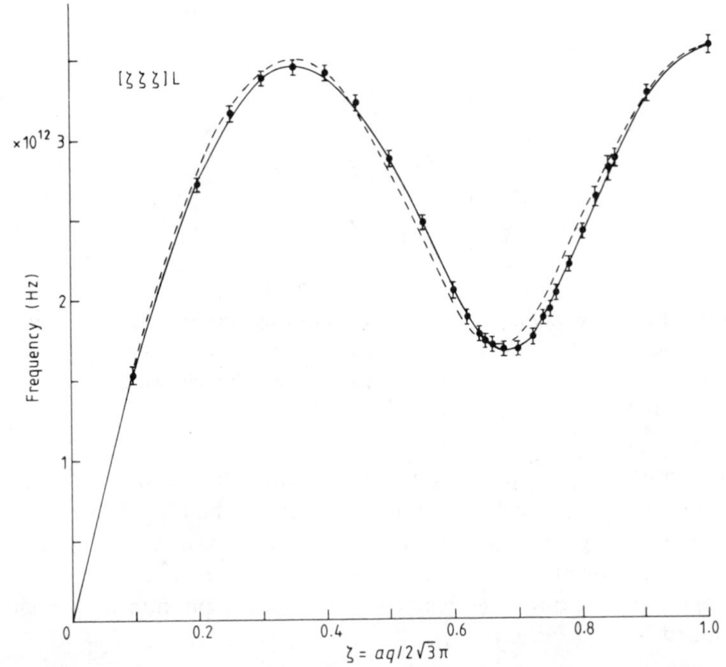

Figure 3.8 Shows Fourier series fits for the [111] branch of phonons in metallic Na of measurements by neutron spectroscopy. A three-term fit is not quite quantitative, but represents a good general description of the phonon dispersion along this symmetry direction. Broken curve, three terms (first and second neighbours); full curve, five terms (fourth neighbour). (From Woods A D B *et al* 1962 *Phys. Rev.* **128** 1112.)

in Pb, the interplanar force constants oscillated in sign. We return to the interpretation of this oscillation in more detail in §4.9.1.

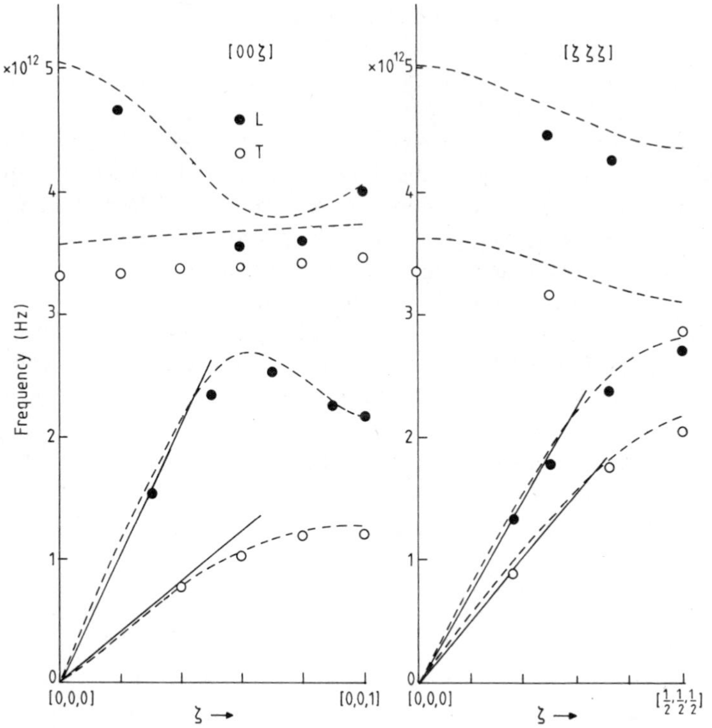

Figure 3.9 Measured dispersion relations of phonons in ionic crystal KBr, in two high symmetry directions ([00ζ], [ζζζ]). Points are for results at 400 K and broken curve is for measurements at 90 K. The straight lines through the origin have the slopes of the appropriate velocity of sound. (From Woods A D B *et al* 1963 *Phys. Rev.* **131** 1025.)

3.6.2 Phonon dispersion curves for KBr

As a further example of the measurement of phonon dispersion relations by neutron inelastic scattering, in figure 3.9 we show results for KBr taken from Woods *et al* (1963). It can be seen that in this material there is a clear separation of acoustic and optic branches. This is readily traced to the large mass difference between K^+ and Br^- (cf §3.4.1).

As regards the theory of these phonon dispersion curves, the earliest treatment (Kellerman 1940) was based on the rigid-ion model, since no account was taken of any possible deformability of the ions as they vibrate. Actually, in KBr, the electronic polarisability α^+ of the K^+ ion is relatively small in comparison with α^- of Br^-, but the rigid-ion model incorporates neither quantity.

In the rigid-ion model, the dynamical matrix $D_{\alpha\beta}$ is separated into two parts:

$$D_{\alpha\beta} = D^{C}_{\alpha\beta} + D^{r}_{\alpha\beta}. \tag{3.50}$$

The term $D^{C}_{\alpha\beta}$ is the long-range Coulomb contribution, while $D^{r}_{\alpha\beta}$ is the short-range repulsive part. In this treatment, the point charge Coulomb contributions embodied in $D^{C}_{\alpha\beta}$ can be handled by a method due to Ewald, which is summarised briefly in the first part of Appendix 3.4.

We shall conclude this section by indicating the usefulness as well as the deficiencies of the rigid-ion results. Firstly, this model predicts a frequency spectrum in which the acoustic branches are in fairly good agreement with the neutron measurements. But the optic branches are too high, sometimes by 50% or more, an example being shown in figure 3.10.

It turns out that the inclusion of polarisation of the ions can remove the major part of these discrepancies and can also provide a realistic mechanism for the interpretation of dielectric properties. It will be convenient to summarise these after we have treated the dielectric properties of polar crystals in some detail in the following chapter.

However, there is a sense in which the rigid-ion model remains a consistent first approximation to more fundamental attempts to calculate phonons in

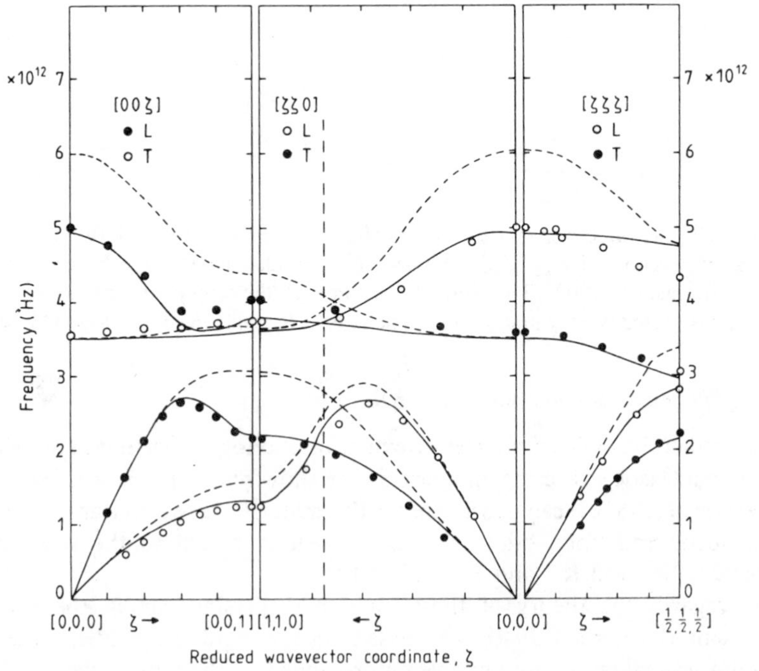

Figure 3.10 Results of rigid-ion model for phonon dispersion at 90 K in KBr. (From Woods A D B *et al* 1963 *Phys. Rev.* **131** 1025.) Broken curves are calculated from the rigid-ion model and full curves from the simple shell model (see §4.6).

ionic crystals. Therefore we have thought it worthwhile in Appendix 3.4 to discuss also how the short-range repulsive contribution $D^{\mathrm{r}}_{\alpha\beta}$ to the dynamical matrix in equation (3.50) can be obtained and the phonon dispersion relations thereby determined within the rigid-ion model.

3.7 Optical absorption due to lattice vibrations

One can expect that the fluctuating electric polarisation in a crystal will be able to interact with electromagnetic radiation and that an absorption spectrum will be found. Such absorption is indeed observed in ionic and in covalent solids, generally in the infrared region.

Analysis of absorption bands is a way of gleaning information about lattice vibrational frequencies at certain points in the Brillouin zone. The argument given below is due to Lax and Burstein (1955).

If $\boldsymbol{P}$ is the polarisation of dielectric, the interaction energy with an electric field $\mathscr{E}$ is $-\boldsymbol{P}\cdot\mathscr{E}$. Since $\mathscr{E}$ is the local field inside the dielectric, it will not be exactly the same as the applied field. If this field is due to an incident light beam, it will have the form $f\exp(\mathrm{i}\,\boldsymbol{k}\cdot\boldsymbol{r}-\omega t)$, where f is the polarisation vector of the electric part of the electromagnetic wave.

One now expands the polarisation as a power series in the atomic displacements $\boldsymbol{u}$:

$$\boldsymbol{P}=\boldsymbol{P}_0+\boldsymbol{P}_1+\boldsymbol{P}_2+\ \ldots . \tag{3.51}$$

$\boldsymbol{P}_0$ represents any static permanent moment which may exist; $\boldsymbol{P}_1$ and $\boldsymbol{P}_2$ are respectively linear and quadratic in the displacements. The absorption coefficient is usefully written as

$$\mu(\omega)=\left[\frac{n}{\epsilon}\left(\frac{\mathscr{E}_1}{\mathscr{E}}\right)^2\right]\frac{4\pi^2\omega}{c}I(\omega) \tag{3.52}$$

where n is the refractive index, ϵ the dielectric constant, $(\mathscr{E}_1/\mathscr{E})$ is the local field correction and

$$I(\omega)=|\langle\psi_{\mathrm{f}}|f\cdot\boldsymbol{P}\mathrm{e}^{\mathrm{i}\boldsymbol{k}\cdot\boldsymbol{r}}|\psi_{\mathrm{i}}\rangle|^2\delta(E_{\mathrm{f}}-E_{\mathrm{i}}-\hbar\omega) \tag{3.53}$$

where ψ_{i}, E_{i} and ψ_{f}, E_{f} refer to wavefunctions and energies of initial and final states. After a short calculation one obtains $I_1(\omega)$ with a factor $\delta(\omega_{\mathrm{TO}}(0)-\omega)$, where $\omega_{\mathrm{TO}}(0)$ is the frequency of the transverse optic (TO) mode in the long-wavelength limit. Thus I_1 consists of a single narrow absorption line at the frequency of the TO mode at $\boldsymbol{k}=0$. This can be shown to be the same as the crystal Raman frequency. Anharmonic effects (cf. §3.8) can be expected to broaden this single phonon line.

Below we will discuss the main qualitative features associated with $I_2(\omega)$ while in Appendix 3.5 a brief summary of the basic equations used to calculate absorption will be given. The argument follows the account given by Elliott and Gibson (1974). A useful focal point for the discussion is the experimental absorption spectrum for GaAs from 0.038 eV to 0.11 eV in

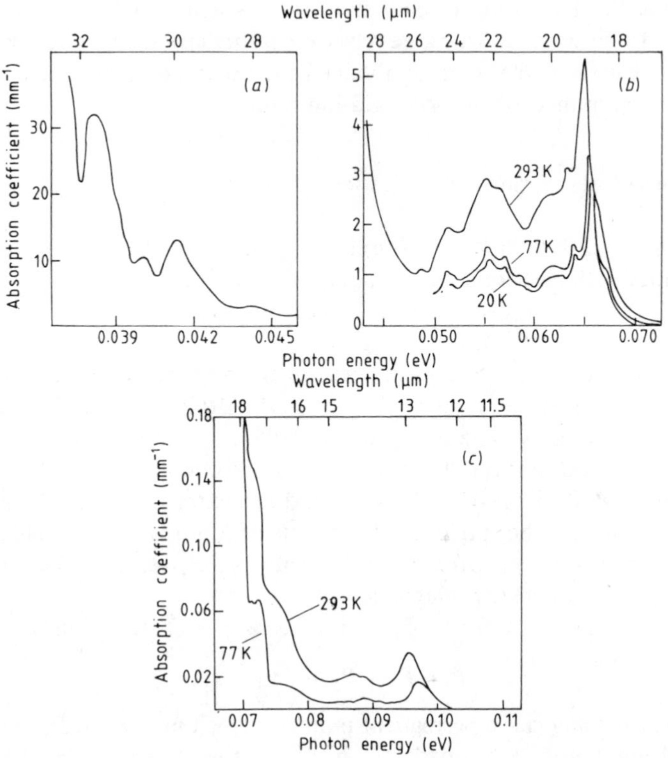

Figure 3.11 Experimental absorption spectrum of GaAs for different temperatures and in different photon energy ranges (redrawn from Cochran *et al* 1961)).

figure 3.11. The absorption, which is seen to be temperature dependent from experiment, is due to processes involving two, three and even four phonons simultaneously.

The probability of a multiphonon process decreases as the number of particles involved increases, but this can be offset by the increase in the number of possible combinations that satisfy the energy and wavevector conservation rules.

The discussion below will merely outline how the theory of two-phonon processes can be developed; higher-order processes introduce no new concepts but are a good deal more complicated to formulate quantitatively.

We note first that if a single photon generates two phonons, the conservation results require

$$\hbar\omega(\boldsymbol{p}) = \hbar\omega(\boldsymbol{q}_1) + \hbar\omega(\boldsymbol{q}_2) \tag{3.54}$$

and

$$\boldsymbol{p} = \boldsymbol{q}_1 + \boldsymbol{q}_2 \tag{3.55}$$

where $\boldsymbol{p}$ is the wavevector of a photon of energy $\hbar\omega(\boldsymbol{p})$. Since the wavelength of the radiation is very much greater than the lattice spacing, $\boldsymbol{p}$ is very much less than a reciprocal lattice vector and generally we can take $\boldsymbol{p} \sim 0$. Thus $\boldsymbol{q}_1 \simeq -\boldsymbol{q}_2$, i.e. the two phonons emitted have almost equal and opposite wavevectors.

The strength of the absorption at a particular frequency will depend mainly on: (a) the density of available phonon states; (b) the phonon distribution; and (c) the matrix element of the transition (cf Appendix 3.5). Taking these three factors in order:

(a) In two-phonon absorption, one is interested in pairs of states satisfying $\boldsymbol{p} = \boldsymbol{q}_1 + \boldsymbol{q}_2$. Since the density of phonon states in $\boldsymbol{q}$-space increases as $\boldsymbol{q}$ increases, two-phonon absorption is favourable for phonons near the Brillouin zone edge. In addition, there are critical points for every branch of the phonon spectrum at the zone edge, where the condition $\nabla_{\boldsymbol{q}}\omega(\boldsymbol{q}) = 0$ is fulfilled. Hence there is a range of $\boldsymbol{q}$ over which the phonon energy is practically constant. Thus in the two-phonon process, the qualitative consideration is that it is phonons near the zone edge that are primarily involved and the sums of the energies of these phonons, in suitable pairs, correspond to the absorption peaks in figure 3.11. A different example in which this is also true is presented in figure 3.12.

(b) The number of phonons of a given energy is, of course, dependent on temperature and this is the feature that determines the observed temperature dependence of the absorption spectrum of GaAs above. The number of phonons $n(\omega)$ of energy $\hbar\omega(\boldsymbol{q})$ is given by the Bose distribution function. The

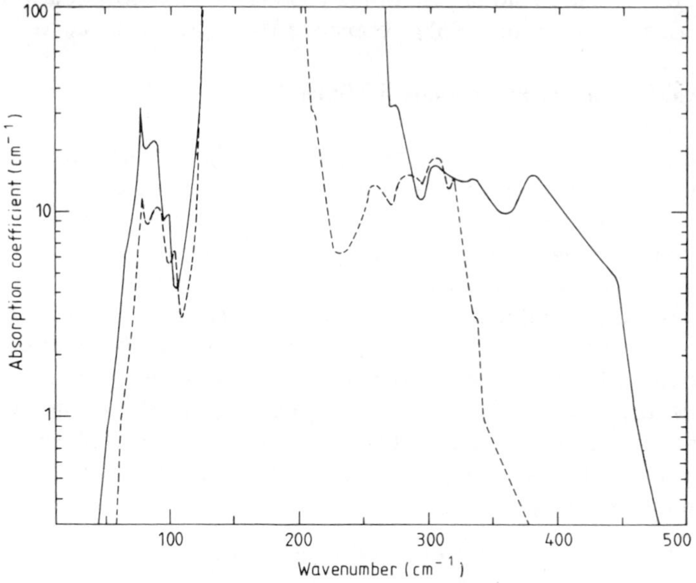

Figure 3.12 Showing two-phonon processes in absorption spectrum. ———, CuCl; – – – –, CuBr.

probability that light will generate an additional phonon of a given energy is proportional to the probability of there being one more phonon, i.e. it is proportional to $n(\omega)+1$. But there is also the possibility of the reverse process, namely that of a phonon being absorbed from the lattice to generate a photon, so that the phonon density decreases. The probability that this will occur is proportional to the equilibrium number of phonons of appropriate energy and hence to $n(\omega)$. If we write the probabilities of phonons of energy $\hbar\omega(\boldsymbol{q}_1)$ and $\hbar\omega(\boldsymbol{q}_2)$ being emitted as $n_1(\omega)+1$ and $n_2(\omega)+1$ respectively, the probability that both phonons are emitted simultaneously is proportional to $[n_1(\omega)+1][n_2(\omega)+1]$ and the net optical absorption is then proportional to

$$\begin{aligned}[n_1(\omega)+1][n_2(\omega)+1]&-n_1(\omega)n_2(\omega)\\ &=1+n_1(\omega)+n_2(\omega)\\ &=\frac{\exp\{\hbar[\omega(\boldsymbol{q}_1)+\omega(\boldsymbol{q}_2)]/k_BT\}-1}{\{\exp[\hbar\omega(\boldsymbol{q}_1)/k_BT]-1\}\{\exp[\hbar\omega(\boldsymbol{q}_2)/k_BT]-1\}}.\end{aligned} \tag{3.56}$$

(c) For certain combinations of phonon pairs, the matrix element of the transition vanishes on symmetry grounds†. Also, the matrix element of allowed transitions may vary with $\boldsymbol{q}$, but the effect of this on the spectrum is usually small compared with the variation in the density of states and can often be neglected.

As already remarked, in order to make clear the way in which points (a) and (c) can be dealt with in a quantitative manner, we have summarised in Appendix 3.5 the equations determining two-phonon absorption which have been used (see Johnson (1965)) to analyse the experimental curves.

3.8 Anharmonicity and phonon lifetimes

In certain problems, it is necessary to transcend the harmonic approximation embodied in equation (3.27) and include the effects of cubic terms in the expansion of the total potential energy in powers of the atomic displacements. Important problems that involve these higher-order terms are thermal expansion, specific heat at high temperatures and thermal conductivity.

Let us begin by considering in some detail the problem of thermal expansion from the standpoint of the phonon theory developed above. Since in that treatment each phonon mode was treated as an independent oscillator, one can express the free energy of the vibrating crystal as the sum over the contributions from each mode. Thus the contribution to the Helmholtz free energy from the lattice vibrations is found from the partition function, determined by the phonon frequencies, as

$$k_BT\sum_{\boldsymbol{q}i}\ln\{2\sinh[\hbar\omega_i(\boldsymbol{q})/2k_BT]\}. \tag{3.57}$$

† In the proximity of such points, the matrix element of the transition will vary with $\boldsymbol{q}$ (cf §2.9.2).

When a cubic crystal expands from the volume V to $V+\Delta V$, the elastic energy expended is

$$\frac{1}{2\kappa_T}\left(\frac{\Delta V}{V}\right)^2 \tag{3.58}$$

where κ_T is the isothermal compressibility (Landau and Lifshitz 1962). The relevant free energy is thus the sum of these contributions, i.e. (3.57) and (3.58). This has to be minimised with respect to volume and one then finds

$$\frac{\Delta V}{V}=\frac{\kappa_T}{V}\sum_{q}[n_i(\boldsymbol{q})+\tfrac{1}{2}]\hbar\omega_i(\boldsymbol{q})\gamma. \tag{3.59}$$

Here $n_i(\boldsymbol{q})$ is the occupation number for the mode $\omega_i(\boldsymbol{q})$ and is given by the appropriate Bose distribution function

$$n_i(\boldsymbol{q})=\frac{1}{\exp[\hbar\omega_i(\boldsymbol{q})/k_BT]-1} \tag{3.60}$$

while γ is defined by the assumption that all ω depend in the same way on volume

$$\gamma \doteqdot -\frac{V}{\omega_j(\boldsymbol{q})}\frac{\partial\omega_j(\boldsymbol{q})}{\partial V}=-\frac{\partial\ln\omega_j(\boldsymbol{q})}{\partial\ln V}=\text{constant}. \tag{3.61}$$

Since $(n_j(\boldsymbol{q})+\frac{1}{2})\hbar\omega_j(\boldsymbol{q})$ is the average energy associated with the $\omega_j(\boldsymbol{q})$ mode, we obtain for the thermal expansion coefficient

$$\left(\frac{\Delta V}{V}\right)=\kappa_T\gamma\langle E\rangle \tag{3.62}$$

where $\langle E\rangle$ is the mean thermal energy density.

Provided that the volume expansion has approximately the same effect on each phonon mode as in equation (3.61) then we find as a consequence

$$\alpha=\frac{1}{V}\frac{\partial V}{\partial T}=\kappa_T\gamma c/V \tag{3.63}$$

where c is the specific heat. This is the Grüneisen relation and γ is known as the Grüneisen constant. It is clear that equation (3.63) is only approximate, but it is an experimental fact that the thermal expansion coefficient has the same temperature dependence as the thermal energy, which follows from equation (3.62).

It must be noted that in the harmonic theory developed above, γ would be identically zero, since $\omega_i(\boldsymbol{q})$ does not depend on volume.

Reiterating the Grüneisen result, since we have seen that one has a Debye T^3 specific heat at low temperatures it follows that the low-temperature thermal expansion α varies as T^4. At high temperatures ($T \gg \Theta_D$), the expansion coefficient becomes independent of T according to the Grüneisen form (3.63).

Two other consequences of anharmonicity are: (a) the deviation of the specific heat c in (3.63) from the Dulong–Petit value at high temperatures; and (b) the finite mean free path of lattice waves. This results in a finite thermal conductivity, though contributions also arise from impurities and imperfections and from the finite dimensions of a crystal. However, for perfect crystals which are not too small, and for temperatures not too low, anharmonicity dominates.

We shall conclude this brief discussion with some remarks of a microscopic nature. In the presence of anharmonicity, one can no longer visualise the system as consisting of non-interacting quasi-particles or phonons. However, provided the anharmonicity is not too strong, we can retain the phonon picture, as long as we now include the possibilities that phonons can collide, split or combine. Figure 3.13 illustrates four processes that can occur. Always one has conservation of momentum, expressed through

$$\boldsymbol{k}_1 + \boldsymbol{k}_2 + \boldsymbol{k}_3 = \boldsymbol{K}, \tag{3.64}$$

$\boldsymbol{K}$ being a reciprocal lattice vector. Actually, to be more precise, the above equation expresses the conservation of crystal momentum. The law of conservation of momentum follows from invariance of the Hamiltonian under infinitesimal translations (Dirac 1957) whereas the law of conservation of crystal momentum expressed in equation (3.64) follows from invariance of the Hamiltonian under translation of a lattice vector. Equation (3.64) shows that the phonon wavevector $\boldsymbol{k}$ can change, but only by a reciprocal lattice vector $\boldsymbol{G}$.

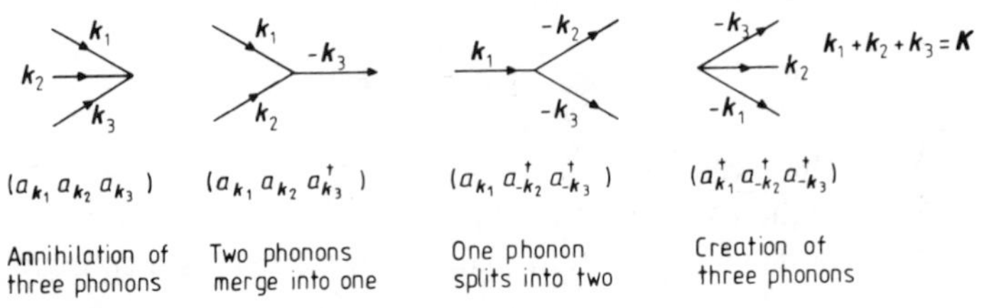

Figure 3.13 Illustrating the effects of including anharmonicity on the picture of independent phonons. ($a^\dagger$ and a are creation and annihilation operators respectively: for details see Jones and March (1973)).

If $\boldsymbol{G} = 0$, we speak of a *normal* process; if $\boldsymbol{G} \neq 0$ we have an *umklapp* process (a name introduced by Peierls meaning the phonon 'flips over'; one may think of it as normal scattering plus Bragg reflexion).

Obviously from the above discussion, phonons in the presence of anharmonicity have finite lifetimes. While we have discussed above an analytical approach to anharmonicity, one can solve numerically the Newtonian equations of motion for atoms on a lattice, interacting by, say, a Lennard-Jones potential. This was done in the molecular dynamics computations of Dickey and Paskin (1969) on solid Kr. These calculations give some valuable information on phonon lifetimes. As an illustration, in figure 3.14 we show the time evolution of a perturbed normal mode at three different temperatures. The reduction of the phonon lifetime as the temperature is increased is apparent. A lifetime τ can

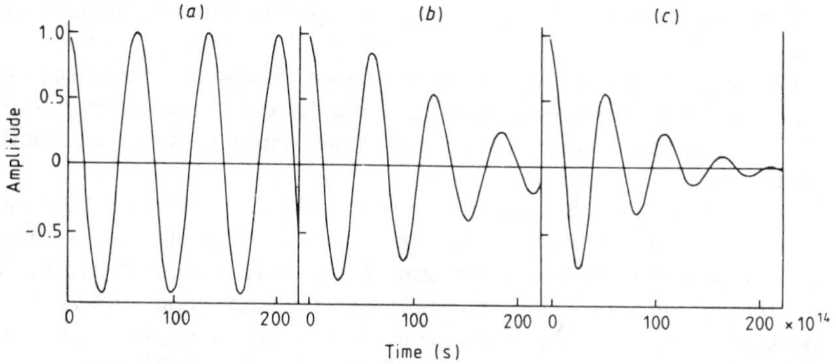

Figure 3.14 Shows time evolution of perturbed normal mode for solid Kr. Results shown were obtained by machine computations and therefore include anharmonicity to all orders: (a) corresponds to temperature of 2.5 K while others are at (b) 91 K and (c) 186 K respectively. (After Dickey J M and Paskin A 1969 *Phys. Rev.* **188** 1407.)

be calculated from the elementary formula for the ratio of the amplitudes A of adjacent maxima:

$$A_1/A_2 = \exp(-t/\tau). \tag{3.65}$$

At high temperatures, τ is found to be of the order of a few periods only.

Finally figure 3.15 shows the temperature dependence of the phonon frequency spectrum $g(\omega)$, the curve (a) being at 2.5 K and the others at 91 K and 186 K.

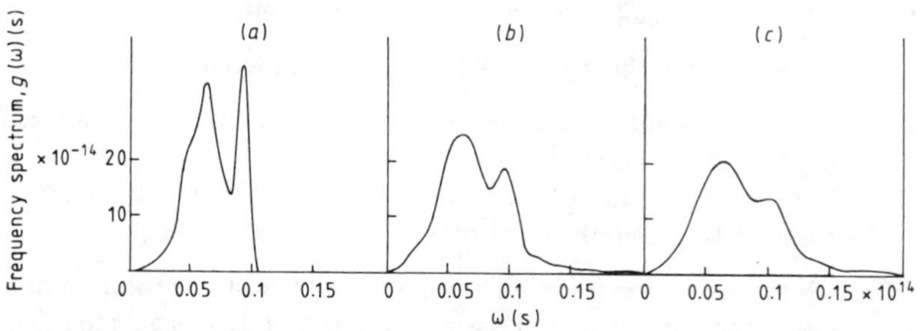

Figure 3.15 Frequency spectrum $g(\omega)$ for solid Kr; curves corresponding to the temperatures in figure 3.14. (After Dickey J M and Paskin A 1969 *Phys. Rev.* **188** 1407.)

3.9 Effect of disorder on atomic dynamics

The discussion of phonons to this point has relied heavily on the periodicity properties of the crystal. However, a problem of considerable interest arises as to the way in which the atomic dynamics is affected by disorder. This is relevant to the properties of glasses, amorphous semiconductors and, though the problem is of a different kind, also to the problem of 'phonons' in liquids (cf Chapter 7).

Because of the lack of translational invariance, the concept of the dispersion relation $\omega(\boldsymbol{k})$ that we have introduced earlier ceases to be valid, because the wavevector $\boldsymbol{k}$ is no longer a 'good quantum number' with which to label the phonon states.

However, one can still define a density of states, or vibrational frequency spectrum $g(\omega)$, and try to calculate this from some specific model. In the general case, this is obviously a very difficult task and the calculations can only be performed on a computer. There is, however, a simple model for which the calculation of $g(\omega)$ can be performed analytically. This is due to Dyson (1953) who studied the vibrational properties of a disordered chain. By this is meant a chain of one-dimensional harmonic oscillators, each coupled to its nearest neighbours by harmonic forces, the inertia of each oscillator or the strength of each spring coupling the masses being a random variable with a prescribed statistical distribution law.

Specifically, suppose the jth particle in the chain has mass m_j, and let its displacement from its equilibrium position be x_j. Throughout this section we are concerned with longitudinal vibrations of the chain. Then if the elastic modulus of the spring connecting particles j and $j+1$ is K_j, the equations of motion are

$$m_j\ddot{x}_j = K_j(x_{j+1}-x_j)+K_{j-1}(x_{j-1}-x_j). \tag{3.66}$$

It is convenient to work with scaled variables $y_j = m_j^{1/2}x_j$ and new force constants $\lambda_1, \lambda_2, \ldots, \lambda_{2N-2}$ given by

$$\lambda_{2j-1} = K_j/m_j, \qquad \lambda_{2j} = K_j/m_{j+1}. \tag{3.67}$$

Dyson works out the theory for two types of disordered chain:

(1) Each of the constants λ_j is an independent random variable with a specified probability distribution function;
(2) Each mass m_j is an independent random variable with a specified distribution function, the constants K_j being fixed and equal.

Though the restriction to one dimension severely restricts the applicability of the results, the great merit of Dyson's work is that he was able to calculate exactly the vibrational frequency spectrum in the presence of disorder in the limit of an infinite chain.

Some of the detailed results he obtained are presented in Appendix 3.6. Here, we shall content ourselves with a comparison of the results obtained for the case of a very disordered chain with those of the perfect linear chain worked out in §3.2.

The quantity plotted in figure 3.16 is the function $M(\mu)$ which gives the number of eigenfrequencies for which $\omega_j^2 \leqslant \mu$. This is obviously related to the frequency spectrum g_2 of the squared frequency (see equation (3.41)) by

$$g_2(\mu) = \mathrm{d}M/\mathrm{d}\mu. \tag{3.68}$$

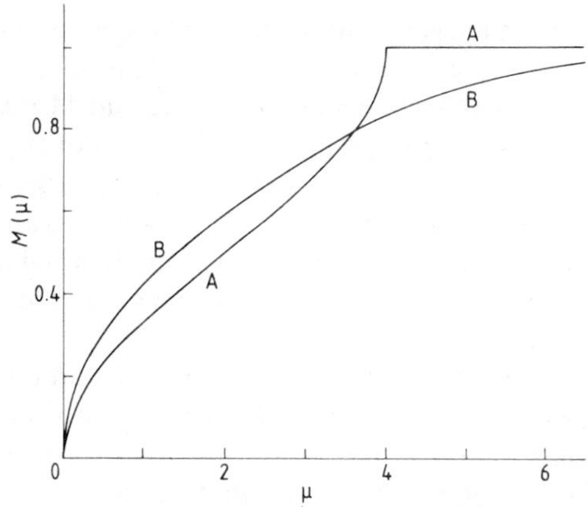

Figure 3.16 Integrated frequency spectrum $M(\mu)$ for Dyson's theory of a disordered linear chain. Curve A is for the perfectly periodic linear chain; there is a Van Hove singularity at $\mu = 4$. Curve B is for the disordered chain; the singularity at $\mu = 4$ now being washed out.

In the case of a perfectly periodic linear chain

$$M(\mu) = \begin{cases} \pi^{-1}\cos^{-1}(1 - \tfrac{1}{2}\mu) & \mu < 4 \\ 1 & \mu > 4 \end{cases} \tag{3.69}$$

as can easily be demonstrated from the results of §3.2. The singularity at $\mu = 4$ is clearly a reflection of the Van Hove singularity in $g_2(\mu)$ (see problem 3.1). As can be seen from figure 3.16, this is completely washed out in the very disordered chain. It is also interesting to note that for $\mu > 4$, $M_{\text{dis}}(\mu)$ is no longer a constant, in contrast to the perfect crystal. This indicates that $g_{\text{dis}}(\mu) = \mathrm{d}M_{\text{dis}}|\mathrm{d}\mu \neq 0$ for $\mu \geqslant 4$. This tailing off of the density of states in regions where $g_2(\mu)$ for the corresponding perfect crystal is zero is a rather general characteristic of disordered systems. In the present example, Dyson has demonstrated that the tailing off is exponential in nature (see Appendix 3.6).

Dean (1972) has reviewed the area of vibrational properties of glasses, following Dyson's pioneering work described above. Another area of considerable interest, both theoretically and experimentally, is that of randomly disordered crystals. Here, in contrast to the problem of glasses, there is an underlying lattice and one has compositional disorder. The presence of the background lattice has allowed a good deal of theoretical progress which is reviewed by Elliott *et al* (1974).

On the experimental side, as for phonons in perfect crystals, the technique of neutron inelastic scattering has been used to study the vibrational properties of such disordered crystals as $Cu_{1-x}Au_x$ (Svensson and Kamitakahara 1971) and $Cr_{1-x}W_x$ (Møller and Mackintosh 1965; see also Cunningham *et al* 1970). In addition, a great deal of experimental information on the optical properties of mixed crystals exists, especially alkali halides, alkaline earth fluorides, III–V and II–VI compounds (see the review by Chang and Mitra 1971); of course, this again contains information related to the vibrational properties of such disordered systems. Thermal properties of disordered systems also give information; as for perfect crystals (see §3.5). However, as an average over part of the vibrational spectrum is inevitably involved, thermal properties are less sensitive to the spectrum of elementary excitations than the neutron or optical studies.

To date, many of the theoretical calculations on disordered systems restrict themselves to the vibrational spectrum, i.e. the density of states. It is therefore worth remarking, in concluding this brief discussion, that one type of experiment that has given important information about the vibrational spectrum is superconducting tunnelling (for some relevant background see Chapter 5). As a specific example, Pb_{1-x} In_x has been investigated at low concentration x by Rowell *et al* (1965) and at large x by Adler *et al* (1966). What emerges is the vibrational density of states, weighted by some unknown (but slowly varying) coupling constant.

Problems

3.1 Use the dispersion relation (3.17) to calculate the number of modes $g(\omega)\mathrm{d}\omega$, lying in the (angular) frequency range between ω and $\omega+\mathrm{d}\omega$.

Notice that the spectrum $g(\omega)$ has singularities (infinities) at the extreme frequencies $\omega = 0$ and $\omega = \omega_{\max}$. What will happen at these frequencies in two and three dimensions?

Finally, discuss qualitatively, in a sentence or two, what will be the effect of introducing disorder by having a distribution of Hooke's law force constants in the springs connecting the atoms (for the reader who wants to study this quantitatively, the results of Dyson summarised in §3.9 should be studied).

3.2 Repeat the calculation in problem 3.1, but with the two different types of masses. Do not try to do all the mathematics this time; merely sketch the main features of the frequency spectrum $g(\omega)$.

3.3 Develop the Debye theory of the specific heat at temperatures high compared with the Debye temperature. Verify directly that the Dulong–Petit law is regained. Will this law hold in the presence of anharmonicity?

3.4 Explain why, when excitons are formed with emission and absorption of phonons, we have an exciton band spectrum with a well defined lower limit, rather than a line spectrum.

3.5 Using the information on the relative force constants in figure 3.8 for Na metal, construct the form of the dispersion relation from the Fourier series (3.49).

3.6 Employing the linear diatomic chain model of §3.3, derive the result (3.35) relating the gap in the frequency spectrum to the maximum frequency and the ratio of the masses.

3.7 For two protons in a high-density electron gas, use electrostatic arguments to derive from the screened potential around a proton the interaction energy $\phi(r)$. Show that at large separations r, $\phi(r) \sim [\cos(2k_F r)]/r^3$ (cf Corless and March 1961). Relate this behaviour to the form of the Lindhard dielectric function in Appendix 2.1.

3.8 In equations (3.23), put $l_A = 2n$, $l_B = (2n+1)$ and show that the ratio of the amplitudes $u_A(0)/u_B(0)$ is (cf equation (3.24))

$$\frac{u_A(0)}{u_B(0)} = \frac{2\gamma\cos kd}{2\gamma - m_A\omega^2} = \frac{2\gamma - m_B\omega^2}{2\gamma\cos kd}. \tag{P3.1}$$

Show that for small k:

(*a*) the negative sign in equation (3.25) gives

$$\omega_-(k) = \left(\frac{2\gamma}{m_A + m_B}\right)^{1/2} kd \tag{P3.2}$$

while equation (P3.1) gives $u_A(0)/u_B(0) = 1$ and

(*b*) the positive sign in equation (3.25) gives

$$\omega_+(k) = \left(2\gamma\frac{(m_A + m_B)}{m_A m_B}\right)^{1/2}\left(1 - \frac{m_A m_B k^2 d^2}{2(m_A + m_B)^2}\right) \tag{P3.3}$$

while equation (P3.1) gives $u_A(0)/u_B(0) = -m_B/m_A$.

Show further that at the zone boundary $k = \pi/2d$, the two solutions are

$$\left.\begin{array}{ll} \omega(k) = (2\gamma/m_A)^{1/2} & u_B(0) = 0 \\ \text{and} & \\ \omega(k) = (2\gamma/m_B)^{1/2} & u_A(0) = 0 \end{array}\right\} \tag{P3.4}$$

with one type of atom stationary in each case.

Finally, show that as $m_A \to m_B$ the gap at $k = \pi/2d$ disappears and the spectrum becomes that of the main atomic chain. [*Hint*. The top branch in figure 3.4 should be folded out to cover the region π/a to $2\pi/a$, where $a = 2d$.]

4

POLARONS AND THE ELECTRON–PHONON INTERACTION

So far, we have treated collective modes of an electron gas, with the ions smeared, and in the previous chapter the collective modes of the vibrating ions have been dealt with. In the present chapter, we shall consider an ionic lattice, such as NaCl, into which we introduce an electron in the conduction band. Considering the effect of this electron as it moves through the crystal lattice, it is clear that, providing it travels sufficiently slowly, the ions can respond to its motion. It will obviously attract positive ions and repel anions.

Let us put this simple physical statement into the phonon language of the previous chapter. We saw there that in the course of a longitudinal optic lattice vibration, positively and negatively charged ions move in opposite directions, the amplitude of this motion varying with position along the direction of oscillation. Therefore, in a variation of this type, alternating regions of positive and negative charge density are set up, in excess of the equilibrium values. As we have already pointed out, an electron introduced into the crystal, being negatively charged, can lower its energy by inducing a polarisation in the lattice, which brings extra positive charge into the region where its probability density is large and which pushes away some negative charge into regions where the electron wavefunction has small amplitude.

The above discussion focuses on the importance of the coupling between the electron introduced into the ionic lattice and the longitudinal optic phonons. This situation is most helpfully considered as one in which, as such an electron moves through the crystal, it carries with it an associated 'polarisation cloud'. The dressed electron, i.e. electron plus lattice polarisation, is a further quasi-particle called a polaron. In this way, the bare electron problem is replaced by the polaron problem, the merit of this picture being that the residual polaron–phonon interaction can be assumed to be small. It is clear that, as mentioned above, the polaron has a lower energy than the electron alone. However, it will have a larger effective mass, due to the fact that it carries with it lattice distortion as it traverses the crystal.

As with the exciton treated in Chapter 2, it is important to distinguish two regimes: the large- and the small-polaron regimes. These are characterised by the size of the deformed region of the lattice due to the presence of the electron. If this deformation extends over many lattice spacings, we have the large-polaron regime. Then one can exploit this fact by replacing the lattice by a continuum, just as was done in the Debye theory of the low-temperature lattice specific heat in the previous chapter. Of course, in the opposite limit in which the extent of the deformed region is of the order of the interionic spacing, the continuum approximation is not appropriate and we have the small-polaron regime.

We now proceed in a little more detail to consider these two limiting cases. The argument given below stems from the work of Landau (1933, see also Austin and Mott 1969).

Denoting the static dielectric constant by ϵ_0, it is clear that at a sufficiently large distance from the electron, the potential energy in which another electron in the crystal finds itself is $e^2/\epsilon_0 r$. However, if the ions could not move, it would be $e^2/\epsilon_\infty r$, ϵ_∞ being the high-frequency dielectric constant (see also § 4.5). Thus the potential energy, $V(r)$ say, due to the displacement of ions is given by

$$V(r) = -\frac{e^2}{r}\left(\frac{1}{\epsilon_\infty} - \frac{1}{\epsilon_0}\right). \tag{4.1}$$

Landau's argument was that because of this potential energy, it is possible for an electron moving through an ionic crystal to be trapped by its polarisation cloud. Of course, if such trapping were to occur it would then profoundly affect the mobility of the electron in the appropriate regime of temperature. We shall take up this point again below.

But for the moment, we assume that the potential $V(r)$ in equation (4.1) is valid beyond a radius r_p say, and that inside this radius the potential is a constant determined by the continuity of V at r_p. Then the problem is reduced to simply that of determining the radius r_p of the potential well, which in turn is a measure of the spread of the electronic wavefunction due to the interaction with the phonons. However, the following argument will show that not all the phonons can effectively interact with the electron. Suppose that, during a period of oscillation of the lattice phonon, say $2\pi/\omega$, the electron travels a distance which is much larger than the phonon wavelength. Then it is clear on physical grounds that the interaction between the electron and such a phonon will be greatly reduced. In other words, the electron–phonon interaction is restricted to those phonons having wavelength λ such that

$$\lambda > \frac{\hbar k}{m_e^*}\frac{2\pi}{\omega} \tag{4.2}$$

where $k = 2\pi/\lambda$ and m_e^* is the effective mass of the electron in the conduction band. The inequality (4.2) follows from the fact that in the interaction with a phonon of wavevector k, the electron acquires a recoil momentum $\hbar k$ and therefore a velocity $\hbar k/m_e^*$. The uncertainty in the momentum of the electrons produces a corresponding uncertainty Δx, in the position of the electron, which can be estimated from the Heisenberg uncertainty principle as $\Delta x \simeq (\hbar/m_e\omega)^{1/2}$. The important conclusion is that the polaron radius r_p is

$$r_p = (\hbar/2m_e^*\omega_0)^{1/2} \tag{4.3}$$

where we have now made the result more precise (cf Fröhlich's (1954a, b) articles on electrons in lattice fields). As already remarked, m_e^* is the band effective mass of the electron, while from the qualitative arguments at the beginning of this chapter it is clear that ω_0 is the average frequency associated with the longitudinal optic phonons.

The above relation (4.3) is valid only if the coupling to the phonons is small and multiphonon processes can be neglected. However, it can still be usefully employed in order to get a qualitative understanding of systems where the coupling is of intermediate strength, which we shall see later is the case for the alkali halides, even though in these materials multiphonon processes are of some relevance.

From relation (4.3), one can evidently estimate the degree of localisation of the polaron (i.e. the extent of the region of lattice deformation). In the case of the small polaron, as anticipated above, r_p is of the order of the interionic distance, while for the Fröhlich (large)-polaron limit the deformation extends over many interionic distances.

The extreme localisation of the small polaron can be expressed in terms of a polaron effective mass which becomes larger and larger as the coupling to the phonons is increased. Clearly, in the case of the large polaron the effective mass of the quasi-particle is still increased from the band effective mass but the increase is much smaller. We shall return to these considerations below when we discuss the mobility of the polaron; however let us return to the question of the polaron energy before discussing transport properties of polarons.

4.1 Energy of small polaron

Returning to the discussion of Landau summarised above, we arrive at the idea of the small polaron, the energy of which can be written as a sum of: (a) the energy required to polarise the medium, i.e.

$$\frac{e^2}{2r_p}\left(\frac{1}{\epsilon_\infty}-\frac{1}{\epsilon_0}\right); \tag{4.4}$$

and (b) the lowering of the potential energy of the electron, i.e.

$$-\frac{e^2}{r_p}\left(\frac{1}{\epsilon_\infty}-\frac{1}{\epsilon_0}\right). \tag{4.5}$$

Of course, such a use of a continuum theory for the small-polaron energy can at best be suggestive, for we have pointed out that only for the large polaron can the continuum model be really justified. However, if we combine equations (4.4) and (4.5) the polaron energy, $-W$ say, is given as

$$-W = -\frac{e^2}{2r_p}\left(\frac{1}{\epsilon_\infty}-\frac{1}{\epsilon_0}\right). \tag{4.6}$$

We shall return to this result later, but we now turn to the other limiting case, that of the large polaron.

4.2 Energy of large polaron

Related to the above discussion, the effective mass of the large polaron, though enhanced from the band effective mass, is not large enough to allow us to neglect the kinetic energy due to its localisation in the potential well (4.1). Such kinetic energy, of magnitude $\hbar^2\pi^2/2m^* r_p^2$, is then an important feature of the problem and we can write

$$-W = -\frac{e^2}{2r_p}\left(\frac{1}{\epsilon_\infty}-\frac{1}{\epsilon_0}\right)+\frac{\hbar^2\pi^2}{2m^* r_p^2}. \tag{4.7}$$

To determine r_p, we minimise this expression with respect to the polaron radius, whence we find

$$r_p = \frac{2\pi^2\hbar^2}{m^* e^2}\left(\frac{1}{\epsilon_\infty}-\frac{1}{\epsilon_0}\right)^{-1}. \tag{4.8}$$

Hence the magnitude of the large-polaron energy is given by

$$W = \frac{e^2}{4r_p}\left(\frac{1}{\epsilon_\infty}-\frac{1}{\epsilon_0}\right). \tag{4.9}$$

The treatment of Fröhlich (1954a) and Allcock (1956) makes a self-consistent calculation of the potential.

4.3 Coupling constant between optical modes and electron

When studying the polaron properties theoretically, one customarily starts from a Hamiltonian H which is the sum of three terms:

$$H = H_e + H_{ph} + H_{e\text{-}ph}. \tag{4.10}$$

Here H_e and H_{ph} are the appropriate Hamiltonians for the non-interacting electron and phonon systems respectively, while $H_{e\text{-}ph}$ describes the interaction between the two. We shall not go into the full explicit form of $H_{e\text{-}ph}$ here, but shall restrict ourselves to a discussion of its importance in different crystals. This can be gauged by introducing a dimensionless quantity α, to which $H_{e\text{-}ph}$ is proportional, and which is a measure of the strength of the electron–phonon interaction.

In order to estimate the coupling constant α, we assume that the electron from which the polaron has originated is spread uniformly over a distance of the order $(\hbar/2m^*\omega_0)^{1/2}$ given by equation (4.3). Then the relevant electrostatic energy in absolute terms, say $|E|$, is given by equation (4.4), with r_p replaced by the result (4.3). The ratio of $|E|$ to the phonon energy $\hbar\omega_0$ is the required coupling constant. A convenient form to use is

$$\alpha = \frac{e^2}{\hbar}\left(\frac{m}{2\hbar}\right)^{1/2}\left(\frac{1}{\epsilon_\infty} - \frac{1}{\epsilon_0}\right)\left(\frac{m^*}{m\omega_0}\right)^{1/2} = 1.4 \times 10^8 \frac{\epsilon_0 - \epsilon_\infty}{\epsilon_\infty \epsilon_0}\left(\frac{m^*/m}{\omega_0}\right)^{1/2} \tag{4.11}$$

where ω_1 is the longitudinal optical mode frequency while m^* is the band mass of the electron.

In table 4.1 we show typical values of the coupling constant α along with ϵ_0 and ϵ_∞, the static and high-frequency dielectric constants respectively. The assumption made, unless m^*/m is known, is that $m^*/m = 1$. Of the cases listed in table 4.1, it is useful to make a gross separation into highly polar substances such as the alkali halides, $\alpha > 3$, intermediate coupling where $\alpha \sim 1.5$, such as the silver halides, and weakly polar crystals such as the compound semiconductors. In a material like PbS, α is small because the optical dielectric constant is almost as large as the static value.

Table 4.1 Typical values of the coupling constant α together with ϵ_0 and ϵ_∞ for different crystals.

Crystal	a(Å)	α	ϵ_0	ϵ_∞
NaCl	5.63	5.5	5.6	2.2
KCl	6.28	5.6	4.7	2.1
Cu_2O	2.46	2.5	10.5	4.0
AgCl	5.55	3.6[a]	12.3	4.0

[a] Including m^*/m, this becomes 1.7.

There are certain materials such as NiO in which one expects to find highly localised polarons. These, as we have indicated, require different treatment (see e.g. Devreese 1972).

4.4 Mobility of small polaron

We have dealt in some detail with the energy of the polaron in the two limiting cases of large and small polarons. We return now to consider the polaron mobility.

We shall focus on the case of small polarons here. While the formation of large polarons, even in the strong coupling limit, does not lead to serious qualitative modification of the transport properties of band electrons, the opposite situation obtains for small polarons.

Holstein (1959) found that for the perfect crystal, small polaron states overlap sufficiently to allow the formation of a polaron band, in which ordinary band conduction can take place. This conduction can then be expected to predominate at low temperatures. However, the width of this polaron band can be shown to decrease exponentially with increasing temperature, and in the vicinity of half the Debye temperature the bandwidth becomes less than the uncertainty in energy due to the finite lifetime of polaron states.

Above this temperature, electrical transport in the polaron band becomes unimportant and the small polaron can be thought of as localised. This is simply the situation we started out from; it is the Landau situation in which the electron is trapped in the local lattice deformation it induces. As we have emphasised, the lattice deformation extends only to nearest neighbours in the small-polaron limit.

Once Landau trapping has occurred, the only way in which the polaron can contribute to conduction is by 'hopping' from one lattice position to an equivalent one. For such an equivalent position to occur, either the lattice round an unoccupied site must deform, the lattice around the polaron must relax, or some mixture of the two must take place. All such lattice deformations require energy in the form of longitudinal optical phonons. At low temperatures, there are few such phonons present and the hopping probability is small. However, at high temperatures the number of phonons increases exponentially and the hopping probability is much larger.

Hopping conduction can be regarded simply as a diffusion of carriers through the lattice, with the assistance of phonons. Since the diffusion constant is connected to the mobility through the Einstein relation, it is apparent that the mobility of a localised carrier conducting by hopping from site to site is of the form

$$\mu = \mu_0 \exp(-W_H/k_B T) \tag{4.12}$$

where W_H is the minimum energy necessary to obtain two equivalent sites (see just below equation (4.13)). The energy W_H can be readily evaluated in terms of the small-polaron binding energy W by noting that for small deviations around any equilibrium configuration the potential must vary quadratically with a suitable defined distortion parameter. A rough argument leading to the correct result is to say, with distortion d, that (cf. Adler 1975)

$$W \propto d^2 \qquad W_H \propto (\tfrac{1}{2}d)^2 + (\tfrac{1}{2}d)^2 \qquad \text{or } W_H = \tfrac{1}{2}W \tag{4.13}$$

the expression for W_H in equation (4.13) coming loosely from the fact that the minimum energy to obtain an equivalent site occurs in a situation in which the region around the carrier distorts 'halfway'; this requiring only a total of half the polaron binding energy since the latter requires a full relaxation of the polaron site. Thus the mobility formula (4.12) becomes

$$\mu = \mu_0 \exp(-W/2k_B T). \tag{4.14}$$

The pre-exponential factor μ_0 can be somewhat dependent on temperature. Thus a factor T^{-1} enters from the Einstein relation

$$\mu = D/k_B T \tag{4.15}$$

relating mobility and the diffusion constant D (see figure 4.1 and also Appel 1968). For the small polaron at high temperatures, Holstein (1959) finds an additional factor $T^{-1/2}$ in D so that μ_0 varies as $T^{-3/2}$. Of course, this is not the dominant variation, the exponential factor itself being the most important dependence on T.

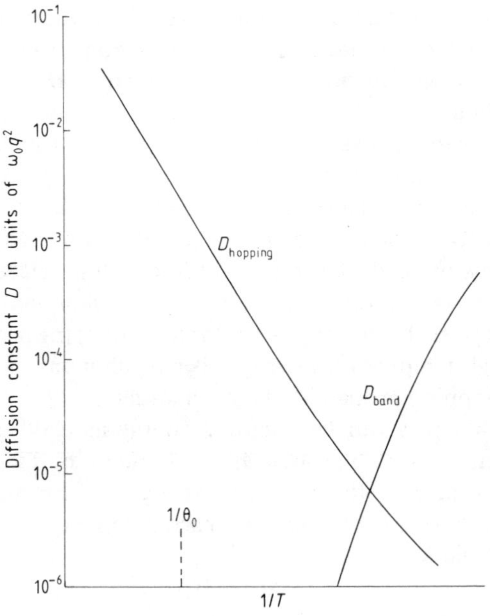

Figure 4.1 Illustrates temperature dependence of mobility for small polarons in hopping and band regimes. Diffusion constant is plotted: it is related to the mobility by the Einstein relation (4.15). Energy band transport occurs at low temperatures, whereas hopping takes place above a transition temperature of the order of $\theta_0 = \hbar\omega_0/k_B$.

It should be stressed here that hopping conduction via localised states is a very different process from band conduction occurring in delocalised states. But since both processes require temperature activation in semiconducting solids, they cannot be distinguished by electrical conductivity measurements alone. For both hopping conduction and semiconduction, for the conductivity σ we expect to find the form

$$\sigma = \sigma_0 \exp(-E_a/k_B T) \tag{4.16}$$

where E_a is the experimentally determined activation energy. The unique feature of hopping conduction is the activation energy for mobility, as in equation (4.12).

This concludes the discussion of the polaron; we turn now to treat the dielectric properties of ionic crystals in a phenomenological way.

4.5 Dielectric function of ionic crystals

In Chapter 3, it was pointed out that optical experiments could be used in ionic crystals to probe the phonon branches (optical) for which $\omega(\boldsymbol{q}) \rightarrow$ constant $\neq 0$ as $\boldsymbol{q} \rightarrow 0$. In this section, we shall derive a precise form for the dielectric constant from which the optical properties can be deduced.

Following Born and Huang (1954), we write down the phenomenological equations for the displacement $\boldsymbol{w}$ of the atoms from their equilibrium positions when they execute a long-wavelength optic vibration. These equations read:

$$\ddot{\boldsymbol{w}} = b_{11}\boldsymbol{w} + b_{12}\boldsymbol{\mathscr{E}} \tag{4.17}$$

and

$$\boldsymbol{P} = b_{21}\boldsymbol{w} + b_{22}\boldsymbol{\mathscr{E}}. \tag{4.18}$$

Here the b_{ij} are phenomenological coefficients† to be identified later, while $\boldsymbol{P}$ is the polarisation vector. This is related to electric displacement $\boldsymbol{D}$ and electric field $\boldsymbol{\mathscr{E}}$ as usual by

$$\boldsymbol{D} = \boldsymbol{\mathscr{E}} + 4\pi\boldsymbol{P}. \tag{4.19}$$

Taking Fourier transforms of equations (4.17) and (4.18) yields

$$-\omega^2\boldsymbol{w}(\omega) = b_{11}\boldsymbol{w}(\omega) + b_{12}\boldsymbol{\mathscr{E}}(\omega) \tag{4.20}$$

$$\boldsymbol{P}(\omega) = b_{21}\boldsymbol{w}(\omega) + b_{22}\boldsymbol{\mathscr{E}}(\omega) \tag{4.21}$$

and $\boldsymbol{w}(\omega)$ can be eliminated to yield a relation between $\boldsymbol{P}$ and $\boldsymbol{\mathscr{E}}$. This takes the form

$$\boldsymbol{P} = \left(b_{22} + \frac{b_{12}b_{21}}{-b_{11} - \omega^2}\right)\boldsymbol{\mathscr{E}} \tag{4.22}$$

which, using equation (4.19), yields for the electric displacement

$$\boldsymbol{D} = \left(1 + 4\pi b_{22} + \frac{4b_{12}b_{21}}{-b_{11} - \omega^2}\right)\boldsymbol{\mathscr{E}}. \tag{4.23}$$

From the definition of the dielectric constant in equation (2.28) one has the frequency-dependent form

$$\epsilon(\omega) = 1 + 4\pi b_{22} + \frac{4\pi b_{12}b_{21}}{-b_{11} - \omega^2}. \tag{4.24}$$

† It can be shown that $b_{12} = b_{21}$; we refer the interested reader to the book by Born and Huang (1954) for the detailed analysis.

This equation can be recast in the form

$$\epsilon(\omega) = \epsilon_\infty + \frac{\epsilon_0 - \epsilon_\infty}{1 - (\omega/\omega_{\mathrm{TO}})^2} \tag{4.25}$$

where we have defined

$$\begin{aligned} 1 + 4\pi b_{22} &= \epsilon_\infty \\ \omega_{\mathrm{TO}}^2 &= -b_{11} \\ \epsilon_0 - \epsilon_\infty &= -4\pi b_{12} b_{21}/b_{11}. \end{aligned} \tag{4.26}$$

Evidently ϵ_0 is the static dielectric constant $\epsilon(\omega = 0)$ while ϵ_∞ gives the response of the system at very high frequencies. 'High' here means frequencies such that the ions are no longer able to respond to the external field, yet low enough for the electronic transitions not to take place.

The meaning of $b_{11} = -\omega_0^2$ can be deduced from equation (4.17) if we split $\boldsymbol{w}$ into longitudinal $\boldsymbol{w}_\mathrm{l}$ and transverse $\boldsymbol{w}_\mathrm{t}$ components. As we know that $\boldsymbol{\mathscr{E}}$ is purely longitudinal, since $\boldsymbol{\mathscr{E}}$ is an irrotational vector satisfying curl $\boldsymbol{\mathscr{E}} = 0$, the transverse part of equation (4.17) then becomes

$$\ddot{\boldsymbol{w}}_\mathrm{t} = b_{11} \boldsymbol{w}_\mathrm{t}. \tag{4.27}$$

This tells us that $\omega_{\mathrm{TO}}^2 = -b_{11}$ has to be identified with the eigenfrequencies of the transverse optic (TO) mode. From this it follows that the dielectric function is resonant at the TO mode frequency.

One could now evaluate the longitudinal optic (LO) mode frequency, by considering the other component of the vector equation (4.17). However, we prefer here to find the LO mode frequency ω_{LO} from the zeros of the dielectric function. The condition $\epsilon(\omega) = 0$ then gives

$$\omega_{\mathrm{LO}}^2 = \omega_{\mathrm{TO}}^2 \epsilon_0/\epsilon_\infty, \tag{4.28}$$

which is the Lyddane–Sachs–Teller relation between the long-wavelength optic mode frequencies.

As in §2.4, the connection to optical properties is readily established. We wish here to focus on the reflectivity coefficient R, which for normal incidence is given in terms of the complex refractive index $n + \mathrm{i}K$ by

$$R = \frac{(n-1)^2 + K^2}{(n+1)^2 + K^2}. \tag{4.29}$$

Using equations (4.24) and (4.29), R can easily be calculated. It is readily shown that R is very large in the region of frequencies between ω_{TO} and ω_{LO}, falling sharply away outside this range.

It is important to emphasise that equation (4.24) for $\epsilon(\omega)$ is only valid in the lower part of the frequency range where $\hbar\omega$ is much less than the electronic transition frequencies. In practice, for alkali halides this means that one has to perform experiments in the infrared region.

4.6 Electronic dipoles and models of phonons beyond rigid ions

The force-constant matrix of lattice dynamics, which can be regarded as directly related to the dynamical matrix of Chapter 3, contains two major classes of contributions: electrostatic interactions and short-range terms associated with the overlap of neighbouring atoms. Following Stoneham (1975), we focus on the dipole approximation, which is the assumption that quadrupole and higher multipole interactions can be ignored.

One very general consequence of the adiabatic, harmonic and electrostatic approximations (which ignore retardation and related effects) is the Lyddane–Sachs–Teller relation discussed above (cf equation (4.28)), which relates the optic mode frequencies to the dielectric constants.

The dipoles whose interaction makes a contribution to the force-constant matrix, and hence to the dynamical matrix, are of several different types (cf Stoneham 1975):

(a) Permanent dipoles associated with any species which can be taken to move rigidly in the lattice motion considered;
(b) Electronic dipoles from the electronic polarisation of the ions;
(c) Displacement dipoles from the rigid movement of charges;
(d) Deformation dipoles, which occur because the electronic and displacement dipoles are not additive; in effect the electronic polarisability is altered by the relative motion of neighbouring ions.

In Chapter 3 we treated the rigid-ion model of lattice dynamics in which electronic polarisation is neglected and only the displacement dipoles are accounted for. As we stressed, such a rigid-ion model puts the optical dielectric constant ϵ_∞ to unity. This model is a natural development of Born's theory of cohesion and in a number of respects describes the lattice dynamics of ionic crystals usefully, as seen in Chapter 3. However, beyond this rigid-ion model, dipole approximations can take on a variety of degrees of refinement, a critical survey being that of Cochran (1971).

We refer here to two other classes of dipole approximations:

(1) The point-polarisable ion model. Here both electronic and displacement dipoles are accounted for. The ionic polarisabilities are normally chosen to yield the optic dielectric constant ϵ_∞ to agree with experiment.

(2) The shell model and related approximations. These models, unlike (1), include electronic, displacement and deformation dipoles; the various types of treatment being reviewed by Cochran (1971). The models include the shell model of Dick and Overhauser (1958) and the deformation dipole model (Szigeti 1950, Hardy 1961, 1962). The common feature of all these models is that the energy of the distorted lattice is written in a general form quadratic in the electronic dipoles and displacements. The rigid-ion limit obviously corresponds to putting the electronic dipoles equal to zero.

The shell model, more specifically, regards each ion as a core of charge $X|e|$ and a shell of charge $Y|e|$, where $(X+Y)|e|$ is just $Z|e|$, the full ionic charge. Roughly speaking the core corresponds to the nucleus and the inner electrons, while the shell includes the valence electrons most influenced by the neighbouring ions. The relative motion of the shell and core, regarded as point charges, gives naturally the electronic dipole moment of the ion. The core has all the ion mass while the shell is assumed massless. There are evidently three types of force constant: (a) shell–shell interactions, which dominate in the short-range interatomic repulsion; (b) shell–core forces, including the interactions within each ion; and (c) core–core terms.

The shell model, and its refinements, have had some success in describing lattice dynamical and dielectric properties of ionic crystals, most notably the less-polarisable ones. The so called breathing shell model (Schröder 1966) includes radial deformations of the ions as an extra degree of freedom and thereby allows some discrepancies between the simple shell model and experiment to be avoided. In particular the longitudinal optic modes in the alkali halides are appreciably improved by the breathing shell model refinement.

4.6.1 Dielectric function in periodic crystals

In concluding this brief discussion of electronic effects in lattice dynamics, it is important to recognise that in the case of a crystal (and in contrast to jellium), the electrons form an inhomogeneous system which has the same periodicity as the lattice. This periodicity will be reflected in the screening characteristics of the crystal and the dielectric function ($\epsilon(k)$ for jellium) will now depend both on wavevector $\boldsymbol{k}$ and on the reciprocal lattice vectors $\boldsymbol{G}$. This leads to the so-called dielectric matrix $\epsilon(\boldsymbol{k}+\boldsymbol{G},\ \boldsymbol{k}+\boldsymbol{G}')$. The elements of this matrix are labelled by the $\boldsymbol{G}$ vectors and in principle the matrix has an infinite number of components. Explicit calculations of this matrix are of course difficult since they involve knowledge of the full energy band structure and the numerical evaluation of many matrix elements.

However, two limiting cases can be considered. The first is relevant to simple (s–p) metals where the distribution of the conduction electrons is only slightly modulated by the lattice. In such cases, the jellium model is a good starting point, and a perturbation expansion in terms of the effective lattice potential is appropriate. Naturally, if such an expansion is valid, the off-diagonal elements of $\epsilon(\boldsymbol{k}+\boldsymbol{G},\ \boldsymbol{k}+\boldsymbol{G}')$ are small. More interesting in the present context is the opposite case of crystals with well-localised electrons. In this case a tight-binding (linear combination of suitably hybridised atomic orbitals) description is appropriate and this allows one to study certain general properties of $\epsilon(\boldsymbol{k}+\boldsymbol{G},\ \boldsymbol{k}+\boldsymbol{G}')$ without resorting to elaborate computations. In particular, the relation between the microscopic $\epsilon(\boldsymbol{k}+\boldsymbol{G},\ \boldsymbol{k}+\boldsymbol{G}')$ and the macroscopic dielectric function has been established, thereby giving theoretical support to the ideas of Clausius and Mossotti.

A particularly useful application of the dielectric formalism is to the calculation of interatomic forces in crystals. Here Sinha (1969) has shown that the shell model described above can be viewed as based on a factorisation property of this dielectric function appropriate to a periodic crystal. He argues that such a factorisation is a consequence of energy bands which can be described by the tight-binding approximation. This approach to lattice dynamics has been treated, for example, by Claesson *et al* (1973), following earlier work by Jones and March (1970, 1973). The article by Sham (1974) is recommended for the reader who wishes to supplement this brief introduction.

4.7 Polaritons in ionic crystals

We are now in a position to discuss the effects that arise when electromagnetic radiation traverses an ionic crystal. If the ionic positions were kept fixed then the only influence of the crystal on the light would be due to the electrons whose only effect is to introduce a uniform dielectric constant ϵ_∞; this reduces the light velocity from c to $c/\epsilon_\infty^{1/2}$. However, as soon as the ions are allowed to respond to the electromagnetic field of the light wave then the lattice vibrations and light become strongly coupled and from their interaction new kinds of collective excitations arise. The dispersion of these new modes can be found, recalling that the equation for the propagation of light in a medium of uniform non-dispersive dielectric constant ϵ is

$$\nabla^2 \mathscr{E} = \frac{\epsilon}{c^2}\frac{\partial^2 \mathscr{E}}{\partial t^2} \tag{4.30}$$

which can be written in reciprocal space as

$$k^2 \mathscr{E} = \frac{\epsilon}{c^2}\omega^2 \mathscr{E}. \tag{4.31}$$

In our case we cannot consider ϵ as a constant but we have to consider the dependence on $\boldsymbol{k}$ and ω. This can be done by writing equation (4.31) in the form (Landau and Lifshitz 1962b):

$$k^2 \mathscr{E}(k, \omega) = \frac{\epsilon(\boldsymbol{k}, \omega)}{c^2}\omega^2 \mathscr{E}(\boldsymbol{k}, \omega). \tag{4.32}$$

In the present problem the dependence of $\epsilon(\boldsymbol{k}, \omega)$ on $\boldsymbol{k}$ can be neglected and we can put $\epsilon(\boldsymbol{k}, \omega) = \epsilon(0, \omega) = \epsilon(\omega)$. In fact while k for light in the visible is of the order of 10^{-3} Å^{-1}, a typical crystal wavevector (say a reciprocal lattice vector) is of the order of 1 Å^{-1}. Therefore the frequencies of light propagating in a dispersive medium can be obtained from the equation:

$$c^2 k^2/\omega^2 = \epsilon(\omega). \tag{4.33}$$

If we now use for $\epsilon(\omega)$ the expression given in equation (4.25), we obtain for the

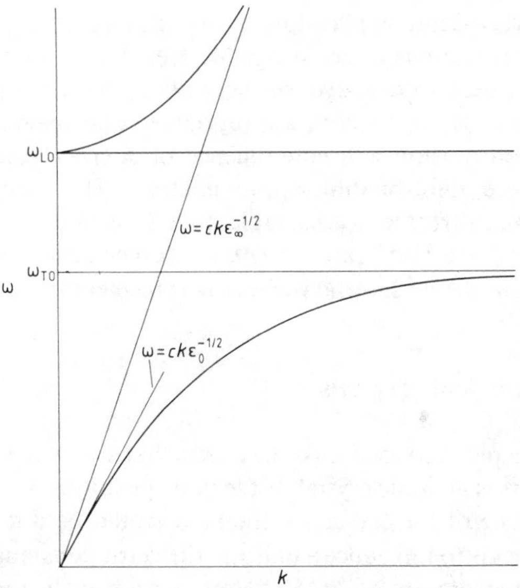

Figure 4.2 Illustrating polariton modes. Straight lines represent limiting cases discussed in the text. Coupling between light mode and phonon mode leads to dispersion curves as shown.

roots of this equation the two branches depicted in figure 4.2. Two limiting regimes may be considered:

(a) $ck \gg \omega_{TO}$. In this case the light frequency ck is so much higher than the lattice frequencies that the coupling between lattice and light vibrations disappears and one finds a light mode $ck/\epsilon^{1/2}$ and a vibrational mode ω_{TO}.

(b) $ck \ll \omega_{TO}$. For these low frequencies, the upper root can be obtained by putting $k = 0$ in equation (4.33) and solving for $\epsilon(\omega) = 0$. As discussed in the preceding section, this gives the longitudinal optic mode frequency ω_{LO}. In this limit the lower branch can be obtained by putting $\epsilon(\omega) \simeq \epsilon_0$ in the right-hand side of equation (4.33). Thus it has the form $ck/\epsilon_0^{1/2}$.

While in these limiting cases one recovers a light mode and a phonon mode, in the intermediate region these are so strongly coupled that the combination of the two has to be considered. This new entity, which for ionic crystals has the dispersion depicted in figure 4.2, is called a polariton. Polaritons can also arise from the interaction of the light with a variety of collective bulk and surface modes and have been observed mainly by means of light scattering (Raman effect).

Finally, we note in figure 4.2 a region of forbidden frequencies between ω_{TO} and ω_{LO}. This corresponds to the region of perfect reflectivity discussed in §4.5.

4.8 Electron–phonon interaction and Kohn anomalies

In the discussion of phonons in the previous chapter, we have not explicitly considered the possibility that when a phonon is excited in a crystal it causes a perturbation of the electronic charge density. This, in turn, will react back on the ions, causing a variation in the interionic potential and therefore in the phonon frequency. If the ion–electron interaction is weak, the response of the electrons to the change in ionic potential due to the vibration can be treated in first-order perturbation theory. This leads to the result that if the change in potential $\delta V(\boldsymbol{r})$ is Fourier analysed as

$$\delta V(\boldsymbol{r}) = \sum_{\boldsymbol{q}} \delta V(\boldsymbol{q}) \exp(\mathrm{i}\boldsymbol{q}\cdot\boldsymbol{r}) \tag{4.34}$$

then the $\boldsymbol{q}$-component of the displaced density $\delta n(\boldsymbol{r})$ is related to $\delta V(\boldsymbol{q})$ by the relation

$$\delta n(\boldsymbol{q}) = -\chi(\boldsymbol{q})\delta V(\boldsymbol{q}) \tag{4.35}$$

where $\chi(\boldsymbol{q})$ is a function of the undistorted electronic state which we assume to be a uniform electron gas. $\chi(\boldsymbol{q})$ in the free-particle approximation can be calculated for systems of different dimensionalities†. The results are shown in figure 4.3, where it is seen that for the one-dimensional electron gas there is a

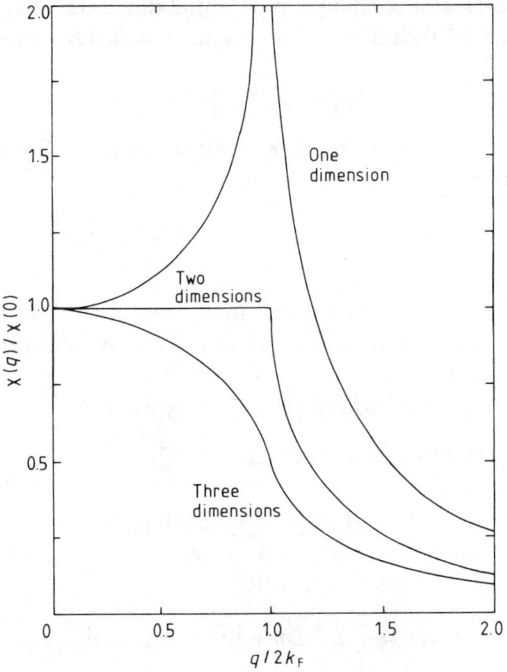

Figure 4.3 Response function $\chi(q)$ of a Fermi gas in one, two and three dimensions.

† Comparison of equations (4.35) and (A2.1.10) shows that in three dimensions $\chi(\boldsymbol{q})$ is the Fourier transform of $j_1(x)/x^2$ (see problem 4.5).

remarkable singularity at $q = 2k_F$. This is present to a lesser degree in two dimensions, while in three dimensions only a kink in $\chi(\boldsymbol{q})$ at $q = 2k_F$ remains.

This different behaviour is not surprising in view of the discussion presented in §2.7, where we have seen that the effect of adding a perturbation is to yield a different asymptotic behaviour of the displaced density $\Delta n = n(\boldsymbol{r}) - n_0$ for different dimensionalities of the electron gas. Thus in one dimension we obtained $\Delta n \sim \sin(2k_F r)/r$ while in three dimensions $\cos(2k_F r)/r^3$ results. From known theorems on Fourier transforms (Lighthill 1958) it follows that different asymptotic behaviour in $\boldsymbol{r}$-space is a reflection of different singularities in $\boldsymbol{q}$-space as can readily be seen from figure 4.3. It should also be added that at non-zero temperature these singularities at $q = 2k_F$ tend to be blurred out and therefore become less important. $\chi(\boldsymbol{q})$ at $T \neq 0$ is referred to in problem 4.5 and will be utilised below.

4.8.1 Peierls distortion

We can now return to the problem of the electron–phonon interaction. Once again, in the formal developments we restrict ourselves to one dimension, which in view of the pronounced singularity at $q = 2k_F$ is also the most interesting case. Here, the change in potential due to the displacement of the atoms from the equilibrium position can be modelled so as to be linear in the displacements:

$$\delta V = \tfrac{1}{2}\gamma[(u_{l+1} - u_l) + (u_{l-1} - u_l)] \tag{4.36}$$

where γ is taken to be a constant in order to simplify the calculations. The Fourier transform of this equation gives

$$\delta V(\boldsymbol{q}) = \tfrac{1}{2}\gamma \sum_N \exp(-\mathrm{i}\boldsymbol{q}\cdot\boldsymbol{r}_l)[(u_{l+1} - u_l) + (u_{l-1} - u_l)] \tag{4.37}$$

where N is the total number of atoms. Finally this form of $\delta V(\boldsymbol{q})$ can be used to calculate the electron density change $\delta n(\boldsymbol{q})$ and the change in energy associated with $\delta V(\boldsymbol{q})$, which is

$$-\tfrac{1}{2}n(\boldsymbol{q})\delta V(\boldsymbol{q}) = -\tfrac{1}{2}\chi(\boldsymbol{q})\delta V(\boldsymbol{q})^2. \tag{4.38}$$

Hence the total Hamiltonian becomes

$$H = H_0 - |e| \sum_{\boldsymbol{q}} \tfrac{1}{2}\chi(\boldsymbol{q})\delta V(\boldsymbol{q})^2 \tag{4.39}$$

where

$$H_0 = \sum_l P_l^2/2m + \tfrac{1}{2}\sum_l K(u_{l+1} - u_l)^2 \tag{4.40}$$

is the Hamiltonian of the unperturbed linear chain that we solved fully in Chapter 3, the eigenfrequencies ($\omega_0(q)$ say) being given by equation (3.17) (with force constant replaced by K_0). The Hamiltonian (4.39) can be solved rather easily and the final result for the eigenfrequencies in the presence of the electron–phonon interaction is (see comment at the end of problem 4.5)

$$\omega_q^2 = \omega_0^2(q)\left(1 - \frac{\gamma^2|e|}{Ka}\sin^2(\tfrac{1}{2}qa)\chi(q,T)\right) \tag{4.41}$$

where we have explicitly shown the dependence on $\chi(q,T)$. At high temperatures, where the singularity in χ at $q = 2k_F$ is smeared out, the effect of the electron–phonon interaction is simply to produce a dip at $2k_F$ in the phonon dispersion curve, as illustrated in figure 4.4. Lowering the temperature enhances the value of $\chi(2k_F, T)$ until a temperature T_c is reached at which

$$1 - \frac{\gamma^2|e|}{Ka}\sin^2(k_F a)\chi(2k_F, T_c) = 0. \tag{4.42}$$

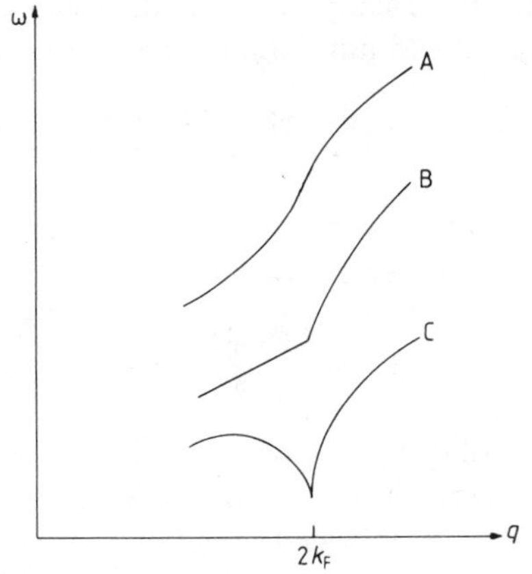

Figure 4.4 Schematic forms of phonon dispersion relation, including Kohn anomaly (dip) in one-dimensional metal. Curve A corresponds to the highest temperature plotted; curves B and C showing the effect of lowering the temperature as represented through equation (4.41).

At this temperature, the phonon dispersion curve touches the axis. This means that it costs no energy to create a phonon of wavelength $q = 2k_F$. And in fact, for temperatures $T < T_c$ a permanent distortion of wavevector $2k_F$ is set up in the system. This transition from undistorted to distorted ground state is called the Peierls transition. A similar mechanism also operates in two and three-dimensional crystals. But there, unless peculiarly favourable circumstances are met, the mechanism will produce only relatively small irregularities (Kohn anomalies) in the phonon dispersion curves.

The above discussion is certainly oversimplified as important features such as those introduced by fluctuations and by electron–electron correlations have been ignored. These are likely to be of particular importance in one-dimensional systems, but will nevertheless not change the qualitative results presented above. Further discussion of the Peierls transition will be deferred to Chapter 8.

4.9 Bohm–Staver formula for velocity of sound in metals

We conclude this discussion of the effect of the electron gas on ionic vibrations in metals by returning to the long-wavelength limit and dealing with the velocity of sound in simple metals.

Consider the vibration of the ions, say in metallic Na, in the background of a uniform distribution of neutralising negative charge; this is the inverted jellium problem. Provided the electron cloud is held fixed in this way, then the characteristic frequency of the ionic vibrations is the ionic plasma frequency. If the ionic mass is M, the ionic charge is ze and the mean ionic density is n_i, then from equation (2.12), applied now to ions, we find

$$\omega_p^{\text{ions}} = \left(\frac{4\pi n_i (ze)^2}{M}\right)^{1/2}. \tag{4.43}$$

Now for a metal of valency z, we have that the mean electron density n_0 is zn_i, and hence equation (4.43) may be written

$$\omega_p^{\text{ions}} = \left(\frac{4\pi n_0 ze^2}{M}\right)^{1/2}. \tag{4.44}$$

But, of course, by holding the electron cloud fixed we have neglected the shielding of the ions by the electrons. In the long-wavelength limit, we can argue from either the general screened Coulomb form (2.13), transformed into wavenumber space (i.e. Fourier transformed), or from the dielectric constant derived in Appendix 2.1, that the ionic charge ze must be screened in such a way that

$$\frac{ze}{k^2} \rightarrow \frac{ze}{k^2+q^2} \quad \text{or} \quad ze \rightarrow \frac{zek^2}{q^2}$$

where q^{-1} is the screening radius and k is the wavenumber. Making this replacement in equation (4.44), we find, as we must, that for long-wavelength phonons in a simple metal like BCC Na the phonon frequency ω is proportional to k. If we eliminate n_0 and q using equations (A2.1.2) and (A2.1.6) in favour of the Fermi velocity v_F then we find

$$\omega = v_s k = \left(\frac{zm_e}{3M}\right)^{1/2} v_F k. \tag{4.45}$$

We see therefore that the velocity of sound v_s in a simple metal is obtained from the Fermi velocity v_F by reducing it by a factor dominated by the term $(m_e/M)^{1/2}$, that is by the square root of the ratio of electron to ion mass (Bohm and Staver 1952, Bardeen and Pines 1955).

4.9.1 Interactions between planes of ions in metallic sodium

As discussed in §3.6.1, for phonons propagating along high symmetry directions, such as a body-diagonal or cube edge in a cubic lattice, the physical motion corresponds to a rigid displacement of those planes of atoms normal to the direction of propagation. As shown by Foreman and Lomer (1957), and set out in §3.6.1, the problem becomes formally equivalent to a linear chain. The successive Fourier coefficients of the dispersion curve are simply related to the interaction of an atom with more and more distant neighbours.

In metals, as summarised in §3.6.1, the experimentally determined interplanar force constants do not decrease monotonically with the interplanar separation, but oscillate in sign. Following Koenig (1964), we consider the force on an ion in the zeroth plane due to a small rigid displacement $\boldsymbol{l}$ of the entire nth plane, all other planes being held fixed. The vector $\boldsymbol{l}$ will be along the direction of propagation for a longitudinal mode and perpendicular to it for a transverse mode. The screening of this displaced plane by the almost-free conduction electrons in Na metal will be equivalent to the screening of a similar plane of dipoles $e\boldsymbol{l}$ by the electron gas.

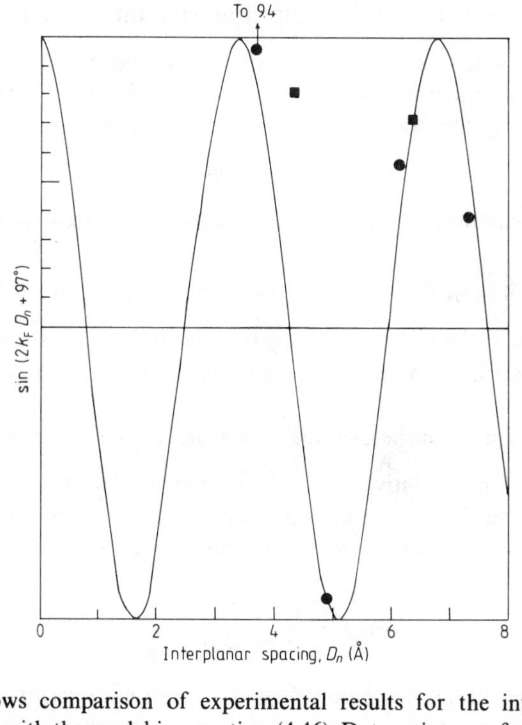

Figure 4.5 Shows comparison of experimental results for the interplanar force constants of Na with the model in equation (4.46). Data points are from Woods *et al* (1962). (From Koenig S H 1964 *Phys. Rev.* **135** A 1693.) Theory curve should be valid beyond about 4.5 Å.

For longitudinal phonons, Koenig shows that the force F_n^{L} is given by

$$F_n^{\mathrm{L}} = 8\pi^3 de\,|\,f_{k_{\mathrm{F}}}(\pi)|\,l \sin(2k_{\mathrm{F}} D_n + \phi)/D_n^2 \tag{4.46}$$

where D_n is the interplanar distance while d is the density of ions in the plane. The factor f is an appropriate scattering amplitude, the detail of which need not concern us here. Koenig took data for interplanar force constants from the work of Woods *et al* (1962) summarised in §3.6.1 that described longitudinal phonons propagating along the $\langle 100 \rangle$ and $\langle 111 \rangle$ directions, adjusted the data for the two directions by dividing by the appropriate density d, multiplying by D_n^2 and then obtaining the best fit to these points with a curve of the form $\sin(2k_{\mathrm{F}} D_n + \phi)$, where $k_{\mathrm{F}} = 0.92 \times 10^8\ \mathrm{cm}^{-1}$, the value for Na. The result is shown in figure 4.5. An analogous equation was obtained for transverse modes.

Problems

4.1 Using perturbation theory, make an estimate of the number of acoustic phonons accompanying a slow electron moving through a polar crystal.

4.2 At very low temperatures, the only states accessible to the polaron are those of small momentum in which any internal degrees of freedom are not excited. Assuming the polaron energy spectrum to be

$$E = E_0 + (p^2/2m^*) + cp^4 \tag{P4.1}$$

show that the partition function Z of the decoupled polaron–phonon system is determined by

$$\ln Z = \ln Z_0 - \beta(E_0 - E_{\mathrm{g}}) + \tfrac{3}{2}\ln(m^*/m) - (15m^{*2}c/\beta) \tag{P4.2}$$

where $\beta = 1/k_{\mathrm{B}}T$. Here Z_0 is the partition function of the entire non-interacting electron–phonon system and E_{g} is the non-interacting ground-state energy.

Could this expression be used to give a definition of the polaron mass?

4.3(a) Making the relatively rough approximation that transverse and longitudinal waves have a common velocity v_{s}, show from elementary Debye theory that the Debye temperature Θ_{D} is given in terms of the atomic volume $\mathscr{V}$, by

$$\Theta_{\mathrm{D}} = \frac{v_{\mathrm{s}}}{\mathscr{V}^{1/3}}\left(\frac{3}{4\pi}\right)^{1/3}\frac{h}{k_{\mathrm{B}}}. \tag{P4.3}$$

(b) Mukherjee (1965) has established the empirical relation

$$\Theta_{\mathrm{D}} = CE_{\mathrm{v}}^{1/2}/\mathscr{V}^{1/3}M^{1/2} \tag{P4.4}$$

in close-packed metals between the energy of formation of a vacancy E_{v}, the

Debye temperature Θ_D and the ionic mass M. Use the Bohm–Staver formula to show that $E_v \propto zE_F$, where z is the valency and E_F is the Fermi energy. (However, the more useful formula is $E_v \sim Mv_s^2$.)

4.4 Estimate the velocity of sound in Na metal, given that the mean interelectronic spacing r_s is $4a_0$, with a_0 the Bohr radius.

4.5 Use the theory given in Appendix 2.1 to show that the function $\chi(\boldsymbol{q})$ in equation (4.35) has a singularity of the type $\ln|1-x/1+x|$, $x = q/2k_F$, in three dimensions.

To deal with time-dependent perturbations we write

$$\delta n(\boldsymbol{q}\omega) = -\chi(\boldsymbol{q}\omega)\delta V(\boldsymbol{q}\omega)$$

which reduces to equation (4.35) when the frequency ω tends to zero. By writing $\chi(q0) \equiv \chi(q)$ generalised the elevated temperature $T \neq 0$ as, (apart from unimportant constants)

$$\chi(q, T) = \sum_{\boldsymbol{p}} \frac{n_{\boldsymbol{q}+\boldsymbol{p}} - n_{\boldsymbol{p}}}{\epsilon_{\boldsymbol{q}+\boldsymbol{p}} - \epsilon_{\boldsymbol{p}}} \tag{P4.5}$$

where $n_{\boldsymbol{p}}$ is the Fermi distribution function for state $\boldsymbol{p}$ and $\epsilon_{\boldsymbol{p}} = p^2/2m$, give an argument for why $\chi(q, T, \omega)$ is obtained from equation (P4.5) by adding $\hbar\omega$ in the denominator.

Note then that in equation (4.41), for $T > T_c$, $\chi(qT)$ should really be interpreted as $\chi(q, T\ \omega = \omega_q)$ which, however, still leads to equation (4.42).

5

SUPERCONDUCTIVITY

The electrical resistance of a pure metal arises because the travelling waves representing conduction electrons moving through the metal are scattered by the vibrating ions.

Kamerlingh Onnes (1913) observed that Hg at 4 K lost all its electrical resistance, and similar behaviour has subsequently been found in many, though not all, metals at low temperatures. The phenomenon is so dramatic that once a superconducting current is established, it has been observed to flow for more than a year, without detectable dissipation.

How is it possible for such a superconducting transition to take place? This problem was tackled phenomenologically by many workers and it was already clear to the London brothers in the 1930's that what one was seeing in the phenomenon of superconductivity was essentially an illustration of a quantum state on a macroscopic scale.

5.1 Electronic specific heat

One early clue towards an understanding of superconductivity was concerned with the electronic specific heat. As remarked in Chapter 3, while the lattice contribution to the specific heat is proportional to T^3 at sufficiently low temperatures, the electronic specific heat of a normal metal is proportional to T. This latter result is easily understood qualitatively. We can say that if E_F is the Fermi energy, only those electrons lying within an energy strip of the order of $k_B T$ of the Fermi level have accessible empty states into which they can be thermally excited. Those electrons in lower energy states than this in the Fermi distribution cannot be excited because of the Pauli exclusion principle. If there are N free electrons in the whole Fermi sea of energy bandwidth E_F, then the number within $k_B T$ of E_F is, in order of magnitude, $Nk_B T/E_F$ and it is only this fraction of electrons which contribute their full classical amount, $\frac{3}{2}k_B T$ per electron, to the internal energy E, which therefore becomes

$$E \sim \tfrac{3}{2}k_B T \frac{Nk_B T}{E_F}. \tag{5.1}$$

Hence the specific heat is given by

$$c_V = \left(\frac{\partial E}{\partial T}\right)_V \propto T. \tag{5.2}$$

However, when we examine experiments on the specific heat of a superconductor, we find it is reduced even more drastically from the classically predicted value. If we express the results in terms of that temperature T_c, usually of the order of a few degrees, below which the electrical resistance vanishes, then we can represent these experimental results as

$$c_V/\gamma T_c = a \exp(-bT_c/T) \tag{5.3}$$

where γT is the low-temperature specific heat of the normal metal, proportional to the density of electronic states at the Fermi surface. The quantities a and b in equation (5.3) are weakly temperature dependent but are about 9 and 1.5 respectively for all superconductors.

The exponentially decreasing specific heat in equation (5.3) is rather direct experimental evidence that there is a scarcity of low-lying excited states of the superconductor; many fewer electrons contributing their classical value $\frac{3}{2}k_BT$ to the internal energy than in the normal state with internal energy (5.1). It is found that when the temperature is raised above T_c the specific heat reverts to the normal behaviour summarised in equations (5.1) and (5.2).

At the critical temperature, there is a discontinuity in the specific heat of about $2.5\,\gamma T_c$. The phase transition is second order (see Chapter 8). One of the tasks of any microscopic theory must be, of course, to explain these observed facts about the specific heat.

5.2 Critical magnetic field and free energies

Kamerlingh Onnes (1914) discovered another important property of superconductors, namely that in a sufficiently strong magnetic field the superconductivity can be destroyed. However, it reappears on removal of the magnetic field. The minimum magnetic field required to destroy the superconductivity depends on the shape and orientation of the specimen. If it is in the form of a long cylinder and is oriented parallel to the field, the transition is sharp. The minimum field which must be applied to destroy the superconductivity in such a case is called the critical field H_c. Its value is independent of the volume of the material but depends on temperature: it is zero at the transition temperature and rises to values varying between a few hundred and a few thousand gauss for most superconductors at $T = 0$. The temperature dependence of H_c for many superconductors is well described by the form

$$\frac{H_c(T)}{H_c(0)} = 1 - \left(\frac{T}{T_c}\right)^2 \tag{5.4}$$

the observed variation being as shown in figure 5.1.

One of the big difficulties which confronted any microscopic theory of superconductivity was the very small difference in free energy between the superconducting (s) and the normal (n) states of a metal. Thus, if F_s and F_n are

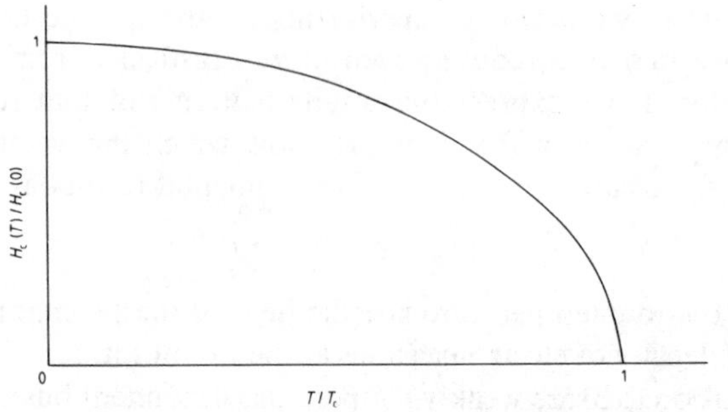

Figure 5.1 Observed variation of critical field H_c with temperature T in a superconductor, T being in units of the superconducting transition temperature T_c.

the free energies per unit volume, then these are related to the critical field H_c discussed above by

$$H_c^2/8\pi = F_n - F_s. \tag{5.5}$$

As already stated, the zero-temperature value of the critical field, H_{c0}, is of the order of a few hundred oersteds and hence a value of around 10^{-8} eV per atom is found for the energy difference at $T = 0$ between the normal and superconducting states.

Bandwidths in metals are known to be of the order of a few electronvolts, while exchange and correlation energies are smaller but of the same order (cf §8.8). Thus it is a remarkable achievement of the theory of Bardeen, Cooper and Schrieffer (1957, see also Schrieffer 1964) that it can interpret successfully such extremely small energies as are involved in the superconducting transition.

5.3 Isotope effect

A striking correlation exists for many non-transition metals between the transition temperature T_c and the isotopic mass M. The relation takes the form

$$T_c M^{1/2} = \text{constant}. \tag{5.6}$$

Since it turns out that T_c is proportional to H_{c0}, a similar relation holds between H_{c0} and M.

This isotope effect was anticipated in a very important paper by Fröhlich (1950). He recognised that it must be via an electron–lattice interaction that the basic mechanism for superconductivity had to be sought.

5.4 Electron–electron interaction via ionic lattice

The specific heat behaviour (5.3), being exponential, could be understood if there was an energy gap in the low-lying electronic excitations. This would be in marked contrast to the normal metallic state, with excited states infinitesimally close to the ground state. Such an energy gap, as anticipated, does indeed lead quantitatively to the specific heat law (5.3), as is demonstrated later by equation (5.30). The gap could arise via binding of electrons, and it would necessitate some attractive interaction which, under certain circumstances, can lead to binding, despite the screened Coulomb repulsion referred to in Chapter 2.

This interaction can arise in a manner which we can usefully describe by referring to the discussion of the polaron in the previous chapter. There we saw how an electron moving through a polar crystal would be surrounded by a cloud of lattice polarisation.

In a metal we observe, in spite of screening connected with the plasmon, that the chosen conduction electron, as it moves through the crystal, will attract the positive ions in its neighbourhood and create a region where there is a local excess of positive charge above its average value. Then, since the ions are massive and move sluggishly compared with the electrons, as the chosen electron passes on it leaves behind it this excess of positive ions. Another electron following up feels itself attracted into this region. In this way, the second electron is effectively coupled to the first, via the lattice.

The experimental discovery of the isotope effect was an essential step in demonstrating that superconductivity was indeed related to an attractive interaction between electrons, mediated by the ionic lattice.

The origin of the attractive interaction between electrons is therefore clear on qualitative grounds. But the question of when the attractive interaction mediated by the phonons can overcome the screened Coulomb potential given in k-space by equation (2.16), which represents electron–electron repulsion, is a quantitative question which remains quite troublesome. We must therefore have a negative term to overcome the positive term in equation (2.16) and this can come, as we have argued, only when one takes the motion of the ion cores into consideration. The lattice deformation around an electron is of course resisted by the same stiffness that results in the elastic behaviour of a solid. It is therefore clear that phonon frequencies must play an important role, from the discussion of Chapter 3 (see Appendices 3.1 and 3.2). We note that, as treated in Chapter 2, when screening by electrons is being considered, the characteristic frequency is then the plasma frequency, which is so high ($\sim 10^{16}$ Hz) that we can assume instantaneous response. However, as will be discussed a little further below, the characteristic phonon frequency is much lower and we shall want to treat a dynamic generalisation of $V(k)$ in equation (2.16), the frequency dependence coming from the phonon response. To press this point a little further, from momentum conservation we can see that if an electron is scattered from a state of wavevector $\boldsymbol{k}$ into one of wavevector $\boldsymbol{k}'$, then the relevant phonon must carry the momentum $\boldsymbol{q} = \boldsymbol{k} - \boldsymbol{k}'$ and the characteristic frequency must be the phonon frequency $\omega_{\boldsymbol{q}}$. As a result it is plausible that the phonon contribution to the screening function must be proportional to

$1/(\omega^2-\omega_q^2)$. Evidently, such a denominator gives a negative sign if $\omega < \omega_q$, corresponding to the physical argument for an attractive interaction given above. For higher frequencies, i.e. electron energy differences larger than $\hbar\omega_q$ the interaction becomes repulsive. This will be relevant to the discussion of the binding of Cooper pairs† and will tell us the cut-off in the matrix elements of the attractive interaction leading to bound Cooper pairs.

A careful treatment of the coupled electron-lattice may be found in the work of Fröhlich (1952) and of Bardeen and Pines (1955). The jellium model has been employed (see e.g. de Gennes 1966), in which crystal structure and hence Brillouin zone effects, and also finite ion-core size have all been neglected. As shown in the book by de Gennes (1966), the above model leads to a dynamic electron–electron interaction which has the approximate form

$$V(k,\omega)=\frac{4\pi e^2}{k^2+q_{\mathrm{TF}}^2}+\frac{4\pi e^2}{k^2+q_{\mathrm{TF}}^2}\,\frac{\omega_k^2}{\omega^2-\omega_k^2} \tag{5.7}$$

The first term on the right-hand side is simply the screened repulsive interaction of equation (2.16), q_{TF}^{-1} being the Thomas–Fermi screening radius given by equation (A2.1.6) of Appendix 2.1, while the second term reflects the phonon-mediated interaction, the denominator having indeed the form anticipated above. This form of the electron–electron interaction is too simplified to afford a criterion for superconductivity, and in particular it reduces to zero for $\omega = 0$, and it is always negative for $\omega < \omega_k$ regardless of the material parameters. What is important for our present purposes is that it does reveal that the phonon-mediated interaction is of the same order of magnitude as the direct Coulomb repulsion term, so that the idea of achieving a net negative interaction matrix element, representing overall electron–electron attraction, is not at all unreasonable. It turns out to be high-frequency phonons which are mainly responsible for the attractive interaction: this fact will be utilised below.

5.5 Cooper pairs

Since electrical resistivity of a metal arises by scattering of electrons at the Fermi surface, we consider a pair of electrons at the Fermi level E_{F}. In the absence of the attractive interaction referred to above, the total energy of the pair is $2E_{\mathrm{F}}$. To see whether binding can occur, we must study the ground state of the two-electron Schrödinger equation:

$$\left(-\frac{\hbar^2}{2m_{\mathrm{e}}}\nabla_1^2-\frac{\hbar^2}{2m_{\mathrm{e}}}\nabla_2^2+v(\boldsymbol{r}_1,\boldsymbol{r}_2)\right)\psi=E\psi. \tag{5.8}$$

† A pair of electrons, with opposed spins, at the Fermi level, which is bound by the lattice mediated electron–electron interaction (see §5.5 for quantitative details).

Here v is to be chosen to simulate the attractive interaction, the physical origin of which was discussed above. For binding it has to be shown that the ground-state eigenvalue is less than $2E_F$. As usual equation (5.8) can be solved by separating the centre of mass motion from the relative motion (see e.g. Pauling and Wilson 1935, p. 113). If the total linear momentum of the pair is zero, the wavefunction $\psi(\boldsymbol{r}_1, \boldsymbol{r}_2)$ depends only on the relative coordinate $\boldsymbol{r} = \boldsymbol{r}_1 - \boldsymbol{r}_2$:

$$\psi(\boldsymbol{r}_1, \boldsymbol{r}_2) = \phi(\boldsymbol{r}). \tag{5.9}$$

We can now expand ϕ in plane waves representing unperturbed excited states

$$\phi(\boldsymbol{r}) = \sum_{|\boldsymbol{k}| > k_F} C_{\boldsymbol{k}} \frac{\exp(\mathrm{i}\boldsymbol{k}\cdot\boldsymbol{r})}{\Omega} \tag{5.10}$$

where Ω is the volume of the system. Substituting (5.10) into equation (5.8), we get

$$\sum_{\boldsymbol{k}} C_{\boldsymbol{k}} \frac{\hbar^2 k^2}{m_e} \exp(\mathrm{i}\boldsymbol{k}\cdot\boldsymbol{r}) + \sum_{\boldsymbol{k}} v(\boldsymbol{r}_1, \boldsymbol{r}_2) C_{\boldsymbol{k}} \exp(\mathrm{i}\boldsymbol{k}\cdot\boldsymbol{r})$$
$$= E \sum_{\boldsymbol{k}} C_{\boldsymbol{k}} \exp(\mathrm{i}\boldsymbol{k}\cdot\boldsymbol{r}). \tag{5.11}$$

Multiplying both sides of (5.11) by $\exp(-\mathrm{i}\boldsymbol{k}\cdot\boldsymbol{r})$ and integrating over $\boldsymbol{r}_1$ and $\boldsymbol{r}_2$ we find

$$\frac{\hbar^2 k^2}{m_e} C_{\boldsymbol{k}} + \sum_{\boldsymbol{k}'} V_{\boldsymbol{k},\boldsymbol{k}'} C_{\boldsymbol{k}'} = E C_{\boldsymbol{k}} \tag{5.12}$$

where

$$V_{\boldsymbol{k},\boldsymbol{k}'} = \frac{1}{\Omega} \int \mathrm{d}\boldsymbol{r}_1 \, \mathrm{d}\boldsymbol{r}_2 \, \mathrm{e}^{-\mathrm{i}\boldsymbol{k}\cdot\boldsymbol{r}} v(\boldsymbol{r}_1, \boldsymbol{r}_2) \mathrm{e}^{\mathrm{i}\boldsymbol{k}'\cdot\boldsymbol{r}} \tag{5.13}$$

is the matrix element of the effective interelectronic potential. Everything now hinges on the choice of $V_{\boldsymbol{k},\boldsymbol{k}'}$ which, following the preceding discussion, we take to be of the form (with $\delta \sim \hbar\omega \sim 10^{-1}$ eV, ω being a typical lattice frequency)

$$V_{\boldsymbol{k},\boldsymbol{k}'} = \begin{cases} -\dfrac{v}{\Omega} & \text{if } E_F \leqslant \hbar^2 k^2/m_e \leqslant E_F + \delta \\ & \text{and } E_F \leqslant \hbar^2 k'^2/m_e \leqslant E_F + \delta \\ 0 & \text{otherwise.} \end{cases} \tag{5.14}$$

This interaction is of course an oversimplification of the real one and it is meant only to simulate in an average sense the main physical effect. Substituting the interaction (5.14) into (5.12) and rearranging, we find

$$C_{\boldsymbol{k}} = \frac{v}{(\hbar^2 k^2/m_e) - E} \frac{1}{\Omega} \sum_{E_F \leqslant \hbar^2 k'^2/2m_e < E_F + \delta} C_{\boldsymbol{k}'}. \tag{5.15}$$

Equation (5.15) can now be solved by summing both sides over the shell of vectors $\boldsymbol{k}$ defined by

$$E_F \leqslant \hbar^2 k^2/m_e \leqslant E_F + \delta. \tag{5.16}$$

This gives the following condition for the existence of a solution:

$$\frac{1}{v}=\frac{1}{\Omega}\sum_{E_F\leqslant \hbar^2k^2/m_e\leqslant E_F+\delta}\frac{1}{(\hbar^2k^2/m_e)-E}. \tag{5.17}$$

Equation (5.17) is solved graphically in figure 5.2 where a solution of energy $E_0 < 2E_F$ is shown to exist no matter how small is the strength of interaction v. Clearly this result cannot be obtained within any perturbation theory. Transforming the sum into an integral over the density of states $\rho(E)$ we find†

$$\frac{1}{v}=\int_{E_F}^{E_F+\delta}\frac{\rho(E')\mathrm{d}E'}{E'-E_0}. \tag{5.18}$$

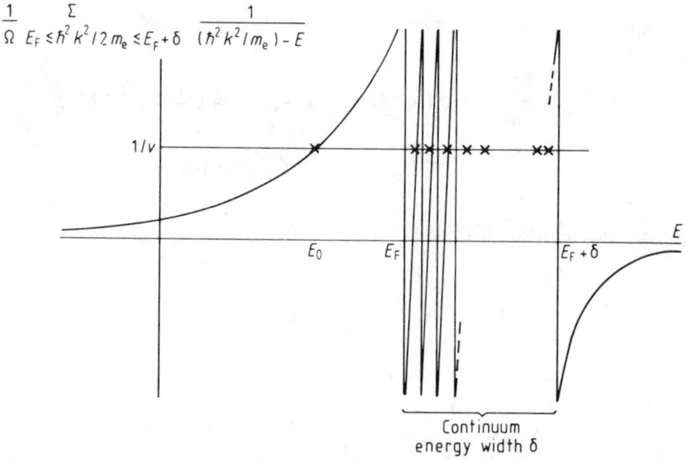

Figure 5.2 Showing binding of Cooper pairs. Crosses between Fermi energy E_F and $E_F+\delta$ correspond to solution of equation (5.17) in continuum. Isolated cross shows bound state with volume-independent energy E_0.

5.5.1 Size of Cooper pair

If we turn again to equation (5.10) we can also calculate the wavefunction of the pair and estimate its extension in real space. In fact

$$C_k = C/[E_0-(\hbar^2k^2/m_e)] \tag{5.19}$$

where C is a constant independent of k. The above equation can be checked by direct substitution into equation (5.12) provided that E_0 is a solution of (5.17). The wavefunction becomes therefore

$$\phi(\boldsymbol{r})=\frac{C}{\Omega}\sum_{\boldsymbol{k}}\frac{e^{i\boldsymbol{k}\cdot\boldsymbol{r}}}{E_0-(\hbar^2k^2/2m_e)}. \tag{5.20}$$

† Assuming δ small ($\sim 10^{-1}$ eV) compared with E_F, the binding energy of the Cooper pair resulting from equation (5.18) is $\sim\delta\exp(-1/\rho(E_F)v)$.

The mean squared radius is obtained by taking the expectation value:

$$\langle r^2 \rangle = \frac{\int \mathrm{d}\boldsymbol{r}\, r^2 |\phi(\boldsymbol{r})|^2}{\int |\phi(\boldsymbol{r})|^2 \mathrm{d}\boldsymbol{r}} \tag{5.21}$$

or in terms of the coefficients $C_{\boldsymbol{k}}$,

$$\langle r^2 \rangle = \frac{\sum_{\boldsymbol{k}} |\nabla_{\boldsymbol{k}} C_{\boldsymbol{k}}|^2}{\sum_{\boldsymbol{k}} |C_{\boldsymbol{k}}|^2}. \tag{5.22}$$

Equation (5.22) can be evaluated and gives

$$(\langle r^2 \rangle)^{1/2} = \frac{2}{\sqrt{3}} \frac{\hbar^2 k_{\mathrm{F}}/m_{\mathrm{e}}}{E_0 - 2E_{\mathrm{F}}}. \tag{5.23}$$

From experiments we know that $E_0 - 2E_{\mathrm{F}} \sim 10^{-4}$ eV and therefore we find for a typical metal ($k_{\mathrm{F}} \simeq 1\,\text{Å}^{-1}$):

$$(\langle r^2 \rangle)^{1/2} \simeq 10^{-4}\,\mathrm{cm}.$$

This means that the extension of a Cooper pair is very large compared to the mean interelectronic separation and the interactions among different pairs become a very important feature of the problem.

Before turning to discuss this point let us dwell for a moment on the spin properties of the bound Cooper pair. Using equation (5.20) and the fact that $C(\boldsymbol{k}) = C(k)$, it is easily shown that ϕ is a function of the modulus only of $\boldsymbol{r}$. Hence the orbital angular momentum of the pair l is equal to zero and the wavefunction is even on the exchange of $\boldsymbol{r}_1 \leftrightarrow \boldsymbol{r}_2$. But the total wavefunction of a many-electron system must be completely antisymmetric on the exchange of the electronic coordinates (space plus spin); therefore the spin part of the wavefunction must be antisymmetric in order to meet this requirement. The antisymmetric combination of two spin-$\frac{1}{2}$ vectors can be obtained by aligning the spins in an antiparallel way in order to form a singlet state. Therefore in a Cooper pair the electronic spins as well as the momenta† of the electrons point in opposite directions.

5.6 Bardeen, Cooper and Schrieffer (BCS) model

The importance of many-body interactions was realised by Bardeen, Cooper and Schrieffer (1957) who were able to describe in a simple model the many-body interactions leading to superconductivity. We will not go into detail because it would be beyond the scope of the present book, but we will content ourselves with the result of their calculations. Not surprisingly, in view of the preceding discussion, the excitations can be described in terms of something

† The total linear momentum is zero.

which closely resembles the Cooper pairs, that is, bound pairs of electrons with opposite spin and momentum. Their dispersion relation can be written in the form, with the normal spectrum measured from E_F,

$$\epsilon(k) = \left[\Delta^2 + \left(\frac{\hbar k^2}{2m_e} - E_F\right)^2\right]^{1/2}. \tag{5.24}$$

The quantity Δ represents the minimum energy required to create an excitation in the system†. This energy gap Δ does not coincide with the binding energy of a Cooper pair that we have discussed above because it contains the effect of the interactions among the pairs. As such it depends on the temperature and it tends to be destroyed at higher temperatures due to the incoherent interference among different pairs. This is depicted in figure 5.3 where the dependence of Δ on T is explicitly shown. Note that, for $T > T_c$, Δ disappears and the system becomes a normal metal.

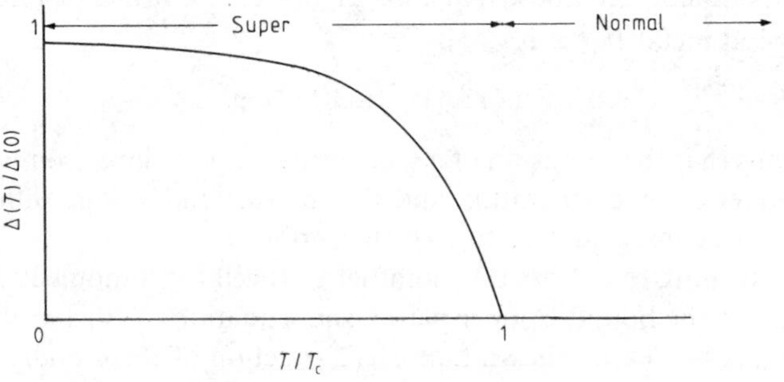

Figure 5.3 Gap parameter Δ against reduced temperature.

The presence of an energy gap in the excitation spectrum of the system induces the striking superconducting properties. In fact, if we establish an electric current in the system it will continue to flow unless some scattering mechanism is present. In a metal the main scattering processes are due to the phonons or to impurities but neither is capable of providing the energy necessary to bridge the gap Δ. Hence the current flows without resistance.

The presence of a gap also explains the behaviour of the specific heat c_V as follows. The total energy of the system can be written as

$$E = \sum_{p} \epsilon(p) n(\epsilon(p)) \tag{5.25}$$

where $n(\epsilon(p))$ is the pair occupation number. At low temperatures the number of excitations is small and n has a Boltzmann distribution:

$$n(\epsilon(p)) \sim \exp\left(-\frac{\epsilon(p)}{k_B T}\right). \tag{5.26}$$

† The argument given here for superconductivity is not valid for gapless superconductors.

Moreover we can neglect the dependence of Δ on temperature and take $\epsilon(\boldsymbol{p})$ as temperature independent. Therefore

$$c_V = \frac{\partial E}{\partial T} \simeq \sum_{\boldsymbol{p}} \epsilon(\boldsymbol{p}) \frac{\partial n(\epsilon(\boldsymbol{p}))}{\partial T}, \tag{5.27}$$

and using (5.26) we evidently find

$$c_V \simeq \sum_{\boldsymbol{p}} \frac{[\epsilon(\boldsymbol{p})]^2}{k_B T^2} \exp\left(-\frac{\epsilon(\boldsymbol{p})}{k_B T}\right). \tag{5.28}$$

In the limit $T \to 0$ we need to consider only the states of lowest energy because all the others have very small weight due to the exponential factor in (5.28). Therefore we expand $\epsilon(\boldsymbol{p})$ around its minimum, $p = p_F$, remembering that $E_F = \hbar^2 p_F^2/2m_e$; when we find from equation (5.24)

$$\epsilon(\boldsymbol{p}) \underset{p \to p_F}{\simeq} \Delta + \frac{1}{\Delta}\frac{\hbar^2}{2} v_F^2 (p - p_F)^2. \tag{5.29}$$

Substituting (5.29) into (5.28) and neglecting higher-order terms, we finally obtain

$$c_V = \exp\left(-\frac{\Delta}{k_B T}\right) \sum_{p} \frac{\Delta^2}{k_B T^2} \exp\left(-\frac{v_F^2}{2\Delta k_B T}(p - p_F)^2\right) \tag{5.30}$$

explicitly showing the exponential dependence on the energy gap Δ.

5.7 Ginzburg–Landau theory and London equation

Though the BCS model offers a convincing explanation of most of the striking properties of superconductors, its detailed application is somewhat complicated. Here we shall therefore employ a semiphenomenological model (see Appendix 5.2 for further discussion) first proposed by Ginzburg and Landau (1950), and subsequently justified on the basis of the BCS model by Gorkov (1959). According to the Ginzburg–Landau treatment, a superconductor whose properties vary slowly in space is uniquely described by a complex order parameter† denoted by $\psi(\boldsymbol{r})$. This order parameter can be used as an effective wavefunction for the macroscopic ground state. Hence the local density of superconducting pairs $n_s(\boldsymbol{r})$, which in turn measures the local degree of superconductivity, is given by

$$n_s(\boldsymbol{r}) = |\psi(\boldsymbol{r})|^2, \tag{5.31}$$

while the current in the presence of an external applied electromagnetic field, represented by a vector potential $\boldsymbol{A}(\boldsymbol{r}, t)$ is given by

$$\boldsymbol{j} = -\frac{e^*}{2m^*}\left[\psi^*(\boldsymbol{r})\left(\frac{\hbar}{\mathrm{i}}\nabla + \frac{2e}{c}\boldsymbol{A}\right)\psi(\boldsymbol{r}) + \mathrm{cc}\right] \tag{5.32}$$

† See especially Chapter 8 on phase transitions.

where cc stands for complex conjugate. Following Cooper pairing ideas and BCS theory $e^* = 2e$ and $m^* = 2m_e$ are respectively the charge and mass associated with a superconducting pair. It is useful to write $\psi(\boldsymbol{r})$ in terms of a phase $\phi(\boldsymbol{r})$ through

$$\psi(\boldsymbol{r}) = [n_s(\boldsymbol{r})]^{1/2}\, e^{i\phi(\boldsymbol{r})}. \tag{5.33}$$

If we further assume that the local density $n_s(\boldsymbol{r})$ varies slowly with respect to $\phi(\boldsymbol{r})$, and can be replaced by a constant n_s, then the equation for the current simplifies to:

$$\boldsymbol{j} \simeq -\left(\frac{2e^2}{m_e c}\boldsymbol{A} + \frac{e\hbar}{m_e}\nabla\phi(\boldsymbol{r})\right) n_s. \tag{5.34}$$

Taking the curl of this equation, the following result, first proposed by London, is obtained:

$$\nabla \times \boldsymbol{j} = \frac{2n_s e^2}{m_e c}\boldsymbol{B} \qquad \boldsymbol{B} = \text{curl}\,\boldsymbol{A}. \tag{5.35}$$

5.8 Meissner effect

One of the important consequences of this equation is that it contains an explanation of an effect discovered by Meissner. This consists in the fact that when a metal becomes superconducting, it cannot be penetrated by the flux lines of a magnetic field provided that the value of this field does not exceed a certain critical value H_c.

This can be understood by combining the London equation with the Maxwell equation

$$\text{curl}\,\boldsymbol{B} = \frac{4\pi}{c}\boldsymbol{j} \tag{5.36}$$

when one obtains

$$\nabla^2 \boldsymbol{B} = \frac{8\pi n_s e^2}{m_e c^2}\boldsymbol{B}. \tag{5.37}$$

To solve this equation naturally involves specifying the boundary conditions. A particularly simple configuration is that in which a magnetic field B is applied parallel to the surface of a semi-infinite superconductor occupying the half-space $z > 0$. In this case one readily obtains the solution

$$B(z) = B\exp(-z/\lambda_L) \qquad \text{for } z > 0, \tag{5.38}$$

where λ_L is the penetration depth given by

$$\lambda_L^2 = \frac{m_e c^2}{8\pi n_s e^2}. \tag{5.39}$$

This tells us that the field cannot penetrate the superconductors to distances greater than λ_L, in agreement with the experimental facts.

5.9 Type-II superconductors: characteristic lengths

Pippard (1953) introduced the idea of coherence length into superconductivity. This is the distance ξ say over which strong correlations exist†. We must expect that if, as in the cases discussed above, the penetration depth λ_L is small compared with the coherence length ξ, a very different situation will exist from that when $\lambda_L/\xi \gg 1$. This is indeed so, and one has to deal differently with the soft or type-I superconductors (λ_L/ξ small) and the hard or type-II ($\lambda_L/\xi \gg 1$).

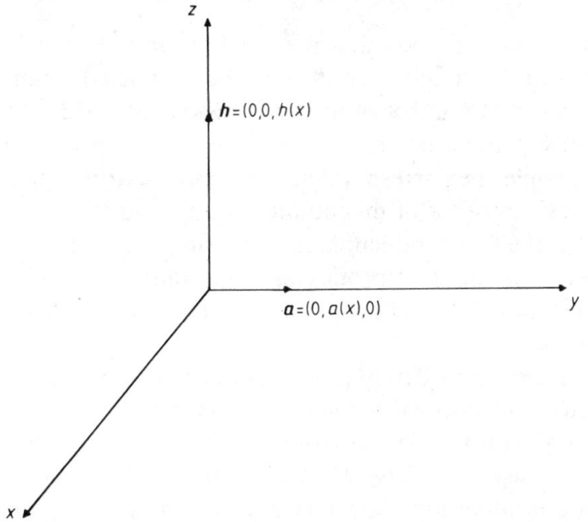

Figure 5.4 Illustrating field and vector potential inside superconductor. ***a*** and ***h*** are dimensionless measures of vector potential ***A*** and magnetic field ***H***, defined precisely in Appendix 5.2, equations (A5.2.3).

In connection with this distinction, it is worth examining briefly the Ginzburg–Landau equations‡ in a situation shown in figure 5.4. Here the superconductor occupies the half-space $x > 0$. A weak applied field, $\boldsymbol{h}_0$ say, exists in the region $x < 0$. Inside the superconductor, the field and vector potential may be described as shown in figure 5.4 where $a'(x) = h(x)$. The Ginzburg–Landau equations then reduce to (with ψ as in equation (A5.2.3))

$$-\frac{1}{K^2}\psi'' + a^2\psi = \psi - |\psi|^2\psi \qquad K \propto \lambda_L^2 \tag{5.40}$$

and (if terms in ψ quadratic in h_0 are neglected) one finds the further equation

$$a'' = a. \tag{5.41}$$

Equation (5.41) has the elementary solution $a = -h_0 e^{-x}$, $h = h_0 e^{-x}$. Thus

† Actually some degree of correlation must extend over the whole superconductor, not merely over the coherence length ξ. Otherwise there would be no quantisation of flux (cf Appendix 5.1) or Josephson effects (cf §5.10).

‡ Specifically equations (A5.2.6) and (A.5.2.7).

$\lambda_L (= 1)$ is a measure of the field penetration into the superconductor and is, in fact, the usual London penetration depth (cf equation (5.39)). This explicit expression for a can now be substituted in equation (5.40) to obtain information about ψ. Expanding the latter as a power series in h_0^2, and discarding h_0^4 terms in equation (5.40), elementary methods yield the result (Douglass and Falicov 1964)

$$\psi = 1 - \frac{K}{(2-K^2)\sqrt{2}}\left(e^{-Kx\sqrt{2}} - \frac{K}{\sqrt{2}}e^{-2x}\right)h_0^2 \qquad (5.42)$$

using the boundary condition that $\psi'(x) = 0$ when $x = 0$.

The dimensionless parameter K can be estimated numerically from equation (A5.2.3) by a number of methods (Goodman 1962). It turns out that usually for soft superconductors, K_0 (the value of K when $T = 0$) is of the order of 0.1 at 0 K, while the corresponding results for hard superconductors are characteristically an order of magnitude greater than this.

When $K \ll 1$, the first exponential in equation (5.42) dominates the second, and ψ increases to its asymptotic value of unity in a distance of order $(K\sqrt{2})^{-1}$. This should, then, be a measure of the coherence length ξ. Thus, we expect $K \simeq \lambda_L/\xi$.

When $K \gtrsim 1$, a rather different physical situation obtains and to understand this it is helpful to introduce in a qualitative way the idea of surface energy (see e.g. F London 1950). The exclusion of an external field H_0 from a superconductor raises the energy per unit volume by an amount $H_0^2/8\pi$. If this were the only consideration, the material would striate into superconducting regions of width $d_s < \lambda_L$, separated from each other by normal regions of width $d_n \ll \lambda_L$. Since $d_n/d_s \ll 1$, maximum advantage is taken of Cooper pairing, while $d_s < \lambda_L$ means (see figure 5.5) that the field would fairly

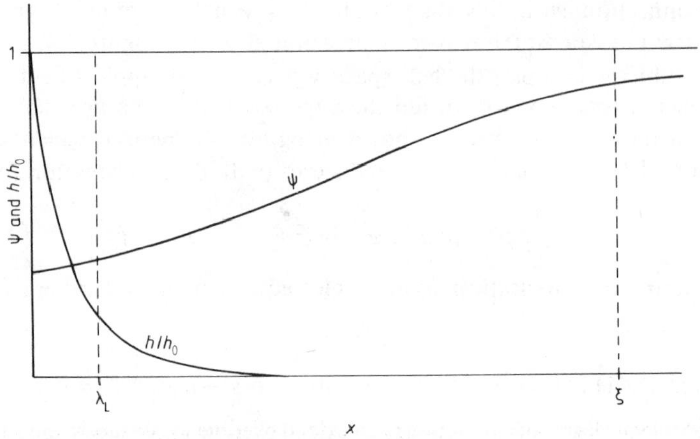

Figure 5.5 Representing solution of Ginzburg–Landau equations in form of equation (5.42) when h_0 is small.

approximate to H_0 everywhere. The reason why this does not usually occur is that there exists a positive surface energy. A rough estimate of this is obtained by observing (figure 5.5) that the boundary reduces the superfluid volume of the metal by a factor of order ξ per unit area, whereas the effective volume not accessible to the magnetic field is obtained by reducing the metallic volume by a factor λ_L per unit area. The surface energy per unit area is thus a factor of order $\xi - \lambda_L$, multiplied by $H_0^2/8\pi$. When $K \ll 1$, $\xi - \lambda_L > 0$ and we have a positive surface energy, whereas, when $K \gtrsim 1$, $\xi - \lambda_L < 0$ and an instability of the kind described above occurs.

Abrikosov has used the Ginzburg–Landau equations to put the above on a firm quantitative basis. The transition value of K is found as $1/\sqrt{2}$ (and furnishes a precise distinction between type-I and type-II behaviour)† and above this value the striations take the form of vortices. It would take us too far from our main theme to develop this further here though when we treat the relation between superconductivity and superfluidity in Appendix 7.2 some further details will be considered.

5.10 Josephson junction

We conclude this discussion by examining what can happen at the junction of two superconductors, between which is sandwiched a thin layer of insulating material: a Josephson junction.

If the two superconductors were completely isolated from each other, the Ginzburg–Landau (GL) equations imply that it would be possible to alter the phases χ‡ in each, independently. This is because, from any chosen solution of the GL equations, other solutions can be found by changing the phase.

On the other hand, in a single superconductor under given external conditions, all phase differences are fixed. But if we consider a Josephson junction in which the thickness of the insulator is gradually reduced to zero, we can expect to go continuously from the properties of the two isolated superconductors to those of a single superconductor. Therefore the free energy of the system must contain a term which depends on the relative phases on the two sides of the barrier.

To examine this in detail, consider two points in a single superconductor, separated by a small distance d say. If the values of the phase χ at the points 1 and 2 are χ_1 and χ_2, then, from the property that the free energy of a superconductor is a function of the momentum of a Cooper pair $\boldsymbol{p}_s$, which in turn is related to the phase by

$$\chi(\boldsymbol{r}) = \boldsymbol{p}_s \cdot \boldsymbol{r}/\hbar \tag{5.43}$$

† Magnetisation versus field curves of type-I and type-II superconductors are shown in figure A5.2.

‡ We now reserve ϕ for difference in phase (see equation (5.48)).

we can write the free energy in the form

$$F\left(\hbar\frac{\chi_1-\chi_2}{d}\right).$$

Hence the current j in the direction 1–2 is given by

$$j=\frac{2ed}{\hbar}\frac{\partial F}{\partial(\chi_1-\chi_2)}=-2e\frac{\partial N_1}{\partial t} \tag{5.44}$$

where N_1 is the number of Cooper pairs in this superconductor.

Now one applies these results to the junction. To do so, assume that the term in the free energy describing the coupling has the form $F(\chi_1-\chi_2)$ where χ_1 and χ_2 are the phases on each side of the insulating slab. But χ is a cyclic variable and a simple form for this coupling term might be

$$F(\chi_1-\chi_2)=\text{constant}\times[1-\cos(\chi_1-\chi_2)] \tag{5.45}$$

yielding a current

$$j=j_{\rm m}\sin(\chi_1-\chi_2). \tag{5.46}$$

This equation shows that for a junction of this kind, tunnelling of a supercurrent occurs up to a maximum value $j_{\rm m}$ (see also problem 5.5).

5.10.1 Junction with voltage across it

We consider briefly a junction with a voltage across it. Cooper pairs on different sides of the barrier then have energies differing by $\Delta E = 2eV$, V being the potential difference between the two sides of the barrier and $2e$ being the charge of a Cooper pair.

Tunnelling through the barrier can then take place only as a virtual process, with associated oscillating currents at a frequency $\nu = \Delta E/h$, that is

$$\nu = 2eV/h. \tag{5.47}$$

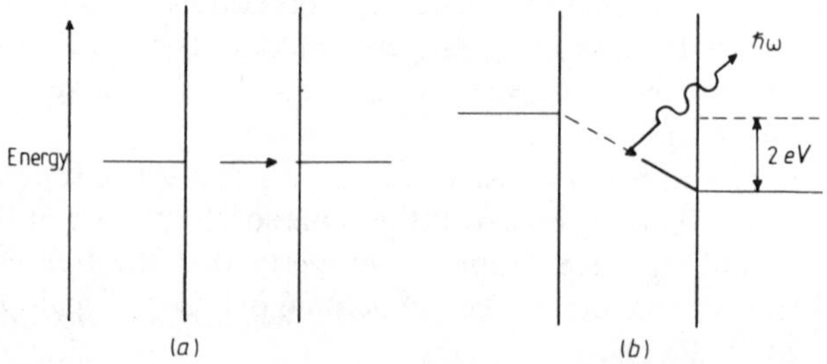

Figure 5.6 Illustrates tunnelling of Cooper pairs through an insulating slab. (a) DC supercurrents at zero voltage; (b) AC supercurrents with applied voltage V, leading to emission of photons. Corresponding to each emitted photon (energy $\hbar\omega$), a Cooper pair tunnels through the barrier. Energy conservation then yields $\hbar\omega = 2eV$, since charge of a Cooper pair is $2e$.

When the interaction between the oscillating currents and the electromagnetic currents is included, one finds that real processes occur, energy being conserved by emission of a photon, as shown in figure 5.6 (Josephson 1964). In this case the radiation is coherent, since every photon comes from an identical process.

Taking into account the fact that the oscillating currents arise from a change in $\phi = \chi_1 - \chi_2$, one finds (see problem 5.5)

$$\dot{\phi} = 2\pi\nu = 2eV/\hbar. \tag{5.48}$$

If V is time independent, then this has solution

$$\chi_1 - \chi_2 = (\chi_1 - \chi_2)_0 + 2eVt/\hbar \tag{5.49}$$

and hence from equation (5.46),

$$j = j_m \sin[(\chi_1 - \chi_2)_0 + 2eVt/\hbar]. \tag{5.50}$$

The current evidently oscillates in time and one can see that a DC voltage induces alternating current.

The effect of RF power on the tunnelling supercurrent of Cooper pairs is also interesting. In his original paper, Josephson (1962) treated this effect and predicted the occurrence of regions of zero slope separated by $\hbar\omega/2e$ in the current–voltage characteristics in the presence of the RF field. To understand how this arises let us write instead of $2eV$,

$$2eV + 2e\mathscr{V} \sin(\omega t)$$

where ω is the angular frequency of the RF power. When $n\omega = 2eV/\hbar, n = 1, 2,$ etc, one will get a DC current as can be seen by averaging over a period $2\pi/\omega$. This has been confirmed experimentally (see e.g. Shapiro 1963).

Problems

5.1 In general there is a non-local relation between current $\boldsymbol{j}$ and vector potential $\boldsymbol{A}$ in a superconductor of the form

$$\boldsymbol{j}(\boldsymbol{r}) = \int K(\boldsymbol{r} - \boldsymbol{r}')\boldsymbol{A}(\boldsymbol{r}')\mathrm{d}\boldsymbol{r}'.$$

Demonstrate that: (a) in the London theory, the Fourier transform of $K(\boldsymbol{r})$ is a constant, independent of $\boldsymbol{k}$; (b) in a non-local treatment, the range of K is determined by the coherence length.

5.2 Flux quantisation in Appendix 5.1 can be treated alternatively by constructing the free energies F_n and F_s of the normal and superconducting phases as functions of the flux Φ. Then to distinguish the states, one must have, because of the Meissner effect (cf §5.8) $\partial F_s/\partial\Phi = 0$.

To implement this, you are given the energy levels of independent electrons in a hollow cylinder (Byers and Yang 1961) as

$$E_n = \frac{1}{2m_e}\left[p_r^2 + p_z^2 + \frac{\hbar^2}{r^2}\left(n + \frac{e}{ch}\Phi\right)^2\right],$$

where p_r and p_z are the momenta in the radial and z-directions while n is an integer. (It is important to think in terms of the level spectrum of pairs of electrons.)

5.3 Give an argument that is plausible to show that to go from a wave function $|\chi\rangle$ for a superconductor to that in which the phase χ is changed to $\chi + \delta\chi$, we must multiply $|\chi\rangle$ by $\exp(iN\,\delta\chi)$, where N is essentially the number (operator) for Cooper pairs.

5.4 In relation to the Josephson junction, argue that if ϕ is the phase, d is the effective thickness of the sheet of flux, $\mathscr{H}$ is the field in the barrier and $\boldsymbol{n}$ is a unit vector normal to the barrier then

$$\operatorname{grad}\phi = \frac{2ed}{\hbar c}(\mathscr{H} \times \boldsymbol{n}).$$

5.5 The equations of motion of the wavefunctions ψ_1 and ψ_2 on either side of the barrier in the Josephson junction are given as (cf Tilley and Tilley 1974)

$$i\hbar\frac{\partial\psi_1}{\partial t} = \mu_1\psi_1 + \gamma\psi_2$$

and

$$i\hbar\frac{\partial\psi_2}{\partial t} = \mu_2\psi_2 + \gamma\psi_1$$

where γ represents the (weak) coupling across the barrier.

Solve these equations by writing $\psi_1 = n_1^{1/2}\,e^{i\chi_1}$, $\psi_2 = n_2^{1/2}\,e^{i\chi_2}$, and show that

$$-\hbar\frac{\partial}{\partial t}(\chi_2 - \chi_1) = \mu_2 - \mu_1$$

and that the current is $j_m \sin(\chi_1 - \chi_2)$.
[NB For voltage difference V between the two sides of the barrier, $\mu_2 - \mu_1 = 2eV$.]

6

MAGNONS

In this chapter we shall be concerned with magnetic materials, especially ferromagnets. We know from the classical work of Weiss that the basic properties of ferromagnets can be explained on the basis of two assumptions:

(a) domains exist;
(b) within one domain, at absolute zero, all the elementary magnets are lined up parallel to one another.

Here we shall be concerned solely with the properties of a single domain.

As referred to in the outline, the basic idea behind constructing an excited state of a ferromagnetic is to start from a perfectly aligned state with all spins parallel. Then we reverse one spin on a particular site, recognising that this misoriented spin could equally well be at any one of the other sites in the lattice. The problem then is to form 'spin waves', in which the misoriented spin propagates through the lattice.

6.1 Dispersion relation for ferromagnetic spin waves

We shall approach this problem in a quite similar manner to that used in dealing with the lattice waves and their quanta, the phonons. As in that case, we shall utilise a linear chain example, though the main result, that $\omega \propto k^2$ for long waves, is valid also in a truly three-dimensional lattice. The energy of a quantum associated with the spin waves, the magnon, is therefore proportional to the square of the wavenumber; like a free electron behaviour $E \propto k^2$. This result, we stress, is true for ferromagnetic materials. As we discuss later, for antiferromagnetic spin waves, $\omega \propto k$ for long waves and we have a situation which is 'phonon-like'.

6.2 Spin–spin interaction

Let us recall briefly the theory of the H_2 molecule (see e.g. Coulson 1954). The total energy E_0 of the two isolated atoms splits in the molecule into two levels given by

$$E_{\pm} = E_0 + \frac{C \pm X}{1 \pm Q^2}. \tag{6.1}$$

Here, the plus sign corresponds to the singlet state, in which the spins of the two electrons are antiparallel; the negative sign to the triplet state, with parallel spins. C and X correspond respectively to Coulomb and exchange terms, while Q is the overlap integral, whose value lies between 0 and 1.

It is easy to show, using this condition, plus $|CQ^2| \ll |X|$, that

$$E_{\text{singlet}} \gtrless E_{\text{triplet}} \qquad \text{for } X \gtrless 0.$$

Calculation of the exchange energy X for the hydrogen molecule gives a negative value; hence the ground state is the singlet, with paired spins (total spin $S = 0$). If X were positive however, as first emphasised by Heisenberg, the triplet state with resultant spin $S = 1$ ($\equiv$ 'ferromagnetic molecule') would be the ground state†.

6.3 Heisenberg Hamiltonian

To get the form of the Hamiltonian of an array of interacting spins, such as in a ferromagnetic crystal, we make the (inessential but convenient) simplification of putting $Q = 0$ in equation (6.1). Also we can write the trivial identity

$$\pm X = X[1 - S(S+1)] \tag{6.2}$$

since this expression gives $+X$ for $S = 0$ and $-X$ for $S = 1$. But we know that the eigenvalues of $\boldsymbol{S}^2$, $\boldsymbol{S}$ being the vector representing the total spin, are $\hbar^2 S(S+1)$ and therefore the energies $E_\pm$ in equation (6.1) can be viewed as the eigenvalues of an operator

$$H = H_0 + H_1 + H_{\text{ex}}, \tag{6.3}$$

where H_0 gives the energy of the isolated atoms, H_1 the Coulomb energy, while H_{ex} gives simply the exchange energy $\pm X$.

We have here the basic idea of the exchange operator. Writing $\boldsymbol{S}$ in terms of the individual spins $\boldsymbol{s}_1$ and $\boldsymbol{s}_2$ of the two electrons we have

$$\boldsymbol{S} = \boldsymbol{s}_1 + \boldsymbol{s}_2 \tag{6.4}$$

and

$$S^2 = s_1^2 + s_2^2 + 2(\boldsymbol{s}_1 \cdot \boldsymbol{s}_2). \tag{6.5}$$

But the quantum number of the individual spins has the value $s = \frac{1}{2}$, and hence s_1^2 and s_2^2 have eigenvalues $\hbar^2 s(s+1) = \frac{3}{4}\hbar^2$. Thus, we obtain for the exchange

† In practice, straightforward calculation of X always yields negative values. Nevertheless, Heisenberg's idea has been extremely fruitful, via the Heisenberg Hamiltonian, derived below.

operator†

$$H_{ex} = -X\left(\frac{1}{2} + \frac{2}{\hbar^2}(\boldsymbol{s}_1 \cdot \boldsymbol{s}_2)\right). \tag{6.6}$$

The first part is constant and uninteresting, but the nub of this expression is the scalar product of $\boldsymbol{s}_1$ and $\boldsymbol{s}_2$, showing very clearly that the corresponding energy depends on the relative orientation of the two spins.

When we have a crystal with spins on each site, then we generalise the above exchange interaction $\boldsymbol{s}_1 \cdot \boldsymbol{s}_2$ between two spins to

$$H_H = -2J \sum_{i,\, j} \boldsymbol{s}_i \cdot \boldsymbol{s}_j \tag{6.7}$$

where the subscript H stands for Heisenberg and where we shall assume that the exchange interactions are sufficiently short range that the sum extends only over all nearest-neighbour pairs. Clearly, as written, the ground state is ferromagnetic if J is positive ($\boldsymbol{s}_i \cdot \boldsymbol{s}_j$ is positive if the spins are parallel): the strength of the exchange interaction is written in the form shown purely for convenience.

We can usefully rewrite H_H once more as

$$H_H = -2J \sum_{i,\, j} (s_i^z s_j^z - \tfrac{1}{4}) - J \sum_{i,\, j} (s_i^+ s_j^- + s_i^- s_j^+) \tag{6.8}$$

which differs from the previous form only through a shift in the zero of energy. Here we have written

$$s_i^{\pm} = s_i^x \pm i s_i^y. \tag{6.9}$$

Now the total z component of spin, M say, is a good quantum number, and if we have N atoms in the chain, with a spin of $\frac{1}{2}$ localised on each, then we can look for eigenstates corresponding to $M = \frac{1}{2}N, \frac{1}{2}N - 1, \frac{1}{2}N - 2, \ldots$. If α represents the spin state with spin $\uparrow$, then the state with $M = \frac{1}{2}N$ is evidently

$$\psi_0 = \alpha_1 \alpha_2 \alpha_3 \ldots \alpha_N. \tag{6.10}$$

This state actually has energy $E_0 = 0$, because of the way we have chosen the zero of energy in the Hamiltonian H_H.

Next we seek states with $M = \frac{1}{2}N - 1$, i.e. with one 'wrong' spin. There are N states with this value of M, corresponding to the N possible atoms to which the $\downarrow$ spin electron, with wavefunction β, can be assigned. The basis states we must therefore use to build the eigenfunctions are:

$$\phi_n = \alpha_1 \alpha_2 \ldots \alpha_{n-1} \beta_n \alpha_{n+1} \ldots \alpha_N. \tag{6.11}$$

Consider now H_H operating on ϕ_n. The first term in H_H is such that $\int \phi_n H_H \phi_m \mathrm{d}\tau$ is zero unless $n = m$, when its value is simply

$J \times$ (number of pairs of antiparallel spins on nearest-neighbour sites) $= 2J$.

† For those folk requiring a little more justification for the argument, we note that all the terms in (6.5) for S^2 commute with one another and with $\boldsymbol{H}_0 + \boldsymbol{H}_1$. Hence s_1^2 and s_2^2 can be replaced by their eigenvalues.

The second term has contributions for which $n \neq m$, and these reflect the fact that the operator changes any $\alpha\beta$ nearest neighbour pair into $\beta\alpha$. The result of all this is remarkably like the phonon treatment, based on the linear chain and after some calculation we get the explicit result

$$H_{\mathrm{H}}\phi_n = 2J\phi_n - J\phi_{n+1} - J\phi_{n-1}. \tag{6.12}$$

From the states ϕ_n we can construct an eigenfunction

$$\psi_n = \sum_n b_n \phi_n \tag{6.13}$$

and then equation (6.12) yields equations for the energy E and the coefficients b_n, namely

$$Eb_n = 2Jb_n - Jb_{n+1} - Jb_{n-1}. \tag{6.14}$$

As with phonons, we seek a wavelike solution representing the β spin (i.e. the misoriented one) propagating through the lattice. If a is the spacing between the atoms in the chain, the desired solution is

$$b_n = \mathrm{e}^{ikna} \tag{6.15}$$

which leads to the energy–wavenumber (dispersion) relation for the spin waves as

$$E(k) = 2J[1 - \cos(ak)]. \tag{6.16}$$

In the long-wavelength limit $k \to 0$, we have here the characteristic quadratic dependence of the energy on k referred to above.

6.4 Measurement of magnon spectrum

The magnon spectrum (given in the model above by equation (6.16)) is accessible to experiment through neutron elastic scattering. Just as for phonons (cf §3.6) we can write the conservation laws for a one-magnon process as

$$\boldsymbol{k}' - \boldsymbol{k} = \boldsymbol{q} + \boldsymbol{G} \tag{6.17}$$

and

$$\frac{\hbar^2}{2M_{\mathrm{n}}}(\boldsymbol{k}'^2 - \boldsymbol{k}^2) = \pm\hbar\omega_{\boldsymbol{q}} \tag{6.18}$$

where $\boldsymbol{q}$ is the magnon wavevector. Again as for phonons, the presence of a reciprocal lattice vector $\boldsymbol{G}$ on the right-hand side of equation (6.17) is a consequence of the periodicity of $\omega_{\boldsymbol{q}}$ in the reciprocal lattice.

To understand how the neutron method may be used to study the spin wave elementary excitations, we start by noting that Bragg reflexions (elastic scattering) enable one to use crystals as spectrometers, first to obtain a monochromatic beam of neutrons and second to separate out neutrons of a definite energy from all those scattered by the target.

For example, the positions of source and analyser, the former giving incident neutrons of a particular energy and the latter counting scattered neutrons of a particular energy, may be fixed while the crystal orientation is varied. There will be a sudden rise in the count of scattered neutrons when equations (6.17) and (6.18) are satisfied simultaneously and the orientation of the crystal will then give the magnon wavevector $\boldsymbol{q}$. The situation in reciprocal space is depicted in figure 6.1, the conditions for satisfying equations (6.17) and (6.18) being indicated.

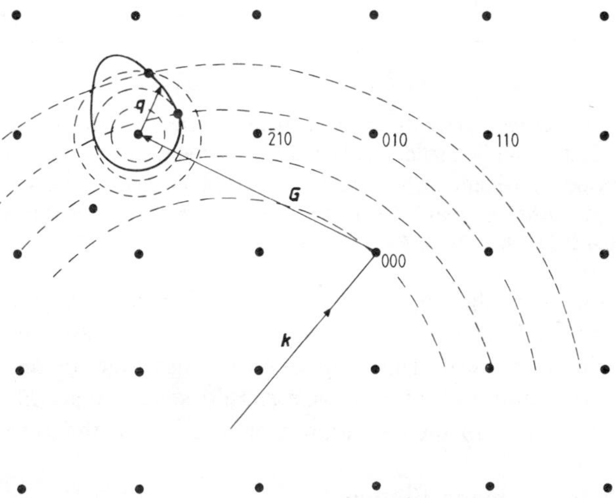

Figure 6.1 Illustrating the simultaneous conservation of energy and momentum in equations (6.17) and (6.18). The inelastic scattering surface is shown as a solid curve: on this curve the conservation requirements are obeyed. (Though the detail is not important for the present purposes, this is the inelastic scattering surface about the point ($\bar{2}10$) in reciprocal space for neutrons of wavevector $\boldsymbol{k}$ incident on a target. The figure is actually applicable to both magnon scattering and phonon scattering (see Appendix A3.3). The broken curves of larger radii represent the constant energy curves for the scattered neutron. Smaller radii curves are the same but for the annihilated phonon.)

In such experiments one finds effects due both to magnons and to phonons. These can be separated in practice as follows. It can be shown that if a magnetic moment $\boldsymbol{\mu}$ is perpendicular to $\boldsymbol{\kappa} = \boldsymbol{k}' - \boldsymbol{k}$ it will not scatter a neutron into the direction of $\boldsymbol{k}'$. Thus application of a magnetic field perpendicular to $\boldsymbol{k}' - \boldsymbol{k}$ results in a reduction of the magnetic scattering by a specimen, whereas the nuclear scattering (reflecting the phonon spectrum) will be unchanged if the two types of scattering are independent. In fact, in the example taken from the experiments of Sinclair and Brockhouse (1960), which is shown in figure 6.2, the phonon peak is enhanced. Figure 6.2 was obtained by varying the crystal orientation as referred to above. The peaks M_1 and M_2 about the [200] direction correspond to the creation of magnons. The peak P corresponds to

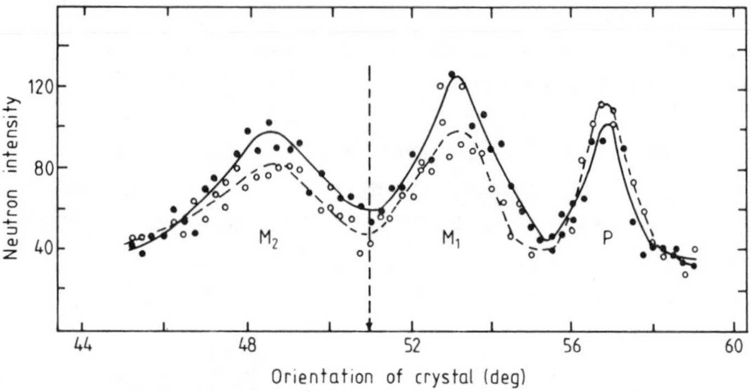

Figure 6.2 Experimental results showing magnon peaks M_1 and M_2 and phonon peak P: •, zero field; ○, applied field. Applied field is perpendicular to $\boldsymbol{k} = \boldsymbol{k}'$. Reduction of magnon peaks is evident on application of magnetic field. However, phonon peak is enhanced, through magneto-vibrational scattering. (After Sinclair R N and Brockhouse B N 1960 *Phys. Rev.* **120** 1638.)

the creation of a phonon. Applying a magnetic field, M_1 and M_2 clearly decrease while P increases. The origin of this increase lies in so called magnetovibrational scattering. The source of this resides in the fact that when analysing the phonon scattering one should strictly take into account the magnetic scattering of the individual atoms as well as the nuclear scattering.

6.5 Spontaneous magnetisation

To see how the magnetisation of a ferromagnet varies with temperature, we next utilise the property that spin wave excitations are bosons. Neglecting any interaction between spin waves†, we obtain the average number of magnons in a state of wavevector $\boldsymbol{k}$ as

$$\langle n_{\boldsymbol{k}} \rangle = \frac{1}{\exp(\hbar\beta\omega_{\boldsymbol{k}}) - 1}. \tag{6.19}$$

where $\beta = 1/k_{\mathrm{B}}T$. If $M_0 = N_\mu$ is the total magnetic moment at $T = 0$, N being the number of spins per unit volume, then

$$M = M_0 - \mu \sum_{\boldsymbol{k}} \langle n_{\boldsymbol{k}} \rangle. \tag{6.20}$$

Replacing the sum over $\boldsymbol{k}$ by an integration through the Brillouin zone of volume Ω_{B} say (see phonon discussion in Chapter 3), we find

$$M_0 - M = \frac{\mu}{8\pi^3} \int_{\Omega_{\mathrm{B}}} \frac{\mathrm{d}\boldsymbol{k}}{\exp(\hbar\beta\omega_{\boldsymbol{k}}) - 1}. \tag{6.21}$$

If we now restrict the argument to low temperatures, then as for the Debye

† The interested reader may consult, for example, Jones and March (1973).

model for the lattice specific heat, only the lowest excited states will be significant and then we can use the quadratic form of the magnon dispersion relation ω for small k for a ferromagnet, namely, for cubic lattices,

$$\omega = 2JSa^2k^2 \tag{6.22}$$

where a is the lattice constant, S the spin and J is the exchange interaction.

We can extend the integration over the whole of space because the Bose function is effectively zero except at small k. By using exactly the same arguments as in Chapter 3 for the Debye specific heat we find

$$M_0 - M \propto T^{3/2} \tag{6.23}$$

which is the famous low-temperature law due to Bloch.

6.6 Low-temperature specific heat

It is of interest to establish what contribution the spin waves in an insulating magnet make to the low-temperature specific heat. To see this, we can evidently write the internal energy as

$$E = \sum_{\boldsymbol{k}} \frac{\hbar\omega_{\boldsymbol{k}}}{\exp(\hbar\beta\omega_{\boldsymbol{k}}) - 1}. \tag{6.24}$$

Again at low temperatures we can integrate over the whole of $\boldsymbol{k}$-space, when we obtain for the energy/unit volume

$$\frac{E}{\mathscr{V}} = \frac{1}{8\pi^3} \int \frac{\hbar\omega_{\boldsymbol{k}}\, \mathrm{d}\boldsymbol{k}}{\exp(\hbar\beta\omega_{\boldsymbol{k}}) - 1}. \tag{6.25}$$

We see by analogy with the calculation of the magnetisation above that the extra factor $\omega_{\boldsymbol{k}} \propto k^2$ alters the temperature dependence by a factor T, yielding

$$E \propto T^{5/2} \tag{6.26}$$

and hence a specific heat $c_V \propto T^{3/2}$.

6.7 Magnetisation–internal energy relation

Having discussed the low-temperature form of the magnetisation and the specific heat associated with spin waves, we want to remark briefly on a relation between magnetisation and internal energy (see e.g. Grout and March 1976).

The specific heat c_V being, of course, positive, it follows that if we write the change in the internal energy from its zero-temperature value as $\Delta E = E(T) - E(0)$, then this is a monotonically increasing function of T. Consequently, we can formally express T as a function $T(\Delta E)$. Thus we can rewrite the magnetisation $M(T)-M(0)=\Delta M(T)$ as a function of ΔE.

6.7.1 Magnetic insulators

For ferromagnetic insulators, the low-temperature form of this relation is readily obtained, since $\Delta M \propto T^{3/2}$ and the spin wave contribution to the specific heat dominates the Debye phonon contribution at sufficiently low temperatures, i.e. $c_V \propto T^{3/2}$, and hence $\Delta E \propto T^{5/2}$. Hence we obtain immediately in the limit $T \to 0$

$$\Delta E \propto (\Delta M)^{5/3}. \tag{6.27}$$

It should be stressed that such a relation has some rather direct physical significance in such insulating materials because both of the quantities ΔM and ΔE are determined by the spin wave excitations.

6.7.2 Metallic ferromagnets

However, the above relation is not correct for metallic ferromagnets, though formally there is still a relation between ΔE and ΔM. The reason is that in metals the specific heat is dominated at sufficiently low temperatures not by the spin wave contribution, but by the linear electronic specific heat due to free-electron excitations. Thus we have (cf equation (5.1))

$$\Delta E \propto T^2 \tag{6.28}$$

whereas ΔM is proportional to $T^{3/2}$ as before, the collective excitations dominating the single-particle (Stoner) excitations in the metal at sufficiently low temperatures. Thus, formally, for the metallic ferromagnet we have

$$\Delta E \propto (\Delta M)^{4/3}. \tag{6.29}$$

In both cases, in the (E, M) plane, it is of some interest that the internal energy approaches its zero-temperature limit with zero slope.

Over the whole temperature range, one can relate E and M from the exact solution of the two-dimensional Ising model†, as set out, for example, by Callaway (1974). In that case the slope $d(\Delta E)/d(\Delta M)$ is finite as $T \to 0$ and therefore the analytic behaviour obviously depends on the dimensionality of the assembly.

6.8 Antiferromagnetism

The phenomenon of antiferromagnetism was explored and described in the pioneering work of Néel (1932, 1936). A variety of substances, including metals, alloys and insulating crystals, exhibit an anomaly in the specific heat at a temperature below which the magnetic susceptibility is smaller than it is

† See problem 6.3 for full details.

above. Néel gave as the explanation of this the tendency of adjacent atoms to have their spins opposite rather than parallel. In that case, one would expect that, at sufficiently low temperatures, the lattice could be subdivided into two sublattices, so that all or most of the spins of one sublattice were mutually parallel, but opposite to those of the other.

A considerable number of the properties of antiferromagnets can then be described in terms of a model like that of Weiss discussed in Chapter 8 for ferromagnets. The major difference is that one has to assume that the force on each spin contains a contribution proportional to the total moment of the sublattice to which the spin belongs, and a term in the total number of the other. Because the tendency is towards opposing spins, the constant of proportionality in the second term must now be negative. The main results are summarised in §8.2.

The difficulty of obtaining eigenstates of the antiferromagnetic system is clearly considerably greater than for ferromagnets. The reason is that quantum mechanical exchange does not permit spins to stay on their sites, but implies that they trade places with one another. This situation is not observable when all the spins are parallel, but clearly has important consequences when they are not.

6.9 Hamiltonian and dispersion relation for two-sublattice model of an antiferromagnet

We now consider the Hamiltonian for a two-sublattice antiferromagnet, in which half the spins have $\langle S^z \rangle \simeq S$ ('up' spins), while the other half have $\langle S^z \rangle \simeq -S$ ('down' spins) and the spins are so arranged that the nearest neighbours of 'up' spins are 'down' spins and vice versa.

It will be assumed that the lattice as a whole is a Bravais lattice and that both the 'up' spin and 'down' spin sublattices are also Bravais lattices.

For the Hamiltonian, one can write

$$H = -\sum_{n,m} J_{nm} \mathbf{S}_n \cdot \mathbf{S}_m - \sum_l \sigma_l H_a S_l^z \tag{6.30}$$

where we have

$$\sigma_l = \begin{cases} 1 & \text{if } \mathbf{S}_l \text{ is an 'up' spin} \\ -1 & \text{if } \mathbf{S}_l \text{ is a 'down' spin.} \end{cases}$$

H_a is the anisotropy field, which defines the preferred direction in the crystal; this has been chosen as the z-direction. The anisotropy field also ensures that the ground state will be such that $\langle S_l^z \rangle \simeq \sigma_l S$. We have assumed that no external field is applied.

It is possible to obtain an approximate solution of the above Hamiltonian by invoking again the concept of spin waves. This can be done by starting from the Heisenberg equations of motion for the spin operators:

$$\dot{S}_l = \frac{-\mathrm{i}}{\hbar}[S_l, H]. \tag{6.31}$$

Substituting the expression (6.30), where H_a has been put equal to zero, and making use of the angular momentum commutation relations we obtain

$$\dot{S}_l = -\frac{1}{\hbar}\sum_m J_{lm}(S_l \times S_m). \tag{6.32}$$

In the case of a linear chain with only near-neighbour interactions, this can be further simplified to

$$\dot{S}_l = -\frac{J}{\hbar}[(S_l \times S_{l-1}) + (S_l \times S_{l+1})] \tag{6.33}$$

Equations (6.33) form a non-linear set and have been solved only in the case $S = \frac{1}{2}$ (see the following paragraph).

In other cases where $S \neq \frac{1}{2}$, one has to resort to a linearisation procedure. This is achieved by assuming that the ground state is not different from the perfectly aligned Néel state described above and then by considering only small deviations from this ideal case. This is not a good approximation for linear chains with a small value of the spins, because there the quantum zero-point fluctuations tend to destroy the perfectly aligned antiferromagnetic state. However, for large values of S these fluctuations become unimportant and our treatment can safely be applied.

We therefore write

$$\begin{aligned}\dot{S}_l \simeq -\frac{J}{\hbar}(\langle S_l\rangle \times S_{l-1} + S_l \times \langle S_{l-1}\rangle \\ + \langle S_l\rangle \times S_{l+1} + S_l \times \langle S_{l+1}\rangle).\end{aligned} \tag{6.34}$$

Then for S_l we take the approximate form

$$|\langle S_l\rangle| \simeq \langle S_z\rangle \simeq (-1)^l S. \tag{6.35}$$

This embodies the assumption that all the spins point in the positive z-direction on one sublattice (l even) and in the opposite direction on the other sublattice (l odd). Equation (6.32) can be further simplified by introducing the linear combinations

$$S_l^{\pm} = S_l^x \pm \mathrm{i} S_l^y \tag{6.36}$$

when one obtains

$$\dot{S}_l^{\pm} = -\mathrm{i}\frac{JS}{\hbar}(-1)^l(S_{l-1}^{\pm} + S_{l+1}^{\pm} + 2S_l^{\pm}). \tag{6.37}$$

Once again we seek propagating solutions of the form

$$S_{2i}^{\pm} = u_1^{\pm} \exp[\mathrm{i}(KR_{2i} - \omega t)] \tag{6.38}$$

$$S_{2i+1}^{\pm} = u_{\downarrow}^{\pm} \exp[\mathrm{i}(KR_{2i+1} - \omega t)] \tag{6.39}$$

which when inserted into equation (6.37) give the following condition on the frequency:

$$\omega^2 = 2\left(\frac{JS}{\hbar}\right)^2 \sin^2(\tfrac{1}{2}qa). \tag{6.40}$$

Equation (6.40) is the desired dispersion relation for spin waves in an antiferromagnet. The most distinctive feature of this $\omega(\boldsymbol{q})$ relation is that the frequency varies linearly with q in the long-wavelength limit; this is a basic characteristic of antiferromagnons.

Knowledge of these elementary excitations permits calculation of the temperature dependence of the sublattice magnetisation, and also of the specific heat, along lines discussed above for the case of ferromagnets.

The qualitative results are that the sublattice magnetisation decreases as $(T/T_c)^2$ for low temperatures. The specific heat, on the basis of the linear ω–q relationship above as $q \to 0$, is quite analogous to the acoustic phonon problem and a T^3 law results.

While the three-dimensional problem remains difficult when we attempt to transcend the simple approximations made above, we conclude this spin wave treatment with a brief summary of a one-dimensional example where progress can be made.

While the three-dimensional problem remains difficult, when one attempts to transcend the simple approximations made above, one-dimensional theory is much more tractable. As an example of a system which, to a first approximation, can be described as an antiferromagnetic chain of N spins with $S = \frac{1}{2}$, results for $CsCoCl_3$ are summarised briefly in Appendix 6.3.

6.10 Spin waves in disordered magnets

The question as to whether spin waves, known to be very well defined collective excitations in both ferromagnets and antiferromagnets remain good concepts in the presence of disorder was tackled by a number of workers in the 1960's. Especially we have in mind the work of Callaway (1963), Callaway and Boyd (1964) and of Murray (1966a, b).

The argument presented by Murray is brief and useful and goes as follows. The spin wave state with $\boldsymbol{k}$ vector equal to zero must be degenerate with the ground state because the total spin commutes with the Hamiltonian. Therefore by continuity arguments, the spin wave energies must tend to zero as k tends to zero. The energy can be expanded as a power series in k, and it must vary as k or as k^2 or as some higher power of k. The only way in which the energy can vary as k is if the square, or higher power of the energy varies as k^2. It can do so only if the equation of motion of the spin deviation operators satisfies a second- or higher-order differential equation (as, for example the position operator does in the corresponding problem for phonons) or if the spectrum is degenerate. This latter situation obtains for spin waves in an antiferromagnet. Since neither of these conditions is satisfied for the Heisenberg ferromagnet, the spin wave energy will vary as k^2. Murray argues that this same situation will hold in a disordered binary alloy, and she verified this by direct calculation.

The above argument can be extended to give information about the way that the scattering cross section of an isolated impurity behaves for spin waves in the limit of small k. The interested reader should refer to problem 6.5 for more information on that point.

For those readers interested in more sophisticated theory of spin waves in disordered systems, Green function methods are utilised by Edwards and Jones (1971) for ferromagnets. They confirm that the concept of a spin wave, in the presence of disorder, is still a good one at low ω and k. They have also discussed spin waves in disordered antiferromagnets (Jones and Edwards 1971).

These arguments are based on short-range interactions; there is considerable interest in exchange interactions in metals, which are long range.

6.11 Indirect exchange interaction between localised spins in a metal

The type of exchange interaction to be discussed below is related to the coupling of localised electron spins via conduction electrons. The specific example we shall use is that of the rare earth metals, where the localised spins are due to f electrons. Before turning to that case, we consider the close relation between the results presented below and the pile-up of electronic charge round an impurity like Zn in Cu, with a positive excess valence of unity.

To be specific, let us introduce a localised spin into a Fermi gas. The effect is to polarise the conduction electrons, in much the same way that in Chapter 2 we saw the introduction of a positive charge caused a piling up of electrons to screen out its field. The analogy goes further, the screening of the test charge led to long-range oscillations in the electron density at large distances, these reflecting the sharp Fermi surface (thus contrast the screening of a charge in a non-degenerate electron gas in §2.5, with that in a degenerate gas with a sharp Fermi edge). In a very similar way, the spin polarisation induced in the conduction electron cloud by a localised spin has long-range oscillations.

For two charges in a degenerate electron gas, one can calculate the interaction energy, in a linear theory, by simply taking one charge as sitting in the electrostatic potential of the other charge, plus its screening cloud. This interaction obviously then oscillates with distance, because of the Friedel oscillations in the potential round a charge discussed in §2.7. In a similar manner one can calculate the interaction energy between localised spins embedded in a Fermi gas and the result is an oscillatory interaction, as first calculated by Ruderman and Kittel and by Yosida (see Kittel 1963 for a detailed derivation).

This type of interaction can explain the observed spin ordering in the rare earth metals where, as already remarked, localised f electrons are coupled via conduction electrons. This spin ordering can occur in two types, which may exist together:

$$\text{type I} \qquad \langle S_n^z \rangle = MS\cos(\boldsymbol{q}\cdot\boldsymbol{R}_n + \phi) \qquad (6.41)$$

$\boldsymbol{R}_n$ being a lattice point. Here $\boldsymbol{q}$ is in fact parallel to the c-axis and this gives a wave-like moment variation along this axis. This structure, to give specific examples, is found to occur in the high-temperature phase of Er and Tm. The second type is:

$$\text{type II} \qquad \langle S_n^x \rangle = M'S\cos(\boldsymbol{q}\cdot\boldsymbol{R}_n) \qquad \langle S_n^y \rangle = M'S\sin(\boldsymbol{q}\cdot\boldsymbol{R}_n) \qquad (6.42)$$

which is a spiral structure, the turn angle between spins in adjacent layers being qc. This is found in the high-temperature phase of Tb, Dy and Ho. The way in which the observed ordering is related to the Ruderman–Kittel interaction has been discussed fully by Elliott and Wedgwood (1963, 1964), to whose work the interested reader is referred.

There is one further, and highly interesting, area which arises as a result of oscillatory Ruderman–Kittel interactions between localised spins in metals. This is the problem of spin glasses. These are magnetic alloys which display some distinctive properties. Despite the fact that one is dealing with random, or almost random impurities carrying magnetic moments in a host like Cu or Ag, magnetic susceptibility and electrical resistivity, to take two examples, follow patterns which show clearly that the magnetic arrangement within the material is a collective feature of the entire magnetic alloy. In particular, the magnetic materials (in the spin glass concentration range which may be from 0.1 to 20 % of magnetic atoms) exhibit a cusp in the magnetic susceptibility, at a temperature T_0 say. Below this temperature the localised moments are frozen in. They have random orientations but are not able to undergo thermal fluctuations. In contrast, above T_0 the moments are free to rotate. The collective behaviour is crucially linked with the long-range oscillatory interaction, the effective magnetic field at any particular impurity being essentially, with N impurities, a superposition of N-1 oscillatory terms with the same wavelength (determined by Fermi surface dimensions, k_F for a spherical Fermi surface) but of arbitrary phase, because each term originates from an impurity atom at a random site. The randomness in the phases limits the coherence, and hence the magnetic order, to a finite range. The reader is referred to the article by Wohlleben and Coles (1973) for details. One important contribution in the theory was that of Edwards and Anderson (1975) who introduced a suitable order parameter q, say, to discuss the cooperative properties. Specifically they chose

$$q = \langle \langle \sigma_i \rangle^2 \rangle_c \qquad (6.43)$$

where σ_i is the spin on the ith site, and the subscript c indicates an average over all configurations of the random system. The other average is the usual thermal average.

Molecular field treatments of the spin glass transition lead to some difficulties, working with the above order parameter q. For example, the interesting work of Sherrington and Kirkpatrick (1975) had a problem associated with the entropy at low temperatures becoming negative. Ways of overcoming this have been proposed by Thouless, Anderson and Palmer (1977). For a fuller discussion the reader may refer to the paper by Khurana and Hertz (1980). Clearly, a lot remains to be done on the theory of spin glasses.

6.12 Kondo effect

A number of alloy systems, for example noble metal solvents containing a dilute concentration of transition metal atoms, exhibit a minimum in the electrical resistivity at low temperatures (see e.g. Coles 1958).

In seeking the origin of this effect, Kondo (1964, 1969) found that when calculated to second order in the exchange interaction J for coupling of the form

$$H = -J\boldsymbol{S}\cdot\boldsymbol{s} \tag{6.44}$$

the amplitude for exchange scattering of a conduction electron spin $\boldsymbol{s}$ at the Fermi surface by a single localised spin $\boldsymbol{S}$ diverges as $\ln T$ when $T \to 0$. The resistivity minimum then follows from the formula

$$\rho = aT^5 + c\rho_0 + c\rho_1 \ln T \tag{6.45}$$

where c is the impurity concentration. Here the different terms arise as follows. The first comes from the phonon-induced scattering by the host lattice, which is assumed proportional to T^5 (cf Ziman 1960), the term $c\rho_0$ includes the temperature-independent result of non-magnetic impurity scattering and the final term is that found by Kondo†. When $J < 0$ (antiferromagnetic coupling), $\rho_1 < 0$ also and one then finds a resistance minimum at

$$T_{\min} = c^{1/5}\left(\frac{|\rho_1|}{5a}\right)^{1/5} \tag{6.46}$$

in reasonable agreement with experiment.

We shall not try to give a quantitative account of the theory of the Kondo effect, though very sophisticated theory now exists (Wilson 1974). Rather, we shall base the brief discussion below on the picture that an electron interacting with a localised spin can cause spin flip. This sudden change in local moment scatters conduction electrons. The problem is somewhat reminiscent of the edge singularity effect discussed in §2.13.

Following Mott (1974), we describe the transition metal impurity at low temperature by a model with integral number of electrons on the ion (e.g. 5 for

† The $\ln T$ term in equation (6.45) can only persist down to some 'Kondo' temperature.

Mn in Cu). There are then $2j+1$ states for the local moment which, in zeroth order, all have the same energy. But the nub of the Kondo low-temperature behaviour is that interaction with the conduction electrons causes the moment to switch between these states, and that this is a slow process. To pass from one of these states to another, the system must go through an intermediate state, in which an electron near the Fermi level moves on to the transition metal ion, a process costing energy E_a say, and then an electron with opposite spin jumps off. The second process is fast but the first is slow. Alternatively, the spin flip can occur by an electron first jumping from the transition metal ion to the Fermi level, and then the second electron jumping back; we let the energy necessary be E_b. The two processes are indicated schematically in figure 6.3. The Kondo frequency† turns out then to be of the form

$$\omega_K = (\Delta/\hbar)\exp(-E_K/\tfrac{1}{2}\Delta) \tag{6.47}$$

where E_K is E_a or E_b, whichever is the smaller. The characteristic energy Δ has to do with the resonance scattering of conduction electrons off the impurity; it in fact measures the energy width over which the scattering of the conduction electrons is strong. This problem is obviously, in total, about dynamic spin polarisation and as such is a more complex problem than we can treat quantitatively with the methods at our disposal in this book.

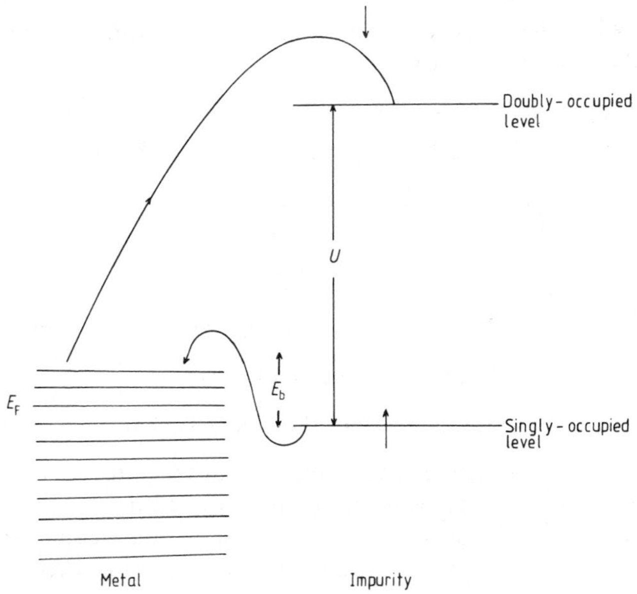

Figure 6.3 Schematic diagram indicating the two processes by which the Kondo effect can occur. U is the intra-atomic Coulomb energy ($=E_a-E_b$).

† One expects the thermal energy corresponding to the Kondo temperature to be $\hbar\omega_K$.

Problems

6.1 Show that the spin lowering operator S^- commutes with S^2 and the Heisenberg Hamiltonian. Relate this to the invariance of the Heisenberg Hamiltonian under rotation.

6.2 Under what situations can the exchange interaction $J(k)$ be regarded as spherical?

6.3 For the two-dimensional Ising model (see e.g. Callaway 1974) the internal energy $E(T)$ turns out to be

$$E(T) = -J\coth(2K)\{1+(2/\pi)[2\tanh^2(2K)-1]K_1(m)\} \qquad \text{(P6.1)}$$

where

$$m = 2\sinh(2K)/\cosh^2(2K) \qquad K = J/k_B T$$

with J the strength of the exchange interaction. $K_1(m)$ is the complete elliptic integral of the first kind.

Also the magnetisation M is given in terms of its zero-temperature value $M(0)$ by

$$M(T) = \begin{cases} M(0)[1-\operatorname{cosech}^4(2K)]^{1/8} & T<T_c \\ 0 & T>T_c. \end{cases} \qquad \text{(P6.2)}$$

Show that the energy as a function of reduced magnetisation $\sigma = M(T)/M(0)$ is explicitly

$$E(\sigma) = -J[1+(1-\sigma^8)^{1/2}]^{1/2} \times (1+(2/\pi)\{2[1+(1-\sigma^8)^{1/2}]^{-1}-1\}K_1(m)) \qquad \text{(P6.3)}$$

where m is given in terms of σ by

$$m = \frac{2(1-\sigma^8)^{1/4}}{(1-\sigma^8)^{1/2}+1} \qquad \text{(P6.4)}$$

(cf Grout and March 1976). Demonstrate that at low temperatures

$$E(T)-E(0) \propto M(0)-M(T) \qquad \text{(P6.5)}$$

and contrast with the result (6.27) for three-dimensional magnetic insulators.

6.4 As discussed in Chapter 1, a ferromagnet at absolute zero is a collection of spins on a lattice, all spins being completely aligned.

As for phonons, the normal modes of oscillation of these spins about their equilibrium directions are characterised by a wavevector $\boldsymbol{k}$. There is only one mode per $\boldsymbol{k}$ vector.

For a simple model of a cubic lattice, the normal mode frequencies are

$$\omega(\boldsymbol{k}) = J[1-\cos(k_x a)-\cos(k_y a)-\cos(k_z a)] \qquad \text{(P6.6)}$$

where J is, as usual, a measure of the strength of the exchange interaction.

Show that ω is quadratic in k for long wavelengths, and calculate the constant of proportionality (the so called spin wave stiffness constant).

Compare the low-temperature specific heat with the Debye specific heat for phonons, both at low temperatures.

6.5 Using the method outlined below (cf Murray 1966a, b), show that the scattering cross section of an impurity for spin waves in a ferromagnet, for small k, is proportional to k^4.

(a) Divide the total Hamiltonian into pure host term H_0 plus a local perturbation V which describes the effect of an impurity.

(b) Write the Schrödinger equations for host and impure solid as

$$H_0|k_0\rangle = d_0 k^2 |k_0\rangle \qquad \text{and} \qquad H|k\rangle = dk^2|k\rangle. \tag{P6.7}$$

(c) Take the scattering cross section

$$\sigma_{\boldsymbol{k}\boldsymbol{k}'} \sim \frac{1}{v_{\boldsymbol{k}}} |\langle \boldsymbol{k}'_0 | V | \boldsymbol{k} \rangle|^2 \rho_{k^2 = k'^2}(E_{\boldsymbol{k}})$$

where $v_{\boldsymbol{k}}$ is the velocity of the spin waves and ρ the density of states.

(d) Use the facts that $v_{\boldsymbol{k}} \propto k$ and that $\rho(E_{\boldsymbol{k}})$ is also proportional to k for small k.

(e) From the Schrödinger equations, prove that

$$d_0 k'^2 \langle \boldsymbol{k}'_0 | \boldsymbol{k} \rangle + \langle \boldsymbol{k}'_0 | V | \boldsymbol{k} \rangle = dk^2 \langle \boldsymbol{k}'_0 | \boldsymbol{k} \rangle. \tag{P6.8}$$

(f) Hence show that

$$\sigma_{\boldsymbol{k}\boldsymbol{k}'} \propto k^4 (d - d_0)^2 |\langle \boldsymbol{k}'_0 | \boldsymbol{k} \rangle|^2 \tag{P6.9}$$

(cf Callaway 1963, Callaway and Boyd 1964).

7

COLLECTIVE EFFECTS IN LIQUIDS

We shall first give a brief summary of the way in which the structure of a liquid can be described. This is most usefully approached by calculating the neutron (or x-ray) scattering from the liquid. Then we shall go on to construct a simple model to describe collective behaviour in a liquid. While the theory is presented below first of all for classical liquids, the most appropriate application of the kind of model we shall use is to describe collective modes in liquid ^{4}He.

7.1 Neutron scattering and liquid structure

Suppose we have a liquid target of N like atoms and an incident monochromatic neutron beam of momentum $\hbar k = h/\lambda$. Since the aim is to explore the liquid structure, the neutron de Broglie wavelength λ must be of the order of the interatomic separation in the liquid, i.e. a few ångströms. This is readily shown to imply that the neutron energy $E_n = \hbar^2 k^2/2M_n$, M_n being the mass of the neutron, is in the thermal energy range. Through its interaction with the liquid, the neutron is scattered into a state having a different momentum $\hbar \boldsymbol{k}'$ say, transferring to the liquid target a momentum

$$\hbar \boldsymbol{q} = \hbar \boldsymbol{k}' - \hbar \boldsymbol{k}. \tag{7.1}$$

The probability for such an event to occur is given by the Fermi golden rule and is thereby proportional to $|\langle \psi | V | \psi' \rangle|^2$, where $|\psi\rangle$ and $|\psi'\rangle$ are the eigenstates of the total system (neutron plus liquid), before and after the scattering event, while V is the neutron–liquid interaction potential. Since neutrons interact only with the nuclei (we exclude here the case of a liquid in which the atoms can carry local magnetic moments, such as in rare earths) and have a wavelength λ which is much larger than the range of the nuclear forces, we can approximate V as a sum of point-like interactions centred on the nuclear positions $\boldsymbol{r}_i$:

$$V = \sum_i f_i \delta(\boldsymbol{r} - \boldsymbol{r}_i). \tag{7.2}$$

In equation (7.2), $\boldsymbol{r}$ is the neutron coordinate while f_i is the scattering length which measures the scattering property of each nucleus. If we assume that all nuclei are identical (in practice we have a mixture of isotopes, with different scattering lengths) then we obtain

$$V = f\rho(\boldsymbol{r}) \tag{7.3}$$

where

$$\rho(\boldsymbol{r}) = \sum_{i=1}^{N} \delta(\boldsymbol{r}-\boldsymbol{r}_i)$$

is the local density of the fluid. Since in the initial and final states the neutron and the liquid do not interact, we may write $|\psi\rangle$ and $|\psi'\rangle$ as a product of the neutron wavefunction, which is a plane wave, and an unperturbed eigenstate

$$\begin{aligned} |\psi\rangle &= \exp(\mathrm{i}\boldsymbol{k}\cdot\boldsymbol{r})|\psi_n\rangle \\ |\psi'\rangle &= \exp(\mathrm{i}\boldsymbol{k}'\cdot\boldsymbol{r})|\psi_{n'}\rangle \end{aligned} \tag{7.4}$$

where $|\psi_n\rangle$ and $|\psi_{n'}\rangle$ are the system wavefunctions before and after the collision. We find therefore (cf Marshall and Lovesey 1971)

$$\begin{aligned} \langle\psi|V|\psi'\rangle &= \langle\psi_n|f\int\exp[-\mathrm{i}(\boldsymbol{k}-\boldsymbol{k}')\cdot\boldsymbol{r}]\rho(\boldsymbol{r})\mathrm{d}\boldsymbol{r}|\psi_{n'}\rangle \\ &= f\langle\psi_n|\rho_{\boldsymbol{q}}|\psi'_n\rangle \end{aligned} \tag{7.5}$$

where

$$\rho_{\boldsymbol{q}} = \int\exp(-\mathrm{i}\boldsymbol{q}\cdot\boldsymbol{r})\rho(\boldsymbol{r})\mathrm{d}\boldsymbol{r} \tag{7.6}$$

is the $\boldsymbol{q}$ Fourier component of the density fluctuations. In order to obtain the total scattered intensity $I(q)$, which is our primary aim, we have now to take the modulus of equation (7.5), sum over the possible final states and average each initial state with its statistical probability p_n. The total scattering probability therefore becomes

$$I(q) = |f|^2 \sum_{n,n'} p_n \langle\psi_n|\rho_{-\boldsymbol{q}}|\psi_{n'}\rangle\langle\psi_{n'}|\rho_{\boldsymbol{q}}|\psi_n\rangle. \tag{7.7}$$

Using the completeness theorem for eigenfunctions, i.e. $\Sigma_{n'}|\psi_{n'}\rangle\langle\psi_{n'}| = 1$ (see e.g. Schiff 1968), we find

$$\begin{aligned} I(q) &= |f|^2 \sum_n p_n \langle\psi_n|\rho_{-\boldsymbol{q}}\rho_{\boldsymbol{q}}|\psi_n\rangle \\ &= |f|^2\langle\rho_{-\boldsymbol{q}}\rho_{\boldsymbol{q}}\rangle = N|f|^2 S(q) \end{aligned} \tag{7.8}$$

where the angular brackets denote the liquid average and we have introduced the definition of the liquid structure factor $S(q)$ through

$$S(q) = N^{-1}\langle\rho_{-\boldsymbol{q}}\rho_{\boldsymbol{q}}\rangle. \tag{7.9}$$

This structure factor is seen to be the correlation between density fluctuations in $\boldsymbol{q}$-space. It is important to note that in equation (7.8) we have separated the terms involving the scattering event ($|f|^2$) from those that describe the unperturbed liquid structure ($S(q)$).

7.1.1 Atomic scattering factor for the case of x-rays

Similar considerations hold for x-ray scattering which is another useful technique for measuring liquid structure. Of course, in this case the x-rays are scattered overwhelmingly by the electrons, nuclear charge scattering being down by a factor involving the ratio of electron to nuclear mass. Thus, one needs the electron density in the atom, say $\rho_e(\mathbf{r})$, in the determination of the intensity. If we denote the atomic scattering factor, which is the Fourier transform of ρ_e, by $f(q)$ then we have explicitly

$$f(q) = \int \exp(\mathrm{i}\mathbf{q}\cdot\mathbf{r})\rho_e(\mathbf{r})\,\mathrm{d}\mathbf{r} \tag{7.10}$$

and the x-ray intensity $I(q)$ is

$$I(q) = N|f(q)|^2 S(q). \tag{7.11}$$

If λ is the x-ray wavelength and 2θ is the angle through which the x-rays are scattered then explicitly

$$q = (4\pi \sin\theta)/\lambda.$$

If one wishes, one can quite properly regard the x-ray intensity equation (equation (7.11)) as the operational definition of the liquid structure factor $S(q)$.

The radial distribution function or liquid pair function $g(r)$ is defined such that if we sit on an atom at the origin $r = 0$, the number of atoms lying at distances between r and $r + \mathrm{d}r$ from it is $4\pi\rho_0 r^2 g(r)\,\mathrm{d}r$, ρ_0 being the average bulk value of the number density of atoms (N/V). Then the structure factor is given by

$$S(q) = 1 + \rho_0 \int [g(r) - 1] \exp(\mathrm{i}\mathbf{q}\cdot\mathbf{r})\,\mathrm{d}\mathbf{r}. \tag{7.13}$$

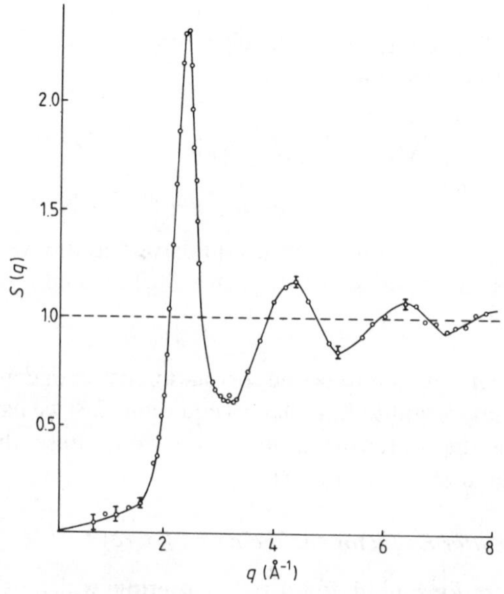

Figure 7.1 Structure factor $S(q)$ of monatomic liquid.

Hence $g(r)$ can be extracted, via equations (7.11) and (7.13), from x-ray scattering.

Actually, since $g(r)$ is really designed to describe nuclear–nuclear correlations, it is best to extract $S(q)$ and hence $g(r)$ from a neutron scattering experiment. However, though x-rays are of course scattered overwhelmingly by electrons, and neutrons by nuclei, we do not need to distinguish between the results of the two experiments for present purposes of getting $S(q)$ or $g(r)$. Schematic forms of $S(q)$ and $g(r)$ for a simple monatomic liquid are shown in figures 7.1 and 7.2.

One other result of considerable importance, for which a fairly elementary justification will be given below, is that as $q \to 0$

$$S(0) = \rho_0 k_B T \kappa_T \tag{7.14}$$

where κ_T is the isothermal compressibility of the liquid.

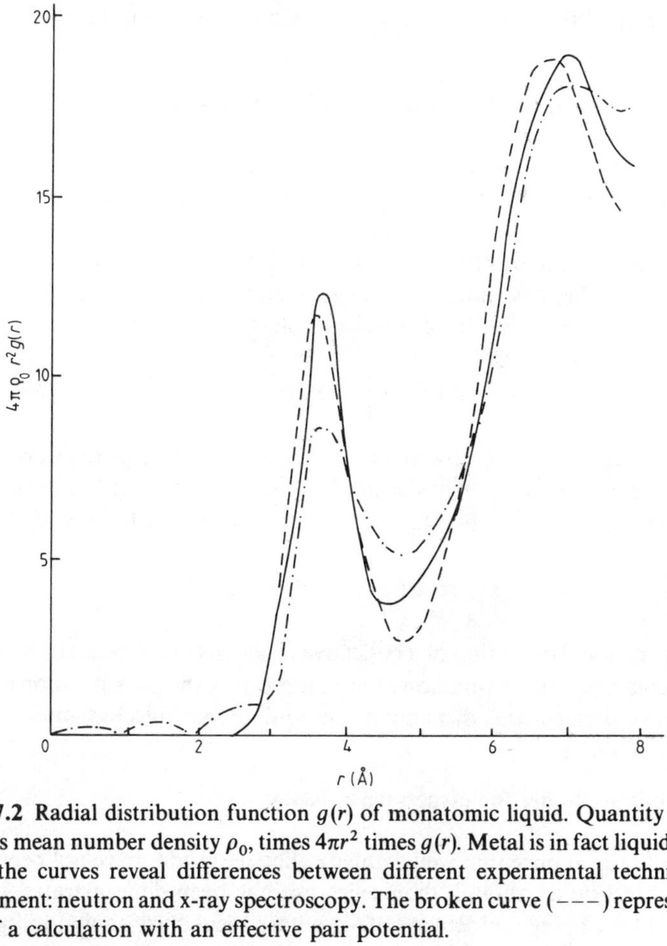

Figure 7.2 Radial distribution function $g(r)$ of monatomic liquid. Quantity actually plotted is mean number density ρ_0, times $4\pi r^2$ times $g(r)$. Metal is in fact liquid Na and two of the curves reveal differences between different experimental techniques of measurement: neutron and x-ray spectroscopy. The broken curve (– – –) represents the result of a calculation with an effective pair potential.

It may be noted that for an ideal gas with equation of state

$$pV = Nk_{\mathrm{B}}T \tag{7.15}$$

$$\kappa_T = \left[-V\left(\frac{\partial p}{\partial V}\right)_T\right]^{-1} = \frac{1}{p} \tag{7.16}$$

and thus from equation (7.14)

$$S(O)_{\mathrm{ideal}} = \frac{N}{V}\frac{k_{\mathrm{B}}T}{p} = 1. \tag{7.17}$$

In marked contrast to equation (7.17), in liquids such as argon, at its triple point $S(0) \sim 0.06$, while for the simpler metals $S(0)$ usually lies in the range 0.01 to 0.03 near the melting point. This major difference between classical liquids and an ideal gas is already reflecting the importance of structure, or put another way, short-range order, in determining the properties of liquids.

7.2 Inelastic neutron scattering and collective modes

We have referred above to the use of neutron scattering to measure the liquid structure factor $S(q)$. But we can get more information than that, for we can measure the probability $S(q, \omega)$, that a neutron incident on the liquid transfers momentum $\hbar q$ and energy $\hbar\omega$ to the liquid. It is not difficult to show that $S(q, \omega)$ is such that if we integrate over all energy transfers, we regain the liquid structure factor $S(q)$, i.e. for a classical liquid

$$S(q) = \int_{-\infty}^{\infty} S(q, \omega)\,\mathrm{d}\omega. \tag{7.18}$$

This relation, often referred to as the zeroth (frequency) moment, can be generalised (see March (1968) should the reader be interested in the proof) to the second moment $\int_{-\infty}^{\infty} \omega^2 S(q, \omega)\,\mathrm{d}\omega$, which turns out to be such that

$$\int_{-\infty}^{\infty} \omega^2 S(q, \omega)\,\mathrm{d}\omega = q^2 k_{\mathrm{B}}T/M. \tag{7.19}$$

An operational definition of a collective mode may now be given by looking for peaks in $S(q, \omega)$ as a function of ω, and plotting the peak position $\omega(q)$ as a function of q to get the 'dispersion relation' of the collective mode.

7.3 Primitive model for dispersion relation

We should say at once that well defined collective modes have not been widely observed in liquids, though their existence has been demonstrated in liquid metal Rb by Copley and Rowe (1974), whose results are shown in figure 7.3.

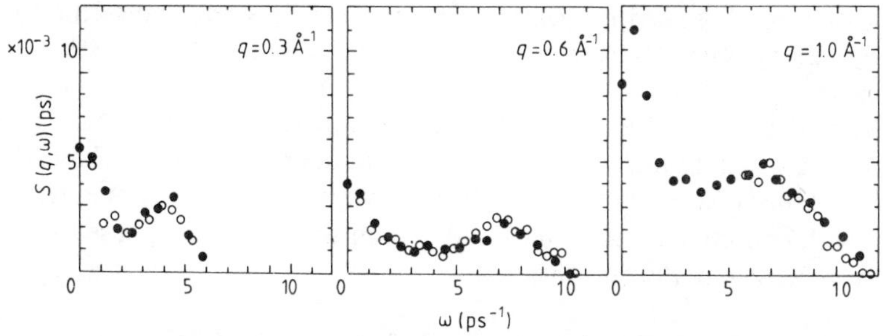

Figure 7.3 Dynamical structure factor $S(q, \omega)$ for liquid metal Rb as measured by neutron inelastic scattering (after Copley and Rowe). Results shown are constant-q plots of $S(q, \omega)$ against ω for three values of q: 0.3, 0.6 and 1.0 in units of $Å^{-1}$. Well defined peaks exist at non-zero frequencies, with peak height frequency increasing with increasing q; they demonstrate the dispersion characteristic of the collective mode. (From Copley J R D and Rowe J M 1974 *Phys. Rev.* A **9** 1656.) Quantity plotted includes the factor $\exp(-\hbar\omega/2k_BT)$. ● = energy gain data; ○ = energy loss data.

They probably also exist in liquid Ni (see Johnson *et al* 1977).

However, let us make the (somewhat oversimplified) assumption that there is a well defined collective mode with dispersion relation $\omega(q)$. Then in a simple classical liquid, we can write first

$$S(q, \omega) = S(q)[\tfrac{1}{2}\delta(\omega - \omega(q)) + \tfrac{1}{2}\delta(\omega + \omega(q))] \tag{7.20}$$

which satisfies the necessary requirements that: (a) it is even in q; (b) it is peaked (actually has infinities) as a function of ω at $\pm\omega(q)$ for a given q, in accord with the above definition of a collective mode; and (c) it satisfies the zeroth moment relation (7.18).

Naturally, even in a liquid like Rb, for which rather well defined collective modes exist, not only the broadening of the side peak but also the presence of a broad central peak around $\omega = 0$ (see § 7.5) has to be taken into account in order to build a quantitative theory of the dynamical structure factor $S(q, \omega)$. But our aim here is to establish the main features of the dispersion relation $\omega(q)$, and this we can now do by inserting equation (7.20) into the second moment relation (7.19). Then we find, with M the atomic mass,

$$\int_{-\infty}^{\infty} S(q, \omega)\omega^2 \, d\omega = S(q)\omega^2(q) = q^2 k_B T/M \tag{7.21}$$

or

$$\omega^2(q) = q^2 k_B T/S(q)M. \tag{7.22}$$

In using such an argument, we have, of course, ignored the contribution of the central peak referred to above to the sum rule (7.19). However, since in equation (7.19), $S(q, \omega)$ is multiplied by ω^2 it appears that within the framework presented here of establishing the main features of the dispersion, the contribution of the central peak to the sum rule does not change the results more than by some quantitative factor which is not important for the present purpose.

We can easily sketch the form of $\omega(q)$ from equation (7.22) since the shape of the liquid structure factor $S(q)$ is as shown in figure 7.1. We shall return to a more detailed discussion of the relation between $\omega(q)$ and $S(q)$ later, when we deal with collective motions in liquid ^{4}He in the superfluid phase.

However, using the fact that

$$\lim_{q\to 0} S(q) = S(0) = \rho_0 k_B T \kappa_T = \gamma k_B T/Mv_s^2,$$

where γ is the ratio of the specific heats c_p/c_V and v_s is the velocity of sound, we find from equation (7.22) that in the long-wavelength limit $q \to 0$,

$$\omega^2 = \frac{q^2}{\gamma} v_s^2, \tag{7.23}$$

which is correct provided γ is near to unity.

Actually, for simple liquid metals, γ at the melting point is only usually 1.1–1.3 and so the above argument is almost quantitative as $q \to 0$. However, for Ar, at the triple point, $\gamma = 2.2$ and the above argument is obviously not appropriate.

It is of interest to press the point a little further and to remark that $\gamma = 1$ is characteristic of a liquid which can be described as independent harmonic oscillators. One then obtains immediately in such a zero-order model not only that $\gamma = 1$ but also that the specific heat $c_V = 3R$. These are reasonable estimates for simple liquid metals and moreover c_V is usually somewhat greater than $3R$. This is again characteristic of an almost harmonic assembly, and the small 'anharmonic' corrections (cf equation (7.27) below) can be related to the structure factor $S(q)$.

7.4 Collective coordinates and thermal properties of liquid metals

Table 7.1 shows the ratio of the specific heats $\gamma = c_p/c_V$ for some liquid metals just above the melting temperature T_m. For those liquid metals with γ near to unity, it will be seen that the values of the specific heat at constant volume c_V are greater than $3R$. Attention is drawn to this because, of course, from the thermodynamic formula

$$c_p - c_V = -T\left(\frac{\partial p}{\partial V}\right)_T \left(\frac{\partial V}{\partial T}\right)_p^2 = \frac{T\alpha^2 V}{\kappa_T}. \tag{7.24}$$

it can be seen that a harmonic theory, for which $\alpha = 0$ (cf the discussion of thermal expansion of a crystal in Chapter 3), will yield $\gamma = 1$. Also, for independent harmonic oscillators, one has $c_V = 3R$.

Table 7.1 Measured specific heat data for some liquid metals just above the melting temperature.

	c_V/R	$\gamma = c_p/c_V$
Na	3.37	1.12
K	3.49	1.11
Rb	3.41	1.15
Zn	3.07	1.25
Cd	3.07	1.23
Ga	3.17	1.08
Tl	2.96	1.21
Sn	3.00	1.11
Pb	2.85	1.20
Bi	3.15	1.15

This circumstance can be exploited (cf Eisenschitz and Wilford 1962) by writing down a suitable Hamiltonian to describe such liquid metals with γ near to 1 and $c_V \gtrsim 3R$ in terms of collective coordinates. The appropriate ones are the density fluctuations $\rho_{\boldsymbol{k}}$, which are defined by (cf equations (7.3) and (7.6))

$$\rho_{\boldsymbol{k}} = \sum_i \exp(\mathrm{i}\boldsymbol{k}\cdot\boldsymbol{r}_i) \tag{7.25}$$

where $\boldsymbol{r}_i$ denotes the position of the ith atom.

The independent oscillator model, taken as the zeroth-order description, then corresponds to the $\rho_{\boldsymbol{k}}$ oscillating independently, with a form

$$\rho_{\boldsymbol{k}}(t) = \rho_{\boldsymbol{k}}(0)\exp(\mathrm{i}\omega_{\boldsymbol{k}}t). \tag{7.26}$$

Eisenschitz and Wilford (1962) therefore develop a Hamiltonian in which, to lowest order, $\rho_{\boldsymbol{k}}$ are introduced in an independent oscillator part H_0 while interactions between the oscillators are incorporated in a perturbation H_1.

While various thermodynamic consequences can be explored, the most interesting of them from the present point of view is the fact that, if H_1 is indeed a perturbation on H_0, then c_V is always increased above $3R$, the value for independent harmonic oscillators.

The correction to c_V can be written down essentially in terms of the liquid structure factor, though a cut-off frequency has to be invoked. As shown by Bratby *et al* (1970), the specific heat c_V then satisfies the inequality

$$c_V \leqslant R\left(3 + \tfrac{1}{2}S_{\max}\sum_{\boldsymbol{k},\boldsymbol{k}'}\frac{(\boldsymbol{k}\cdot\boldsymbol{k}')^2}{N^2k^2k'^2}\right) \tag{7.27}$$

where $S_{\max}$ is the maximum of $S(k)$ in the range $0 < k < k_0$, with $k_0 = (18\pi^2 N/V)^{1/3}$, and for K the experimental structure factor was used to obtain an estimate of the correction as $\sim R$ or less. The result is consistent with the experimental value of $3.5R$ in Table 7.1.

It would appear that the conditions which lead to a rather good approximation using a zeroth-order model of independent density fluctuations are:

(a) Relatively soft cores. This restricts the considerations to those metals with sp electrons, but no d electrons.
(b) Long-range interactions. This suggests that the Friedel oscillations should not be rapidly damped out, which means liquid metals with relatively long electronic mean free paths.

The merit of the above collective coordinate approach in explaining some of the results in Table 7.1 is made clear when one notes that even a simple pair potential theory of c_V (or c_p) leads to three- and four-atom correlation functions entering the theory (see e.g. March and Tosi 1976, p. 80). It turns out that the final results for c_V (or c_p) come from a marked cancellation between large individual terms and this makes the above collective coordinate approach the more attractive.

We caution the reader that in a liquid like Ar, which near its triple point has a ratio of specific heats γ around 2.2, a theory like the above which gives $\gamma = 1$ in zeroth order is of no utility. This reflects the fact that an assembly of billiard balls is, of course, a very anharmonic assembly. The force law of Ar, characterised by a hard core and a short-range tail, means that a density fluctuation model of independent oscillators will fail.

7.5 Hydrodynamics and collective modes in classical liquids

Although the methods described above can be usefully employed in order to understand some of the qualitative features of collective modes in liquids, a quantitative theory often requires very heavy calculations.

There is, however, one limiting case in which simple results can be obtained. In fact, for frequencies $\omega \ll 2\pi/\tau$, where τ is the relaxation time, and wavenumber $q \ll 2\pi/a$, a measuring the interparticle separation, the phenomenological equations of hydrodynamics can be employed to describe the system. For simplicity, we shall neglect for the time being the coupling with thermal fluctuations. Then one needs only to consider the two continuity equations:

$$\frac{\partial \rho(\boldsymbol{r}, t)}{\partial t} + \nabla \cdot \boldsymbol{p}(\boldsymbol{r}, t) = 0 \tag{7.28}$$

and the conservation law of momentum

$$\frac{\partial \boldsymbol{p}(\boldsymbol{r}, t)}{\partial t} + \nabla \cdot \boldsymbol{\Pi}(\boldsymbol{r}, t) = 0. \tag{7.29}$$

Here $\rho(\boldsymbol{r}, t)$ is the mass density, $\boldsymbol{p}(\boldsymbol{r}, t) = \rho\boldsymbol{v}(\boldsymbol{r}, t)$ the momentum density and $\boldsymbol{\Pi}(\boldsymbol{r}, t)$ the stress tensor†. In the hydrodynamic limit, this takes the Navier–Stokes form (Landau and Lifshitz 1959)

$$\Pi_{\alpha\beta}(\boldsymbol{r}, t) \simeq \delta_{\alpha\beta}p(\boldsymbol{r}, t) - \eta\left(\frac{\partial v_\alpha}{\partial x_\beta} + \frac{\partial v_\beta}{\partial x_\alpha}\right) - \delta_{\alpha\beta}(\zeta - \tfrac{2}{3}\eta)\nabla\boldsymbol{v}(\boldsymbol{r}, t) \qquad (7.30)$$

where $p(\boldsymbol{r}, t)$ is the local pressure and ζ and η the bulk and shear viscosities respectively. Substituting for $\boldsymbol{\Pi}$ in equation (7.29) yields

$$\frac{\partial \boldsymbol{p}(\boldsymbol{r}, t)}{\partial t} + \nabla p(\boldsymbol{r}, t) - \eta\nabla^2 v - (\zeta + \tfrac{1}{3}\eta)\nabla(\nabla v) = 0. \qquad (7.31)$$

We can now express $\boldsymbol{v}$ in terms of $\boldsymbol{p}$ and split $\boldsymbol{p}$ into longitudinal (irrotational) and transverse (divergence free) parts. Thus we find for the transverse component $\boldsymbol{p}_{\mathrm{T}}$

$$\left(\frac{\partial}{\partial t} - \frac{\eta}{\rho}\nabla^2\right)\boldsymbol{p}_T(\boldsymbol{r}, t) = 0 \qquad (7.32)$$

and a longitudinal equation

$$\left(\frac{\partial^2}{\partial t^2} - D_{\mathrm{L}}\nabla^2\frac{\partial}{\partial t}\right)\langle\rho(\boldsymbol{r}, t)\rangle = \nabla^2 p(\boldsymbol{r}, t) \qquad (7.33)$$

where $D_{\mathrm{L}} = (\tfrac{4}{3}\eta + \zeta)/\rho$, and use has been made of the continuity equation.

For small disturbances from equilibrium, provided we neglect thermal fluctuations, we can write

$$\delta p(\boldsymbol{r}, t) \simeq \left(\frac{\partial p}{\partial \rho}\right)_{\mathrm{T}} \delta\rho(\boldsymbol{r}, t) \qquad (7.34)$$

and equation (7.33) becomes

$$\left(\frac{\partial^2}{\partial t^2} - D_{\mathrm{L}}\nabla^2\frac{\partial}{\partial t}\right)\langle\rho(\boldsymbol{r}, t)\rangle = c_T^2\nabla^2\langle\rho(\boldsymbol{r}, t)\rangle \qquad (7.35)$$

with c_T, the isothermal speed of sound, given by

$$c_T^2 = \left(\frac{\partial p}{\partial \rho}\right)_{\mathrm{T}}.$$

The reader will recognise the difference in type of equations (7.32) and (7.35). The first is purely diffusive, while the second describes sound wave propagation with a damping term proportional to D_{L}. The corresponding solutions in $\boldsymbol{k}$-space take the form

$$\boldsymbol{p}_{\mathrm{T}}(\boldsymbol{k}, t) \simeq \boldsymbol{p}_{\mathrm{T}}(\boldsymbol{k}, 0)\ \exp(-k^2\eta t/\rho) \qquad (7.36)$$

and

$$\rho(\boldsymbol{k}, t) \simeq \rho(\boldsymbol{k}, 0)\cos(c_T k t)\exp(-k^2 D_{\mathrm{L}} t). \qquad (7.37)$$

† For the reader not familiar with these aspects of hydrodynamics equations (7.32) and (7.35), with solutions (7.36) and (7.37) embody the essence of this discussion.

This means that, once established, a long-wavelength transverse excitation can only relax exponentially towards equilibrium while a longitudinal excitation, or equivalently a density fluctuation, will act as a damped oscillator. We can rephrase this by saying that for the transverse mode the eigenvalue is purely imaginary, with value

$$\omega_{\mathrm{T}} = -\mathrm{i}k^2\eta/\rho \tag{7.38}$$

while in the longitudinal case there are two frequencies

$$\omega_{\mathrm{L}} = \pm c_T k - \mathrm{i}k^2 D_{\mathrm{L}} \tag{7.39}$$

as follows directly from equation (7.37).

Inclusion of thermal fluctuations into the above treatment leads to a new, pure imaginary, relaxation mode

$$\omega = -\mathrm{i}k^2 K/\rho c_p \tag{7.40}$$

where K is the thermal conductivity and c_p the specific heat at constant pressure. Also, the isothermal sound velocity c_T should be substituted by the adiabatic sound velocity c_S. Furthermore, D_{L} is changed into D'_{L} say, given by

$$D'_{\mathrm{L}} = D_{\mathrm{L}} + \frac{K}{\rho c_p}\left(\frac{c_p}{c_r} - 1\right) \tag{7.41}$$

due to the fact that heat diffusion processes can contribute to the sound wave damping.

Of these four modes, only those that are coupled to the density (see equation (7.35)) can appear in the scattering function $S(\boldsymbol{k}, \omega)$. Therefore, as there is no coupling in this limit between transverse and longitudinal modes, only three peaks will appear in $S(\boldsymbol{k}, \omega)$: a central one (Rayleigh peak) due to the heat relaxation mode centred on $\omega = 0$, and two side peaks (the Brillouin doublet) centred at frequencies $\omega = \pm c_S k$. The shapes of these peaks are approximately Lorentzian, with half-widths as indicated in figure 7.4.

These features of the spectrum, displayed in figure 7.4 have been confirmed by light scattering experiments. Since these experiments are concerned with k-values of the order of 10^{-3} Å^{-1}, we are clearly in a regime where the above hydrodynamic theory will be valid.

As we move to regions of higher k and ω, the hydrodynamic theory will eventually cease to apply, and also one must use neutron scattering rather than light scattering, to measure $S(\boldsymbol{k}, \omega)$. However, extrapolating from the hydrodynamic results, one would expect the modes to broaden rapidly and therefore eventually to merge into one broad central peak.

This situation obtains in rare gas liquids such as Ar or Ne, where the experiments performed have demonstrated that the side peaks are only resolved for small values of q.

But in the case of liquid metallic Rb, in contrast these peaks are still in evidence at values of q lying well outside the hydrodynamical regime. In order

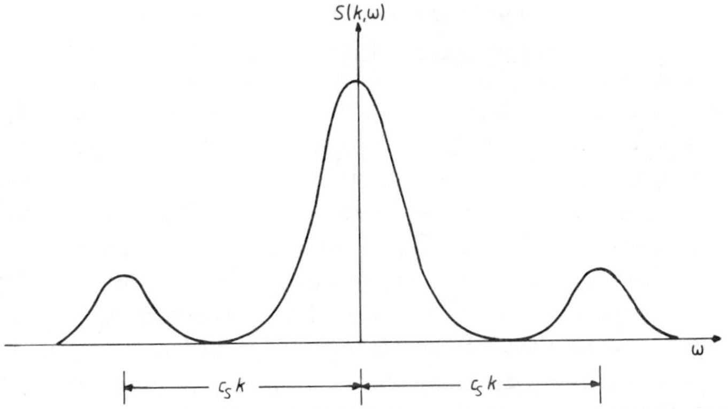

Figure 7.4 Showing central Rayleigh peak and Brillouin doublet centred on $\omega = \pm c_S k$. The half-width of the Brillouin doublet peaks is determined by D_L' in equation (7.41). The Rayleigh peak at $\omega = 0$ has a width determined by heat diffusion.

to emphasise the difference between the insulating liquid Ar and the conducting liquid Rb, we refer again to the dispersion relation displayed in figure 7.3. As already discussed, this can be understood in terms of the very different nature of the interatomic forces in Ar and in Rb, and gives further confirmation of the picture of the simple liquid metals as 'harmonic' liquids.

7.6 Optical modes in ionic liquids

Let us consider an ionic liquid, such as would be obtained by melting an alkali halide crystal. Since such a fluid will obviously consist of cations and anions, we now require three static structure factors to describe cation–cation, anion–anion and cation–anion correlations. The generalisation of the neutron scattering formula (7.8) to this case then takes the form

$$I(q) = f_+^2 S_{++}(q) + f_-^2 S_{--}(q) + 2f_+ f_- S_{+-}(q). \tag{7.42}$$

By changing the scattering lengths, which can be effected by changing the isotopic composition of the liquid, and repeating the neutron measurements on three isotopically different samples, one can extract the three partial structure factors $S_{++}(q)$, $S_{--}(q)$ and $S_{+-}(q)$, which in turn lead to the pair correlation functions in $\boldsymbol{r}$-space between cations, anions and the cation–anion cross-correlations respectively. The $S_{ij}(q)$ are defined as

$$S_{ij}(q) = \frac{1}{n} \langle \rho_{\boldsymbol{q}}^i \rho_{\boldsymbol{q}}^j \rangle \qquad i, j = +, - \tag{7.43}$$

where n is the number of molecules in the liquid. With this definition one can form the following linear combinations:

$$S_{NN} = \tfrac{1}{2}(S_{++} + S_{--} + 2S_{+-}) \tag{7.44}$$

$$S_{QQ} = \tfrac{1}{2}(S_{++} + S_{--} - 2S_{+-}) \tag{7.45}$$

$$S_{NQ} = \tfrac{1}{2}(S_{++} - S_{--}) \tag{7.46}$$

which respectively give the correlations between fluctuations in total number $\rho_q^N = \rho_q^+ + \rho_q^-$, total charge $\rho_q^Q = \rho_q^+ - \rho_q^-$ and the cross-correlations.

By forming these new structure factors, one finds a very revealing picture of the local order in the liquids as can be seen from the example shown in figure 7.5 for the case of NaCl. Thus S_{NQ} is small everywhere and can usually be neglected. $S_{NN}(q)$ is seen to be very smooth, reflecting the large degree of disorder in the position of the particles. But in contrast, $S_{QQ}(q)$ is sharply peaked as a consequence of the strong charge ordering imposed by the constraint of the local neutrality. Due to this characteristic feature of the relative arrangement of ions, we may expect that these liquids will be capable of sustaining collective excitations of the optic mode type. In such motions, as we saw for crystals in Chapter 3, one species moves in phase opposition relative to the other, causing a charge separation.

If such modes do indeed exist, they should be observable as side peaks in the dynamic generalisation of $S_{QQ}(q)$, namely $S_{QQ}(q, \omega)$, which describes the dynamics of the charge density fluctuations. Unfortunately, it is very difficult to extract information on $S_{QQ}(q, \omega)$ from inelastic neutron scattering (cf

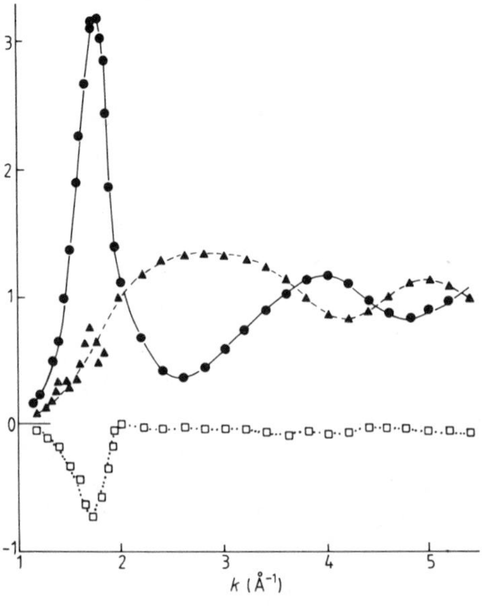

Figure 7.5 Local order in molten NaCl at 1148 K: results obtained by neutron experiments using isotopes. Charge-number correlation functions are shown: ●, $S_{QQ} = \frac{1}{2}(S_{++} + S_{--} - 2S_{+-})$; ▲, $S_{NN} = \frac{1}{2}(S_{++} + S_{--} + 2S_{+-})$; □, $S_{NQ} = \frac{1}{2}(S_{++} - S_{--})$. (From Parrinello M and Tosi M P 1979 *Riv. Nuovo Cim.* **2** No. 6.)

Copley and Dolling 1978). One has therefore to rely, for where optical experiments are no longer relevant, on computer simulation. Such calculations show that $S_{QQ}(q, \omega)$ exhibits a well defined side peak due to the presence of propagating optic modes. These modes show a negative dispersion, as in solids, and appear as a distinct feature in the spectrum up to wavenumbers $q < 2\pi/a$, with a as usual the mean interparticle separation.

Caution should be exercised here, in the sense that the results of such computer experiments do not give a decisive demonstration that optic modes exist in real ionic melts, since there are substantial uncertainties in the form of their interaction potentials, the input data in the computer simulation studies.

From the above discussion, it follows that the only experiments available at the time of writing are infrared measurements. From these, the form of $S_{QQ}(k, \omega)$ as $\boldsymbol{k} \to 0$ can be deduced. The only alkali halide for which data are available is LiF, its spectrum being shown in figure 7.6. This exhibits one central peak, due to dissipative processes, and two side peaks due to optical modes. The occurrence of optical modes for $\boldsymbol{k} \to 0$ is not a general feature of molten salts. In fact, in systems like the silver halides, optical modes are so heavily damped that they only appear as a persistent tail in the high-frequency range of the spectrum.

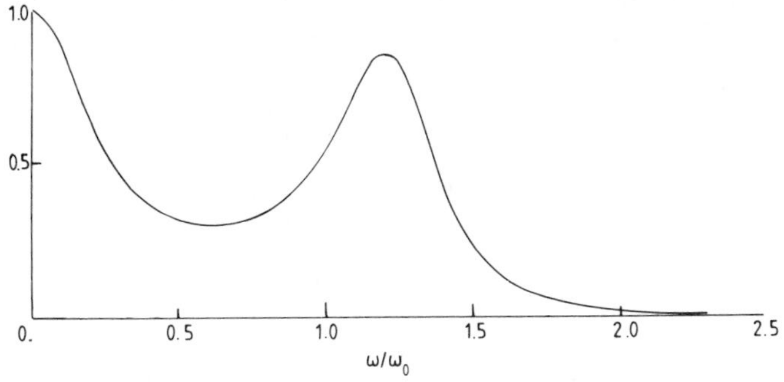

Figure 7.6 Charge fluctuation spectrum, obtained from optical data on molten LiF. (From Parrinello M and Tosi M P 1979 *Riv. Nuovo Cim.* **2** No. 6.) ($\omega_0 = 530\,\text{cm}^{-1}$)

7.7 Quantum liquids

We turn from the classical liquids discussed above to the case of quantum liquids. All important in this discussion is the relation between the thermal de Broglie wavelength and the interparticle spacing. If the thermal de Broglie wavelength h/p_{th} is very much less than the interparticle separation, then nothing is lost in letting Planck's constant tend to zero, that is in passing to the classical limit. The other point that should be emphasised is that it is only in the light atoms in the condensed state that the zero-point energy is sufficiently large to cause the condensed phase to remain liquid down to the lowest temperatures so far achieved, at atmospheric pressure. Therefore one has to worry about quantal effects in liquid helium and here exciting manifestations of collective behaviour arise. As we shall see, these turn out to depend sensitively on the statistics. We shall begin the discussion with liquid ^{4}He.

7.8 Liquid ^{4}He

Having a total nuclear spin of zero, ^{4}He obeys Bose–Einstein statistics. Although it is not possible to neglect the interactions between ^{4}He atoms, in their liquid state, we consider first what the collective properties of liquid ^{4}He would be if it were an ideal Bose–Einstein gas.

7.8.1 Condensation in a Bose–Einstein gas

Consider an assembly of non-interacting bosons. At absolute zero, all particles will occupy the lowest energy (momentum) state. What is important is that a substantial fraction of the particles can remain in the ground state when the temperature is raised somewhat. If we write the Bose–Einstein distribution function in the usual form:

$$n(\epsilon, T) = \frac{1}{\exp[(\epsilon-\mu)/k_B T]-1} \tag{7.47}$$

then the occupancy of the lowest state $\epsilon = 0$ is given by

$$n(0, T) = \frac{1}{\exp(-\mu/k_B T)-1}. \tag{7.48}$$

But, as remarked above, when $T = 0$ the occupancy of the lowest state is simply the total number of particles N, i.e.

$$n(0, 0) = N = \lim_{T\to 0} \frac{1}{\exp(-\mu/k_B T)-1}. \tag{7.49}$$

It may be noted here that as $T \to 0$ the chemical potential μ must always be lower in energy than the lowest state (here $\epsilon = 0$), in order that the occupancy of every state shall never be negative.

Because of this finite occupancy of the ground state, it is helpful to write

$$N = \sum_j n_j = N_0(T) + N_{exc}(T) \tag{7.50}$$

where N_0 is evidently the number of particles in the lowest energy state while N_{exc} is the number in the excited states.

Now suppose that the number of allowed energy states in the range E to

$E+\mathrm{d}E$ is $g(E)\mathrm{d}E$. Then we can write

$$N = N_0 + \int_0^\infty g(E)n(E,T)\mathrm{d}E. \tag{7.51}$$

For N free particles in a volume V, we have explicitly (cf Appendix 2.1) for particles with zero spin

$$g(E) = \frac{V}{4\pi^2}\left(\frac{2M}{\hbar^2}\right)^{3/2} E^{1/2} \tag{7.52}$$

where M is the mass of the particles. Since $g(E) = 0$ at $E = 0$, it is essential that we treat the ground state $E = 0$ separately as above. Then using the explicit forms (7.47) and (7.52) for $n(E,T)$ and $g(E)$ respectively in equation (7.51) we find $N_{\mathrm{exc}}(T) \propto T^{3/2}$† and if we define a degeneracy temperature such that

$$N_{\mathrm{exc}}(T_0) = N \tag{7.53}$$

then we can write, for $T < T_0$, the approximate equation

$$N_0(T) = N[1-(T/T_0)^{3/2}]. \tag{7.54}$$

This result shows that the number of particles in the ground state, N_0, is macroscopic for $T < T_0$. T_0 is readily shown to be given by

$$T_0 = \frac{2\pi\hbar^2}{Mk_{\mathrm{B}}}\left(\frac{N}{2.612V}\right)^{2/3} \tag{7.55}$$

where use has been made of the value of the definite integral arising from

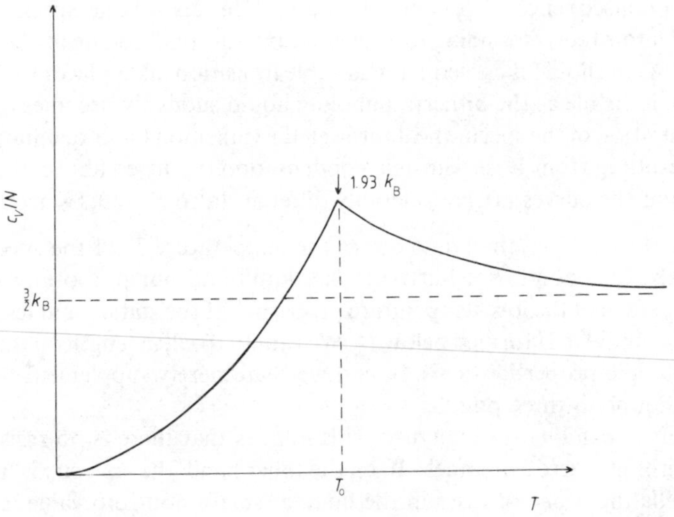

Figure 7.7 Specific heat c_V for ideal Bose gas. For $T < T_0$, $c_V \propto T^{3/2}$.

† For low temperatures, in which case from equation (7.49), $\mu/k_{\mathrm{B}}T \to N^{-1}$.

equation (7.51), i.e.

$$\int_0^\infty \mathrm{d}x \frac{x^{1/2}}{e^x - 1} = 1.306\pi^{1/2}. \tag{7.56}$$

The internal energy as a function of temperature is readily constructed by a similar argument to that used for $N(T)$ above. The result for the specific heat c_V derived by differentiating this internal energy with respect to T is shown in figure 7.7.

7.8.2 Properties of liquid ^{4}He

At this point it will be useful to summarise a number of the properties of helium.

(i) The potential curve for the interaction between He atoms has, as essential characteristics: (a) a hard repulsive core, with atomic diameter about 2.7 Å; and (b) a weakly attractive van der Waals tail.

(ii) Helium gas liquefies at 4.2 K and then remains liquid down to the lowest temperatures studied (for pressures up to about 25 atmospheres). The reasons for this exceptional behaviour are: (c) the small atomic mass, which results in a large zero-point energy; and (d) the relatively weak van der Waals attraction which, combined with (c), makes any lattice configuration unstable.

It turns out that other possible candidates do not fulfil both the necessary requirements. Thus, the other inert gases have weak van der Waals interactions, but are too massive. Hydrogen, on the other hand, satisfies (c) but the intermolecular forces are too strong.

As a consequence of (c) and (d), liquid ^{4}He has a large specific volume (46 Å^3/atom) corresponding to an average interatomic spacing of about 3.6 Å.

(iii) As the liquid is cooled, a remarkable transition takes place at 2.2 K. This change is visible as the hitherto bubbling liquid suddenly becomes quiescent. The variation of the specific heat through the transition looks qualitatively like that resulting from Bose–Einstein condensation discussed above, though the details of the curves are considerably different (also $T_0 = 3.2$ K from (7.55)).

Liquid He II. He I, the liquid above the temperature T_λ of the specific heat anomaly, or λ-point, is a fairly normal liquid and our purpose below is to explain some of the low-temperature properties of the static (or at least slowly moving) fluid He II formed below T_λ. We cannot dwell at length on the various macroscopic properties of He II, and we shall merely supplement the above discussion with three points:

(i) He II exhibits superfluidity. This implies that there is no resistance to flow through narrow channels. If, on the other hand, the viscosity is measured by oscillating a pile of discs in the fluid, a (small) non-zero value is found.

(ii) He II has an exceedingly large thermal conductivity. The rate of heat transport is not governed by the customary temperature gradient theory and is so large that it has been described as a thermal superconductor (Lynton 1959).

It is this property that explains the boiling liquid becoming quiescent below the λ-point. In He I, heat is carried away by the bubbling process just as in any other normal liquid near its boiling point. As soon as He II is formed, this mechanism is no longer necessary.

(iii) Well defined temperature waves may be propagated through the fluid. In all essential respects, these are analogous to ordinary sound waves. This phenomenon of second sound was predicted by Landau (cf F London 1961) and subsequently demonstrated experimentally by Peshkov (1946).

7.8.3 Structure and excitations

Landau (1947) suggested that the energy spectrum of He II should have the form sketched in figure 7.8. There is a phonon-like region at small momentum p, with a linear relation $\epsilon(p) = cp$ with c the speed of sound. In the vicinity of the minimum in the curve of figure 7.9, at a characteristic momentum $p_0 \sim 1/a$ where a is the mean interatomic spacing, $\epsilon(p)$ can be expanded in a series in $p - p_0$ and takes the form†

$$\epsilon(p) = \Delta + (p - p_0)^2/2\mu \tag{7.57}$$

where Δ is the energy of the minimum and μ is a constant used to define the curvature in the region of the minimum. Henshaw and Woods (1961) observed

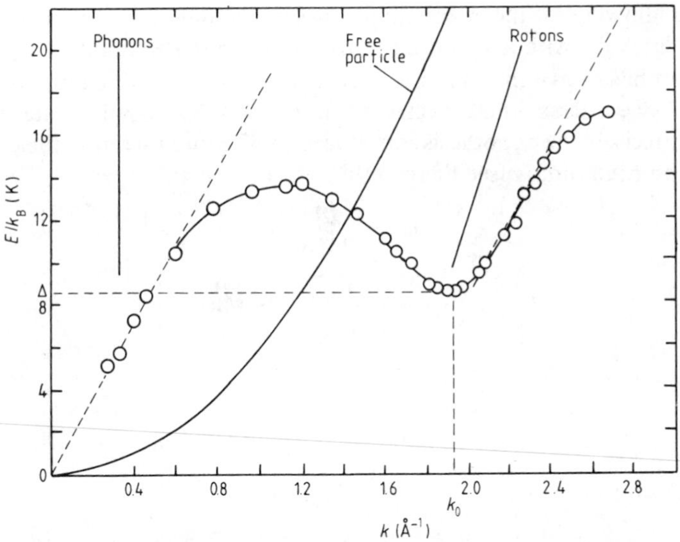

Figure 7.8 Landau curve of elementary excitations in He II.

† Landau originally postulated a 'roton' excitation as shown in figure 7.8, requiring a certain minimum energy Δ for its existence, because phonons alone could not explain the large specific heat as T was raised to about 1 K (see equation (7.66) below).

that $\Delta/k_B \simeq 8.6$ K, $p_0/\hbar \simeq 1.91$ Å^{-1} and $\mu \simeq 0.16\,M$, where M is the mass of a ^{4}He atom.

The form of this energy spectrum of the elementary excitations was explained by Feynman (1953). He chose to relate the spectrum, not to the force law of He atoms, but to the observed structure of the liquid, i.e. to the structure factor $S(k)$ discussed in §7.1. Rather than give his detailed argument (see March *et al* 1967 for a summary of this) let us focus on the nature of the scattering function $S(k, \omega)$, from which the structure factor $S(k)$ is obtained by integration over all energy transfers (compare the classical result (7.18)).

A useful starting point is again to consider the ideal Bose–Einstein gas. It is straightforward to get the scattering function for this non-interacting system at $T = 0$, the result being simply

$$S(k, \omega) = \delta\left(\omega - \frac{\hbar k^2}{2M}\right) \tag{7.58}$$

M being the mass of a ^{4}He atom. Equation (7.58) in fact describes single-particle excitations from the initial state of zero momentum to an excited state with momentum $\hbar k$. For a free particle, the change in energy in such a transition, $\Delta E = \hbar\omega$, is evidently

$$\hbar\omega = \hbar^2 k^2/2M. \tag{7.59}$$

From what we said above, these results cannot be transferred to ^{4}He liquid, without allowing for the interactions, since in a liquid, each particle at a given instant interacts with a significant portion of the remaining fluid. However, Feynman has shown that it is reasonable, as a first approximation, to assume that at $T = 0$ a Bose liquid spectrum is dominated by a single excitation line. This is precisely the hypothesis introduced in §7.3 and therefore the expected excitation spectrum is (see figure 7.9)†

$$\hbar\omega = \frac{\hbar^2 k^2}{2MS(k)}. \tag{7.60}$$

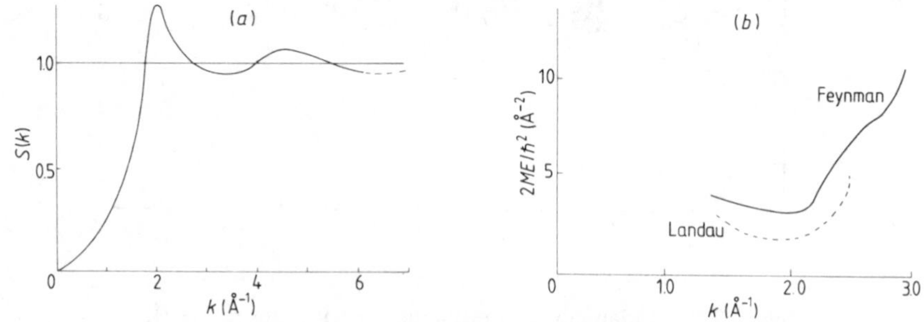

Figure 7.9 (a) Schematic form of measured structure factor for liquid ^{4}He; (b) schematic form of excitations given by equation (7.60).

† The argument is as in §7.3 except that the classical moment relation (7.19) is replaced at $T = 0$ by $\int_0^\infty \omega S(k,\omega)\mathrm{d}\omega = \hbar k^2/2M$ with $S(k, \omega) = S(k)\delta(\omega - w(k))$.

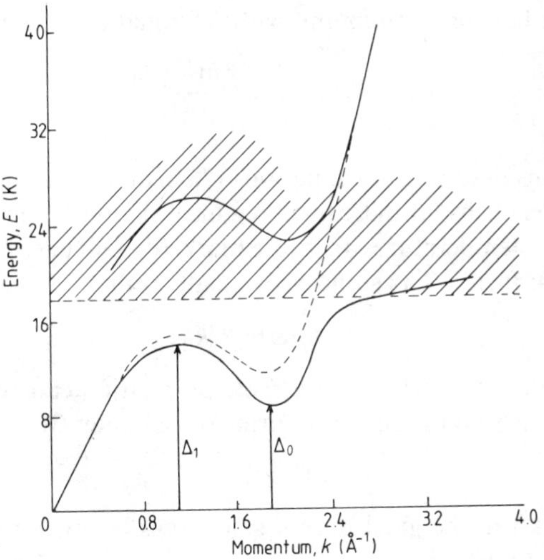

Figure 7.10 Measured excitations in liquid ^{4}He. Note upper broader excitation branch in addition to form qualitatively like that predicted by equation (7.60). Shaded region involves roton–roton interactions.

For free particles, $S(k) = 1$ since the radial distribution function $g(r) = 1$, and then the above equation reduces to equation (7.59).

Equation (7.60), however, turns out to be only partially in agreement with experiment. In fact the experimental findings while consistent with a sharp resonance of the qualitative form predicted by equation (7.60), also reveal an upper, broader excitation branch (see figure 7.10). Thus, the hypothesis of a single excitation is too restrictive and has to be modified by making allowance for the interaction between different modes.

Based on the above findings as to the nature of the spectrum, one can write the dynamical structure factor as

$$S(k, \omega) \simeq Z(k)\, \delta(\omega - \omega(k)) + S^{\mathrm{II}}(k, \omega). \tag{7.61}$$

Here $\omega(k)$ is the dispersion relation of the lower branch, $Z(k)$ defines the strength of the single mode excitation and $S^{\mathrm{II}}(k, \omega)$ describes the multimodes (eg multiphonon processes). The situation is similar to that of a crystal where one can distinguish between one-phonon and multiphonon processes. Before turning to examine $S^{\mathrm{II}}(k, \omega)$, let us dwell on the form of $\omega(k)$. Three regions can be usefully distinguished, extending the previous discussion:

(i) A linear region near $k = 0$. Here $\omega(k) = ck$, and the form of $Z(k)$ in the phonon regime is qualitatively similar to one-phonon scattering in a solid (cf Appendix 3.3).

(ii) A region having a maximum, around which $\omega(k)$ can be represented by

$$\omega_{\max}(k) = \Delta_1/\hbar - \frac{\hbar(k - k_1)^2}{2\mu_1}. \tag{7.62}$$

(iii) A region having a minimum, with (cf equation (7.57))

$$\omega_{\min}(k) = \Delta_0/\hbar + \frac{\hbar(k-k_0)^2}{2\mu_0} \tag{7.63}$$

respectively.

The excitations described by equations (7.62) and (7.63) are called maxons and rotons respectively. Of these three regimes, obviously the thermodynamic properties at low temperatures are more markedly influenced by the two low-lying regions. In fact, if T is sufficiently small so that

$$T < \hbar\omega(k_0)/k_{\mathrm{B}} \tag{7.64}$$

then only the phonon part of the dispersion relation can contribute to the specific heat, which takes the usual form (see Chapter 3)

$$c_V^{\mathrm{ph}} \propto T^3. \tag{7.65}$$

This is in contrast to the ideal Bose–Einstein gas behaviour where there is a $T^{3/2}$ dependence (cf §8.5).

At higher temperatures, however, the rotons can also be excited and their contribution to the specific heat will take the form

$$c_V^{\mathrm{rot}} \propto \exp\left(-\frac{\hbar\omega(k_0)}{k_{\mathrm{B}}T}\right). \tag{7.66}$$

The interplay between phonon and roton contributions has indeed been observed experimentally.

Now we must turn to discuss the multimode contribution $S^{\mathrm{II}}(k, \omega)$. The theory of this has been given by Pitaevskij (1959, 1966) and later by Zawadowski (1978). As it is rather complicated, we limit ourselves to remarking that $S^{\mathrm{II}}(k, \omega)$ is expected to be usefully approximated by

$$S^{\mathrm{II}}(k, \omega) \propto \int f(\boldsymbol{k}, \boldsymbol{q})\, \delta(\omega(\boldsymbol{q}) + \omega(\boldsymbol{k}+\boldsymbol{q}) - \omega)\, \mathrm{d}\boldsymbol{q} \tag{7.67}$$

where $f(\boldsymbol{k}, \boldsymbol{q})$ is in general rather a complicated function. It should be noted that $S^{\mathrm{II}}(k, \omega)$ has the same structure as in two-phonon processes (see A3.5 and §3.7). We can expect that the most important contributions to the integral come from the regions of $\boldsymbol{q}$-space where

$$\nabla_{\boldsymbol{q}}(\omega(\boldsymbol{q}) + \omega(\boldsymbol{k}+\boldsymbol{q})) = 0. \tag{7.68}$$

When $\boldsymbol{k} \simeq 0$, as, say, in a Raman experiment, then the largest contribution to the integral will come from maxons and rotons, for both of which $\nabla_{\boldsymbol{q}}(\omega(\boldsymbol{q})) = 0$. However, an accurate examination of the Raman spectrum and of the neutron scattering experiments has led to the conclusion that the data cannot be interpreted completely without allowing for some coupling between the modes. This is particularly relevant in the roton case, as it has been shown that an attractive interaction between the rotons leads to a roton–roton bound state (see Zawadowski (1978) for further details).

Finally, we shall present an argument due to Landau, which allows one to explain the superfluid property of liquid ^{4}He below a certain transition temperature. This property is its ability to flow in a capillary tube without apparent viscosity.

7.8.4 Landau's explanation of superfluidity

We consider the fluid to flow along a capillary tube with velocity $\boldsymbol{v}$. Its initial energy is the sum of its internal energy E_0 and its kinetic energy $\frac{1}{2}Mv^2$. Friction with the walls means that the energy must decrease from its original value of $E_0+\frac{1}{2}Mv^2$. As we are dealing with a quantum system, all the exchanges of energy can be made through excitations of quanta $\hbar\omega(k)$ with $\omega(k)$ as depicted in figure 7.9(b). If the system was at rest, the creation of an elementary excitation $\hbar\omega(k)$ would lead to a variation of energy $\hbar\omega(k)$ and momentum $\hbar k$. As the fluid as a whole is in motion, we must apply a Galilean transformation and we find that the energy E of the moving system with one excitation $\hbar\omega(k)$ is

$$E = \hbar\omega(k) + \hbar\boldsymbol{k}\cdot\boldsymbol{v} + \tfrac{1}{2}Mv^2 + E_0. \tag{7.69}$$

The creation of one excitation is energetically favourable then if

$$\omega(k) + \boldsymbol{k}\cdot\boldsymbol{v} < 0.$$

For a given $\boldsymbol{k}$, $\omega(k)+\boldsymbol{k}\cdot\boldsymbol{v}$ is a minimum only if $\boldsymbol{k}$ and $\boldsymbol{v}$ are antiparallel and the condition for the creation of an excitation of wavevector $\boldsymbol{k}$ is

$$\omega - kv < 0 \quad \text{or} \quad v > \omega/k.$$

With a spectrum like the one in figure 7.11 this condition is satisfied for any k provided that $v > v_{\mathrm{m}}$ where v_{m} is the slope of the straight line depicted in this figure. Hence we conclude that for $v < v_{\mathrm{m}}$ the above inequality is never satisfied and the system is superfluid because no process in which its energy is diminished is possible.

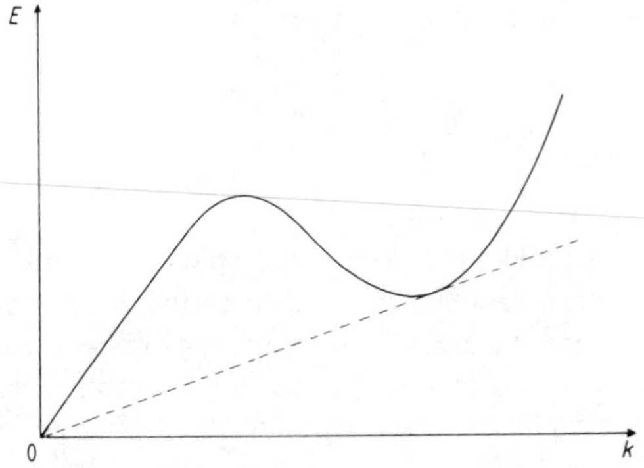

Figure 7.11 Shows excitation spectrum in liquid ^{4}He in schematic form, together with a line of slope v_{m} referred to in text after equation (7.69).

In fact, the value for v_m obtained in this way is wrong by several orders of magnitude. This is because we have neglected the possibility of creating other kinds of excited states such as, for instance, vortices†. Their statistical weight in determining the equilibrium properties of the system is in fact negligible, but they become important in any discussion of the hydrodynamics.

7.8.5 *Elementary excitations in very thin superfluid ^{4}He films*

We shall conclude this discussion of liquid ^{4}He with some brief comments on elementary excitations in very thin superfluid ^{4}He films (cf Rutledge *et al* 1978).

It turns out that the concepts of surface phonons and surface rotons allow an interpretation of various properties of such films. Quantum hydrodynamics can be reformulated using surface quantities, to describe surface phonons. The sound velocity can then be expressed, with M the bare ^{4}He mass, as

$$c^2 = 3A\sigma_0/M(a+\sigma_0)^4 \tag{7.70}$$

where A and a are constants entering the formula for the van der Waals binding of the film to the substrate, while σ_0 is the average superfluid surface density of the film.

Experiments on such films reveal an excitation with an energy gap which one can regard as a surface roton. An intuitive way to view this is as consisting of a bound pair of vortices of opposite circulation, with the surface roton being the smallest or most tightly bound pair allowed by quantum mechanics. In parallel with the bulk roton dispersion relation, one may write a phenomenological dispersion relation in the surface roton region as

$$\hbar\omega_k = \Delta + \frac{\hbar^2(k-k_0)^2}{2\mu} \tag{7.71}$$

but for further details we must refer the reader to the account of Rutledge *et al* (1978).

In Appendix 7.2 we discuss the relation between superfluidity and superconductivity and also treat the quantisation of circulation (cf equation (A7.2.8)).

7.9 Liquid ^{3}He

In contrast to ^{4}He, ^{3}He has a nuclear spin of $\frac{1}{2}$ and therefore obeys Fermi–Dirac statistics, its degeneracy temperature being only a few degrees kelvin. Below this temperature, it behaves in many respects as a normal Fermi gas, until one reaches very low temperatures ($\sim$ 1 mK). Then it undergoes a transition to another phase, to be discussed later. For instance, one finds that

† See the discussion in Appendix 7.2 where vortices in He II are referred to in relation tc vortices in type-II superconductors.

for temperatures less than 100 mK, the specific heat c_V is approximately proportional to T, as is to be expected for a Fermi gas (cf equation (5.2)). This behaviour is, at first sight, somewhat surprising in view of the presence of strong interactions that are expected to lead to substantial departures from ideality. The explanation of these unexpected properties has been given by Landau. He hypothesised that in a Fermi liquid the net effect of interactions was to screen each atom by a cloud of surrounding particles. As utilised throughout this book, the entity of atom plus screening cloud is a quasi-particle. Once one works with quasi-particles rather than with bare particles, one can describe the system as a gas of quasi-particles. In the Landau theory, a crucial point is that the quasi-particles obey Fermi statistics, and thus at low temperatures we can describe the low-lying excited states as due to the excitation of quasi-particles across the Fermi surface. Thus the same argument that was used in § 5.1 can be applied to this case to demonstrate that $c_V \propto T$ at low T. There is one major difference to be emphasised between quasi-particles and bare particles, namely that the quasi-particles will not have the same mass as the He atoms. In fact, because of the interactions, at atmospheric pressure the effective mass m^* of the ^{3}He quasi-particle turns out to be almost three times that of a ^{3}He atom. The effective mass can in fact be extracted from the experimental value of c_V/T, which for $T \to 0$ depends on m^* (cf Chapter 5).

The quasi-particle picture is only valid if the momentum and energy of the quasi-particle are not very different from P_F and E_F. In fact, as we move to

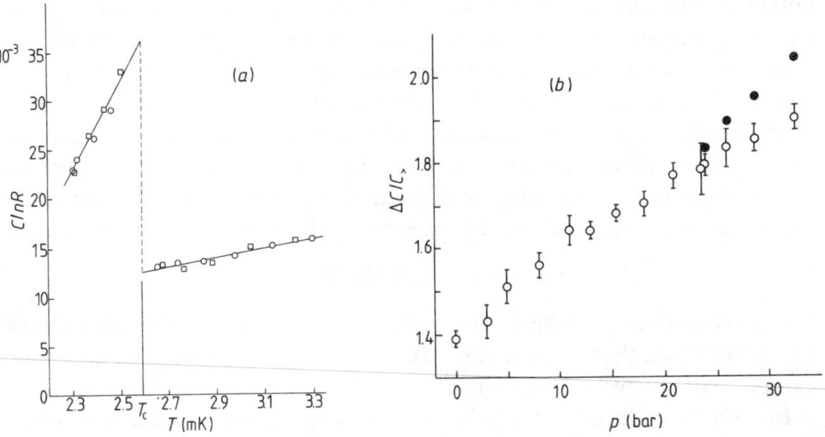

Figure 7.12 Anomalies in specific heat in superfluid phases of liquid ^{3}He. (a) Molar specific heat as a function of temperature of liquid ^{3}He at 33.4 bar pressure (from Webb R A, Greytak T J, Johnson R T and Wheatley J C 1973 *Phys. Rev. Lett.* **30** 210). (b) The pressure dependence of the specific heat discontinuity at T_C: ○, B phase; ●, A phase. (From Alvesalo T A, Haavasoja T, Manninen M T and Soinne A T 1980 *Phys. Rev. Lett.* **44** 1076.)

regions in the excitation spectrum involving quasi-particles of different energy and momentum, more complicated theories have to be applied.

For some time, it was thought that ^{3}He continued to behave as a normal Fermi liquid down to $T = 0$. However, with the enormous progress in the field of cryogenics, it has proved possible to cool liquid ^{3}He down to temperatures of the order of 1 mK. As already stated, at this temperature the system undergoes a phase transition to what is termed a superfluid phase. Investigation of the phase diagram as a function of pressure has also shown that two superfluid phases, termed A and B, are possible. These are signalled, as discussed in the following chapter, by anomalies in the specific heat at the critical temperatures. This is illustrated in figure 7.12, where two phase transitions can be identified.

7.9.1 Angular momentum state of Cooper pairs

The theory that has been invoked in order to explain ^{3}He superfluidity parallels closely that of Bardeen, Cooper and Schrieffer for superconductivity, which was discussed in Chapter 5. There, we stressed that a Fermi liquid is unstable towards the formation of Cooper pairs, provided that some attractive interaction, no matter how weak, is present in the system. In ^{3}He, a natural source of attractive interaction is provided by the van der Waals forces.

However, two effects tend to reduce their influence, and to complicate the naive picture of Cooper pairs bound by van der Waals attraction. The first is the presence of the other particles, which screen the direct interparticle pair potential. The second comes from the repulsive part of the potential, which obviously reduces the effect of the attractive forces. Thus, if some effective attractive interaction is present in the system, this can lead to the formation of a bound Cooper pair only if the resulting wavefunction is sufficiently small at short distances where the interaction becomes repulsive. An almost automatic way of achieving this result can be found by building Cooper pair wavefunctions of total angular momentum different from zero†. In fact, for such a state, the wavefunction can be written in terms of the spherical harmonics Y_{lm} as

$$F(\boldsymbol{r}) = Y_{lm}(\theta, \phi) f(|\boldsymbol{r}|). \tag{7.72}$$

This wavefunction vanishes at the origin for $l \neq 0$. The experimental evidence suggests, in fact, that in the case of ^{3}He in the A and B phases one gets bound Cooper pairs of p($l = 1$) character.

In order to understand the difference between the two phases, we have now to consider the spin part of the Cooper pair wavefunction, which has hitherto been ignored. For $l = 1$, $F(\boldsymbol{r})$ is obviously antisymmetric in the exchange of coordinates $\boldsymbol{r} \to -\boldsymbol{r}$. Thus, in order to comply with the Pauli principle, the spin part of the wavefunction has to be symmetric. There are only three symmetric

† The ensuing discussion follows the lines of Leggett's article in the book edited by Ruvalds and Regge (see Zawadowski (1978)).

spin states than can be built from two particles, each with spin $\frac{1}{2}$. These correspond to the three possible components 1, 0, -1 of a state of total spin angular momentum $S = 1$, and are:

$$S_z = \begin{cases} 1 & |\uparrow\uparrow\rangle \\ 0 & \frac{1}{\sqrt{2}}(|\uparrow\downarrow\rangle + |\downarrow\uparrow\rangle) \\ -1 & |\downarrow\downarrow\rangle \end{cases} \tag{7.73}$$

where the notation is self-explanatory. Thus the total wavefunction, which is the product of space and spin parts, can in general be written as

$$F(\boldsymbol{r}, \alpha, \beta) = F_{\uparrow\uparrow}|\uparrow\uparrow\rangle + F_{\uparrow\downarrow}\frac{(|\uparrow\downarrow\rangle + |\downarrow\uparrow\rangle)}{\sqrt{2}} + F_{\downarrow\downarrow}|\downarrow\downarrow\rangle. \tag{7.74}$$

Different forms of $F(\boldsymbol{r}, \alpha, \beta)$ have been postulated for the A and B phases.

In the A phase, one assumes that there exists a direction in the spin space $\boldsymbol{d}$, which we may take to lie in the z-direction, such that along this direction $S_z = 0$. Thus in this phase, $F(\boldsymbol{r}, \alpha, \beta)$ becomes

$$F(\boldsymbol{r}, \alpha, \beta) = F(\boldsymbol{r})\frac{|\uparrow\downarrow\rangle + |\downarrow\uparrow\rangle}{\sqrt{2}}. \tag{7.75}$$

Clearly the frame of reference in the spin space is immaterial. Had we chosen $\boldsymbol{d}$ to lie on the x,y plane, then $F(\boldsymbol{r}, \alpha, \beta)$ would have taken the form

$$F(\boldsymbol{r}, \alpha, \beta) = F(\boldsymbol{r})(|\uparrow\uparrow\rangle + e^{i\phi}|\downarrow\downarrow\rangle) \tag{7.76}$$

with the angle ϕ depending on the direction of $\boldsymbol{d}$ in the x,y plane.

The B phase instead would correspond in an atom to a $^3P_0$ state and the corresponding wavefunction reads:

$$F(\boldsymbol{r}, \alpha, \beta) = F(\boldsymbol{r})\left(Y_{1,-1}|\uparrow\uparrow\rangle + Y_{1,0}\frac{(|\uparrow\downarrow\rangle + |\downarrow\uparrow\rangle)}{\sqrt{2}} + Y_{1,1}|\downarrow\downarrow\rangle\right). \tag{7.77}$$

Clearly, in both phases, two privileged directions can be identified: one in the spin space corresponding to say the direction along which the component of the spin is zero and the other in real space $\boldsymbol{l}$ corresponding to say the direction along which the orbital angular momentum has component 1.

In the absence of any other interaction, these two directions would be uncorrelated. In reality, the interaction between $\boldsymbol{l}$ and $\boldsymbol{d}$ is provided by the small dipolar interaction (V_d) between the nuclear magnetic moments:

$$V_d = \mu_n^2 \frac{\boldsymbol{\sigma}_1 \cdot \boldsymbol{\sigma}_2 - 3\boldsymbol{\sigma}_1 \cdot \hat{\boldsymbol{r}}_{12}\boldsymbol{\sigma}_2 \cdot \hat{\boldsymbol{r}}_{12}}{r_{12}^3} \tag{7.78}$$

where μ_n is the nuclear magnetic moment and $\boldsymbol{\sigma}$ is a spin operator, its components being Pauli spin matrices. This might appear to be very small; it is in fact of the order of 10^{-7} K at the distance of closest approach of two He atoms. However, Cooper pairs are Bose particles and as $T \to 0$ they all tend to occupy the same quantum state. Therefore in the ground state they all have the same direction of $\boldsymbol{l}$ and $\boldsymbol{d}$. Thus, the importance of V_d is enhanced by the coherence of the state and has macroscopic effects. The expectation value of V_d can be calculated using the forms (7.75) and (7.77) for the wavefunctions†. The result one obtains for the dipolar energy in the A phase is

$$\langle V_d \rangle = -\tfrac{3}{5} g_d (\boldsymbol{d} \cdot \boldsymbol{l}) \tag{7.79}$$

and in the B phase:

$$\langle V_d \rangle = \tfrac{4}{5} g_d (\cos\theta + 2\cos^2\theta) \tag{7.80}$$

where

$$g_d = \tfrac{1}{2}\gamma^2 \hbar^2 \int \frac{1}{r^3} |F(\boldsymbol{r})|^2 \mathrm{d}\boldsymbol{r}, \tag{7.81}$$

γ being the ratio of spin angular momentum to spin magnetic moment. Hence in the A phase $\boldsymbol{d}$ is parallel to $\boldsymbol{l}$ while in the B phase $\cos\theta = -\tfrac{1}{4} (\theta \simeq 104°)$. These interactions between orbital and spin momentum have been investigated by means of nuclear magnetic resonance (NMR). In fact when a varying external field is applied, which tends to vary the direction of the spins and hence that of $\boldsymbol{d}$, $\langle V_d \rangle$ is expected to vary. Thus, an additional force is exerted on the spins, which may change the position and the shape of the NMR line. A variety of collective excitations have been predicted, and in part observed, in superfluid ^{3}He. These correspond to different ways of breaking the complex order present in the system.

7.9.2 Spin wave dispersion in superfluid ^{3}He

As an example, we refer to the experiments by Osheroff *et al* (1977) on spin wave dispersion in superfluid ^{3}He. They employed a resonance method in order to excite standing waves in thin layers of ^{3}He. These were formed between twenty sets of quartz plates and an external magnetic field $\boldsymbol{H}$ was applied in the plane of these plates. In the bulk liquid, $\boldsymbol{d}$ would tend to align perpendicular to $\boldsymbol{H}$; hence the spin direction was fixed. In the neighbourhood of the surface, the boundary condition tended to align $\boldsymbol{l}$ (say $\boldsymbol{l}$ perpendicular to the wall); hence because of the dipolar energy $\boldsymbol{d}$ was also fixed there too. In this way, the direction of the spin is caused to vary smoothly in a precisely determined manner between the plates.

Spin waves now consist of coherent deviations from this equilibrium situation. They can be excited by applying an electromagnetic field in the radio frequency range. By examining the resonant absorption of energy, one can measure the spin-wave frequency. The corresponding wavelength of these

† The A phase is described by $\boldsymbol{d}$ and $\boldsymbol{l}$, the B phase is characterised by a rotation angle θ and a rotation axis which relates spin and orbital space.

excitations, that are obviously standing wave excitations, is fixed by the condition that the thickness of the layer must be an integral number of half wavelengths.

In the experiments of Osheroff *et al*, a series of absorption peaks was observed, corresponding to the fundamental and higher modes of the spin waves being excited. They occur roughly at the frequencies predicted theoretically. Though quantitative discrepancies remain between observation and theory, there can be little doubt that Osheroff *et al* have indeed observed collective excitations in the form of spin waves in superfluid ^{3}He.

7.10 Structure factor and radial distribution function of an electron liquid

We shall conclude this discussion of quantum fluids by referring to the structure of an electron liquid. While we shall primarily mean here the one-component degenerate plasma (jellium), we shall briefly refer to a more realistic model of the electrons in a molten metal like Al at the end of this discussion.

Let us start from a determinant of plane waves $e^{i\boldsymbol{k}\cdot\boldsymbol{r}}$, for $|\boldsymbol{k}| < k_F$, i.e. the total electronic wavefunction Ψ has the form

$$\Psi = \det |e^{i\boldsymbol{k}\cdot\boldsymbol{r}}|. \tag{7.82}$$

If we now form the pair correlation function $g(r)$ as

$$g(|\boldsymbol{r}_1 - \boldsymbol{r}_2|) \equiv g(r) \propto \int \Psi^*(\boldsymbol{r}_1, \boldsymbol{r}_2, \ldots, \boldsymbol{r}_N)\Psi(\boldsymbol{r}_1, \boldsymbol{r}_2, \ldots, \boldsymbol{r}_N)\mathrm{d}\boldsymbol{r}_3 \ldots \mathrm{d}\boldsymbol{r}_N, \tag{7.83}$$

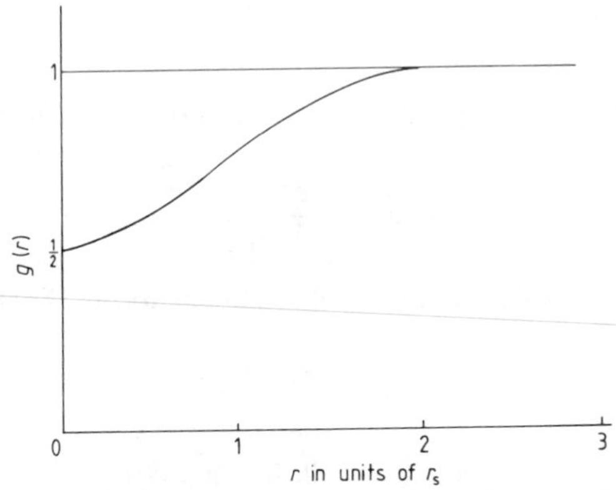

Figure 7.13 Exchange (Fermi) hole around an electron in a degenerate electron gas, given by equation (7.84). Note that though there are oscillations according to equation (7.84) at large r, these are not revealed to graphical accuracy. The unit of r is the mean interelectronic separation.

we find after some calculation that

$$g(r) = 1 - \frac{9}{2}\left(\frac{j_1(k_F r)}{k_F r}\right)^2 \tag{7.84}$$

where j_1 is the first-order spherical Bessel function given by $j_1(x) = (\sin x - x\cos x)/x^2$. $g(r)$ is plotted in figure 7.13. It is the so called Fermi hole surrounding an electron and arises solely from statistical correlations.

If we use equation (7.13) relating $g(r)$ to the structure factor $S(k)$, then, after a short calculation, we find that

$$S(k) = \begin{cases} \dfrac{3k}{4k_F} - \dfrac{1}{16}\left(\dfrac{k}{k_F}\right)^3 & k < 2k_F \\ 1 & k > 2k_F. \end{cases} \tag{7.85}$$

The interest now is to enquire how this structure factor is modified by collective effects arising from the long-range Coulomb interaction.

The answer can be obtained by introducing the direct correlation function $c(r)$ of Ornstein and Zernike†. This is defined from the total correlation function $h(r) = g(r) - 1$ as

$$h(r) = c(r) + \rho_0 \int h(\boldsymbol{r}')c(\boldsymbol{r}-\boldsymbol{r}')\mathrm{d}\boldsymbol{r}'. \tag{7.86}$$

The idea is to split the total correlation function $h(r)$ into a part due to direct interactions between particles and a part arising from indirect effects; the latter being the convolution in equation (7.86). For electrons, at large r, $c(r) \sim e^2/r$, the direct Coulomb interaction. Since, as defined, $c(r)$ is dimensionless, e^2/r must be divided by a characteristic energy of the electron plasma, which is in fact the zero-point plasmon energy $\frac{1}{2}\hbar\omega_p$, ω_p as usual denoting the plasma frequency. Thus, asymptotically at large r,

$$c(r) \sim \frac{e^2/r}{\frac{1}{2}\hbar\omega_p} \tag{7.87}$$

which in $\boldsymbol{k}$-space evidently yields, as $k \to 0$ (Fourier transform relating large r to small k),

$$c(k) \sim \frac{e^2/k^2}{\frac{1}{2}\hbar\omega_p}. \tag{7.88}$$

But the Ornstein–Zernike definition of $c(r)$ in equation (7.86) leads in Fourier transform to

$$c(k) = \frac{S(k) - 1}{S(k)} \tag{7.89}$$

and since $c(k) \sim 1/k^2$ as $k \to 0$ it follows that $S(k) \sim k^2$ near $k = 0$. The

† The use of $c(r)$ in the theory of the liquid–gas critical point will be illustrated in Chapter 8 (see especially Appendix 8.3).

coefficient of the k^2 term is easily calculated from the above and one obtains

$$S(k) \sim \frac{\frac{1}{2}\hbar\omega_p}{4\pi\rho e^2}k^2 \tag{7.90}$$

showing, by comparison with equation (7.85) for the Fermi hole, that the collective plasma behaviour cancels out the k term at small k and introduces the k^2 term given by equation (7.90). Thus, through its zero-point energy, the plasmon strongly alters the structure factor of the ground state of the interacting electron fluid.

Assuming knowledge of this structure factor for the jellium model with interactions, Cusack *et al* (1976) have introduced electron–ion interactions to discuss the structure factor of the electron liquid in molten metals, and in particular in Na and in Al. For Na, they were able to conclude that if one could measure $S(k)$ for the conduction electrons, the electron–ion effects are so weak that one would be studying, in effect, the jellium model at the electron density appropriate to liquid metal Na ($r_s \sim 4\,a_0$). For Al, however, with a stronger electron–ion interaction, one should observe deviations from the jellium electron–electron structure factor. In case these considerations may seem purely academic, Egelstaff *et al* (1974) have proposed a way to extract the electron–electron structure factor from a combination of x-ray, neutron and electron scattering experiments on the same molten metal. The analysis of Egelstaff *et al* (1974) suggests that the oscillations in $g(r)$ for the electrons, while small compared with those in $g(r)$ for the ions, in fact extend out to larger values of r for the electrons than for the ions. The implication is that the short-range order in the electronic assembly extends further than for the ions. This area, combined with that of electron crystallisation to be discussed in the next chapter, promises to be a fruitful one for future studies, especially further scattering experiments.

Problems

7.1 The model of Feynman and Cohen (1956) for the structure factor $S(k)$ of liquid ^{4}He at elevated (low) temperatures, but for long waves, treats the liquid as a continuous compressible medium.

If $\rho(\boldsymbol{r}, t)$ is the number of atoms per unit volume in such a medium, we define the normal coordinates $\rho_{\boldsymbol{k}}$ by (cf equation (7.6))

$$\rho_{\boldsymbol{k}} = \int \rho(\boldsymbol{r}, t)\exp(\mathrm{i}\boldsymbol{k}\cdot\boldsymbol{r})\mathrm{d}\boldsymbol{r}. \tag{P7.1}$$

Then the energy E, assuming the $\rho_{\boldsymbol{k}}$ vary harmonically, is given by

$$E = \frac{1}{2}\sum_{\boldsymbol{k}}\left(\frac{M}{Nk^2}\right)\left(\dot{\rho}_{\boldsymbol{k}}\dot{\rho}_{\boldsymbol{k}}^* + \omega_k^2\rho_{\boldsymbol{k}}\rho_{\boldsymbol{k}}^*\right) \tag{P7.2}$$

where $\omega_k = v_s k$, v_s being the velocity of sound.

You are now given that the structure factor $S(k)$ is the expectation value of $N^{-1}|\rho_{\boldsymbol{k}}|^2$. Use the virial theorem for a harmonic oscillator (kinetic energy equals potential energy on average) and equation (P7.2) to show that

$$S(k) = \frac{\langle E_k \rangle}{Mv_s^2} \tag{P7.3}$$

where M is the mass of a ^{4}He atom and $\langle E_k \rangle$ is the average energy of the oscillator representing sound of wavenumber k.

Hence:

(a) Calculate $\langle E_k \rangle$ at $T = 0$ and hence show that

$$S(k) = \frac{\hbar k}{2Mv_s} \qquad \text{at small } k. \tag{P7.4}$$

(b) Using the result that at elevated temperatures the probability of finding the oscillator representing phonons of wavenumber k in its nth excited state is proportional to $\exp(-E_n/k_B T)$, show that

$$\begin{aligned} S(k) &= \frac{\hbar k}{2Mv_s} \coth\left(\frac{\hbar v_s k}{2k_B T}\right) \\ &\simeq \left(\frac{Mv_s^2}{k_B T}\right)^{-1} + \frac{\hbar^2 k^2}{12Mk_B T} + \ldots \qquad k \text{ small.} \end{aligned} \tag{P7.5}$$

Finally show that $S(0)$ is related to the compressibility, in accordance with equation (7.14).

8

PHASE TRANSITIONS

Our aim in this final chapter is to provide an introduction to phase transitions. There has been great progress in theory in this area since around 1960. However, some of the theory is highly technical and we shall therefore focus in the main on phenomenological treatments. In the latter part of the chapter though, modern trends in relation to quasi-one-dimensional conduction, structural instabilities and electron crystallisation will be considered.

First, though, it will be helpful to review molecular field treatments of magnets, both ferromagnets and antiferromagnets.

8.1 Weiss molecular field theory of ferromagnetism

We shall begin this discussion of phase transitions by recalling the molecular field theory of ferromagnetism due to Weiss.

Weiss recognised that if we start with an assembly of N dipoles, each carrying moment μ, then in order to explain the occurrence of a cooperative phenomenon (i.e. ferromagnetism below a certain temperature) the simplest assumption to make is that the total field acting on the dipoles is the sum of two parts:

$$B_{\text{tot}} = B_{\text{ext}} + B_{\text{int}} \tag{8.1}$$

where the subscripts have obvious meanings. He then made the assumption that B_{int} is directly proportional to the magnetisation M, i.e.

$$B_{\text{int}} = \lambda M \tag{8.2}$$

the internal or molecular field λM having its basic origin in the interaction between the dipoles†.

8.1.1 Paramagnetic behaviour

To obtain the paramagnetic behaviour of such an assembly, one can now start out from the classical Langevin theory of a paramagnetic gas, in which there is a balance between the tendency of the applied field B_{ext} to align the moments parallel to the field and the disordering randomising influence of thermal agitation. Then the mean moment $\bar{\mu}$ in such a classical (i.e. no quantisation of direction) assembly of spins (dipoles) is given by

$$\bar{u} = \frac{\int \mu \cos\theta \, \mathrm{d}N}{\int \mathrm{d}N} = \mu \overline{\cos\theta} \tag{8.3}$$

† Of course, we know from Chapter 6 that in a material like Fe, the interaction arises from quantum mechanical exchange and is electrostatic in origin.

where $\mathrm{d}N$ is the number of dipoles making an angle between θ and $\theta + \mathrm{d}\theta$ with B_{ext}. Now from the Boltzmann law of classical statistics we have for the number $\mathrm{d}N$ of dipoles with energy $E = -\mu B_{\text{ext}} \cos\theta$, in solid angle $\mathrm{d}\Omega$,

$$\frac{\mathrm{d}N}{\mathrm{d}\Omega} = \text{constant} \times \exp\left(-\frac{E}{k_{\mathrm{B}}T}\right) = \text{constant} \times \exp\left(\frac{\mu B_{\text{ext}} \cos\theta}{k_{\mathrm{B}}T}\right) \tag{8.4}$$

and hence, with $\mathrm{d}\Omega \equiv 2\pi \sin\theta \, \mathrm{d}\theta$

$$\mathrm{d}N = \text{constant} \times \sin\theta \exp\left(\frac{\mu B_{\text{ext}} \cos\theta}{k_{\mathrm{B}}T}\right)\mathrm{d}\theta. \tag{8.5}$$

Inserting (8.5) into (8.3) and integrating over θ from 0 to π we find the well known Langevin result

$$\overline{\cos\theta} = \frac{\bar{\mu}}{\mu} = \coth x - \frac{1}{x} \equiv \mathscr{L}(x). \tag{8.6}$$

where $x = \mu B_{\text{ext}}/k_{\mathrm{B}}T$. The form of the magnetisation M in units of the saturation magnetisation $N\mu$ is shown in figure 8.1. Obviously, on physical grounds, $M \propto B_{\text{ext}}$ for small fields, and tends to the saturation value $N\mu$ for

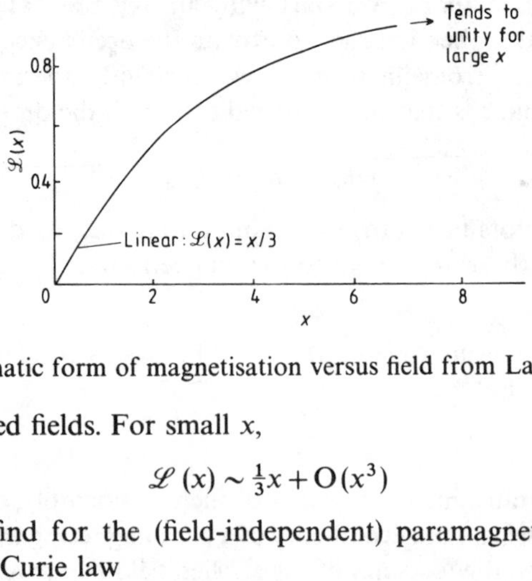

Figure 8.1 Schematic form of magnetisation versus field from Langevin form (8.6).

very large applied fields. For small x,

$$\mathscr{L}(x) \sim \tfrac{1}{3}x + \mathrm{O}(x^3) \tag{8.7}$$

and hence we find for the (field-independent) paramagnetic susceptibility $\chi = M/B_{\text{ext}}$ the Curie law

$$\chi = \frac{N\mu^2}{3k_{\mathrm{B}}T}. \tag{8.8}$$

Into the above Langevin argument, we now introduce the interactions by replacing B_{ext} by B_{tot} according to equations (8.1) and (8.2). Again working in the field-independent limit we find

$$\frac{M}{N\mu} \simeq \tfrac{1}{3}\left(\frac{\mu B_{\text{tot}}}{k_{\mathrm{B}}T}\right) = \frac{\mu B_{\text{ext}}}{3k_{\mathrm{B}}T}(1 + \lambda\chi)$$

or

$$\chi = \frac{N\mu^2}{3k_B T}(1 + \lambda\chi). \tag{8.9}$$

Solving for χ leads therefore to the Curie–Weiss law:

$$\chi = \frac{N\mu^2}{3k_B(T - T_c)} \qquad T > T_c \tag{8.10}$$

where

$$T_c = \frac{\lambda N\mu^2}{3k_B}.$$

This shows that as the temperature falls to a critical value T_c the susceptibility diverges.

We now turn to the region of lower temperatures, to see whether the Weiss assumptions can lead under suitable conditions to spontaneous magnetisation, i.e. to $M \neq 0$ in the absence of any external field (i.e. $B_{ext} = 0$).

8.1.2 Ferromagnetic ordering

With $B_{ext} = 0$, we therefore return to equation (8.6) for the magnetic moment $M = N\bar{\mu}$ and write

$$\frac{M}{N\mu} = \mathscr{L}\left(\frac{\mu\lambda M}{k_B T}\right), \tag{8.11}$$

using the total field in equation (8.1) with $B_{ext} = 0$. Clearly it is equation (8.11) which must be solved to obtain the spontaneous magnetisation M as a function of temperature T.

A graphical solution shows rather clearly what is happening. Thus, we write

$$\eta = \mu\lambda M / k_B T \tag{8.12}$$

and

$$M/N\mu = \mathscr{L}(\eta) \tag{8.13}$$

and we can regard equations (8.12) and (8.13) as simultaneous equations which can be solved by finding the intersection of the straight line $M/N\mu = \eta k_B T/N\mu^2\lambda$ with the plot of equation (8.13), as shown in figure 8.2.

It is clear that for T less than a certain limiting temperature, an intersection occurs. The limiting temperature is obtained when the straight line $M/N\mu = \eta k_B T/N\mu^2\lambda$ is tangential at the origin with the Langevin function $\mathscr{L}(\eta)$. From equation (8.7) this means

$$\frac{1}{3} = \frac{k_B T}{N\mu^2\lambda} \tag{8.14}$$

and we regain the temperature T_c of equation (8.10). Thus, spontaneous magnetisation exists in the Weiss theory below this temperature T_c.

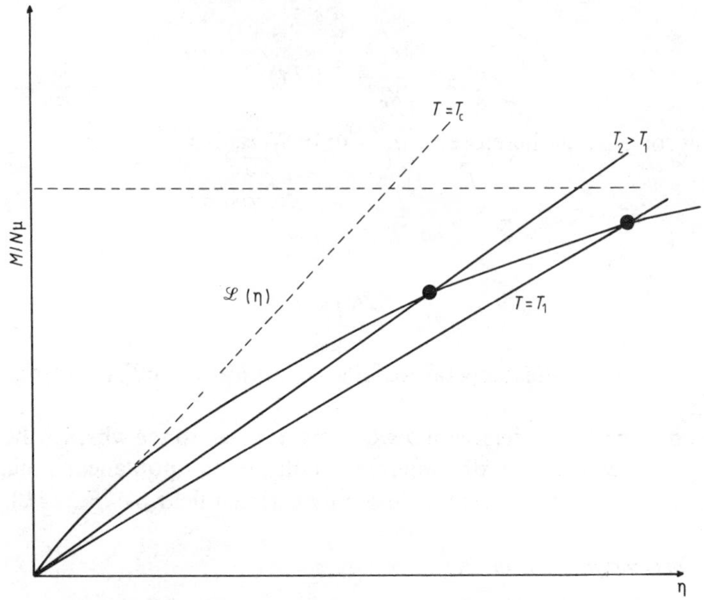

Figure 8.2 Construction to obtain spontaneous magnetisation in Weiss theory. $\mathscr{L}(\eta)$ is Langevin function (essentially the quantity plotted in figure 8.1). Line labelled $T = T_c$ is tangential to $\mathscr{L}(\eta)$ at origin. (Straight lines represent equation (8.12) for three different temperatures $T \leqslant T_c$.)

8.1.3 Molecular field theory and critical exponents

When we solve equations (8.12) and (8.13) simultaneously the plot of $M/N\mu$ against T/T_c is as shown in figure 8.3. Experimental results for Fe and Ni are also shown: there is semi-quantitative agreement. One thing the reader will immediately notice is that the theory curve is not steep enough around $T/T_c = 1$.

If we use the next higher order approximation to equation (8.7), namely

$$\mathscr{L}(\eta) \sim \tfrac{1}{3}\eta - \tfrac{1}{45}\eta^3 + \mathrm{O}(\eta^5), \tag{8.15}$$

then we readily find that the behaviour of the magnetisation just below the critical (Curie) temperature T_c has the form

$$\frac{M}{N\mu} \propto (T_c - T)^{1/2}. \tag{8.16}$$

In practice, it is found that very near to T_c

$$\frac{M}{N\mu} \propto (T_c - T)^{\beta} \tag{8.17}$$

where $\beta \simeq 0.33 \pm 0.03$. This fits the experimental results, and β is referred to as a critical exponent. This shows that the Weiss theory of M will give a curve which is not as steep as the experimental results near $T/T_c = 1$, as noted above.

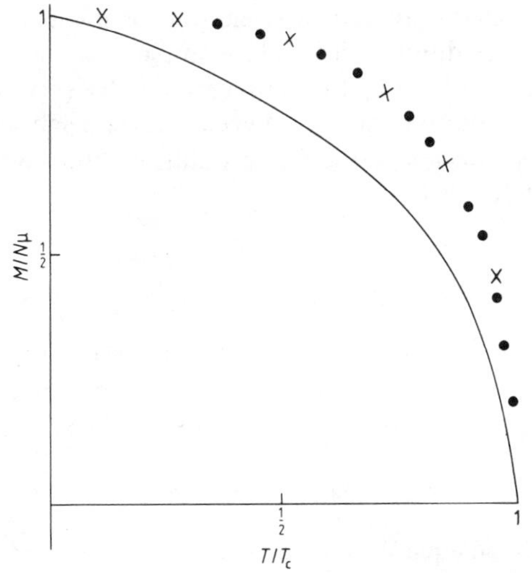

Figure 8.3 Temperature dependence of spontaneous magnetisation of ferromagnet in Weiss theory (full curve). Crosses denote experimental results for iron; full circles are for nickel.

A second critical exponent γ is defined by writing the field-independent paramagnetic susceptibility for $T > T_c$ as

$$\chi \propto (T - T_c)^{-\gamma} \tag{8.18}$$

and the exponent γ comes out from experiment to be around 4/3, whereas molecular field theory gives $\gamma = 1$, as seen by comparing equations (8.10) and (8.18).

Thus, while the molecular field idea gives a semi-quantitative account of the phase transition from the ferromagnetic to the paramagnetic state, a richer theory is required to give a full account of the critical exponents. Modern theories of critical phenomena supply just this (see Domb and Green 1972, 1974). Since the spontaneous magnetisation M goes to zero as $T \to T_c$ according to equation (8.17) with positive β and is always zero in the paramagnetic region above T_c, this transition is second order†. Though we have seen that molecular field theory is inaccurate in its values of the critical indices, it correctly predicts the order of the transition.

8.2 Weiss field form of two-sublattice model of an antiferromagnet

As remarked previously, the simplest situation in antiferromagnetism arises when the lattice of paramagnetic ions can be divided into two interpenetrating

† To be contrasted with first-order transitions, to be treated later in this Chapter, in which there would be a discontinuous drop to zero at $T = T_c$.

sublattices A, B such that all nearest neighbours of an ion on sublattice A lie on sublattice B. This condition is satisfied by simple cubic (SC) and body-centred-cubic (BCC) lattices, but not by the face-centred cubic (FCC) lattice. If the only interactions are antiferromagnetic between nearest neighbours one can write for the magnetisation above the Curie point, within the framework of the Weiss theory (cf Kittel 1956, 1963)

$$TM_{\mathrm{A}} = C'(H - \lambda M_{\mathrm{B}}) \tag{8.19}$$

$$TM_{\mathrm{B}} = C'(H - \lambda M_{\mathrm{A}}). \tag{8.20}$$

Here C' is the Curie constant of one sublattice and the effective field on sublattice A has been written as $H - \lambda M_{\mathrm{B}}$. For positive λ, this represents antiferromagnetic interactions between A and B. Forming the total magnetisation M,

$$M = M_{\mathrm{A}} + M_{\mathrm{B}}, \tag{8.21}$$

by adding the above equations, one obtains

$$TM = 2C'H - C'\lambda M \tag{8.22}$$

which yields the susceptibility $\chi = M/H$ as

$$\chi = \frac{2C'}{T + C'\lambda} \tag{8.23}$$

or

$$\chi = \frac{C}{T + \theta}, \tag{8.24}$$

where $C = 2C', \theta = C'\lambda$. The transition temperature T_{c} is by definition that temperature below which each sublattice possesses a magnetic moment in the absence of an external field.

The transition temperature T_{c} is thus to be obtained from equations (8.19) and (8.20) for $H = 0$, and the condition that these equations have a non-trivial solution is that the determinant of the coefficients of the unknowns $M_{\mathrm{A}}, M_{\mathrm{B}}$ should be zero; namely

$$\begin{vmatrix} T & \theta \\ \theta & T \end{vmatrix} = 0 \tag{8.25}$$

or $T_{\mathrm{c}} = \theta$.

This Weiss model therefore predicts that the transition temperature T_{c} should equal the constant θ in the Curie–Weiss law. Experimental results are collected in table 8.1 and show that there is a spread of values of θ/T_{c} between about 1.5 and 5.

If one includes next-nearest neighbour interactions (Néel 1948) and allows for more general kinds of sublattice arrangements (Anderson 1950, Luttinger 1951) values of θ/T_{c} of the observed magnitude may be obtained.

Having discussed phase transitions in magnets by molecular field theory, we

Table 8.1 Transition temperature T_c compared with the constant θ in the Curie–Weiss law for antiferromagnets.

Substance	Paramagnetic ion lattice	T_c (K)	θ(K)	θ/T_c
MnO	FCC	122	610	5.0
MnF_2	BC rectangular	72	113	1.57
FeF_2	BC rectangular	79	117	1.48
$FeCl_2$	hexagonal layer	23.5	48	2.0

turn next to treat another type of phase transition, namely the liquid–gas critical point. In spite of the very different physical assemblies involved in the ferromagnet and in the fluid transitions, we shall find that important common features exist.

8.3 Liquid–gas critical point

The most obvious definition of the critical point is that point at which the isotherm has a point of inflexion (see figure 8.4):

$$\left(\frac{\partial p}{\partial \rho}\right)_T = 0; \qquad \left(\frac{\partial^2 p}{\partial \rho^2}\right)_T = 0. \tag{8.26}$$

From this, it follows immediately that the isothermal compressibility κ_T

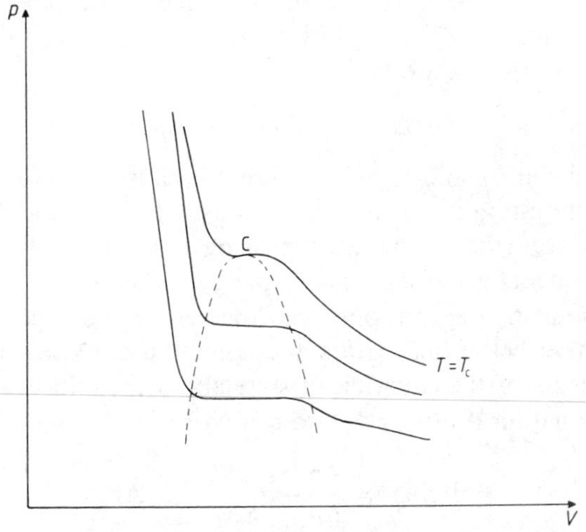

Figure 8.4 p, V isotherms of a substance as obtained from experiment. The tangent to the critical isotherm $T = T_c$ at point C is horizontal and C is also a point of inflexion. C is the critical point. The horizontal portions of the two lower curves correspond to two-phase regions of liquid–vapour equilibrium.

diverges at the critical point. Usually, the form of κ_T is taken as

$$\kappa_T \sim (T - T_c)^{-\gamma} \tag{8.27}$$

measured at the critical density ρ_c, T_c being the critical temperature. For insulating liquids, experiment indicates that $\gamma \sim 1.1$. As we discuss below, this is near to the value for a fluid described by the equation of state of van der Waals. We have anticipated the common features with the ferromagnetic transition by writing the critical exponent as γ in equation (8.27) just as in equation (8.18) for the susceptibility.

Also, we can write for the difference between liquid (l) and gas (g) densities

$$\rho_l - \rho_g \sim (T_c - T)^{\beta} \tag{8.28}$$

where again the exponent has been written the same as for the magnetisation M in equation (8.17). Experiment indicates that for insulating fluids $\beta = 0.35 \pm 0.02$.

8.3.1 Fluid described by the van der Waals equation of state

Let us now turn to examine how such critical behaviour can arise, by studying the model of the van der Waals equation of state, namely

$$\frac{p}{k_B T} = \frac{\rho}{1 - b\rho} - \frac{a\rho^2}{k_B T}. \tag{8.29}$$

The critical values of pressure, density and temperature will be denoted by p_c, ρ_c and T_c respectively. We now proceed to find the critical exponents for κ_T and $\rho_l - \rho_g$ in equations (8.27) and (8.28) by expanding about the critical point. Writing $\Delta p = p - p_c$, etc we find

$$\Delta p = a_1 \Delta T + b_1 \Delta T \Delta\rho + d(\Delta\rho)^3 + \dots \tag{8.30}$$

where the terms in $\Delta\rho$ and $(\Delta\rho)^2$ are absent because of the conditions (8.26).

Now it is interesting that although the constants a_1, b_1 and d can, of course, be determined from the parameters appearing in the van der Waals equation of state, the exponents γ and β do not depend on them.

These exponents depend only on the assumption (unfortunately, as discussed further below, inaccurate, though still useful) that one can make a Taylor expansion of the equation of state about the critical point.

Dividing equation (8.30) by $\Delta\rho$ we can write

$$d(\Delta\rho)^2 = \frac{\Delta p}{\Delta\rho} - a_1 \frac{\Delta T}{\Delta\rho} - b_1 \Delta T. \tag{8.31}$$

In the limit $\Delta\rho$ tends to zero we find that

$$\frac{\Delta p}{\Delta\rho} \to \left(\frac{\partial p}{\partial\rho}\right)_{T = T_c} = 0 \tag{8.32}$$

and

$$\frac{\Delta T}{\Delta \rho} \to \left(\frac{\partial T}{\partial \rho}\right)_p = -\frac{(\partial p/\partial \rho)_{T=T_c}}{(\partial p/\partial T)_{\rho=\rho_c}} = 0 \tag{8.33}$$

since the ratio $\Delta p/\Delta T$ tends to a finite constant at the critical point. Thus equation (8.31) takes the form

$$\Delta \rho = \pm \left(\frac{a_1}{d}\right)^{1/2} (\Delta T)^{1/2}. \tag{8.34}$$

It follows that $\beta = \frac{1}{2}$ in this van der Waals case. This equation (8.34) predicts further that the coexistence curve is symmetrical about $\rho = \rho_c$ in the neighbourhood of the critical point (see P8.8 for details, including the 'rule of equal areas').

The value $\beta = \frac{1}{2}$ is not in quantitative agreement with the measured value $\beta = 0.35 \pm 0.02$ quoted above for insulating fluids. The Taylor expansion assumed (which is certainly valid for the van der Waals equation of state) is clearly therefore an oversimplification for insulating fluids. Nevertheless, we see from the van der Waals equation of state the correct general behaviour for $\rho_l - \rho_g$ as $T \to T_c$, though the model is quantitatively in error.

One can also calculate $(\partial p/\partial \rho)_T$ and then one obtains

$$\left(\frac{\partial p}{\partial \rho}\right)_T \sim b_1 \Delta T + 3d(\Delta \rho)^2. \tag{8.35}$$

Along the critical isochore $\Delta \rho$ is zero and

$$\kappa_T = \frac{1}{\rho}\left(\frac{\partial \rho}{\partial p}\right)_T = \frac{1}{a_1 \rho_c}(\Delta T)^{-1}, \qquad T \to T_c^+. \tag{8.36}$$

Similarly, along the coexistence curve, equation (8.34) applies and one has

$$\kappa_T = \frac{1}{2a_1 \rho_c}(\Delta T)^{-1}, \qquad T \to T_c^-. \tag{8.37}$$

One has that the critical exponent $\gamma = 1$, but it should be noted that the coefficients of $(\Delta T)^{-1}$ in equations (8.36) and (8.37) differ by a factor of 2. The exponent $\gamma = 1$ is in reasonable agreement with experiment. It should be noted that, as for the van der Waals description of the critical point, the molecular field theory of ferromagnetism has already been shown (in equation (8.10)) to give $\gamma = 1$.

We shall return to the critical exponents briefly below, when we make a fuller comparison between the critical behaviour of ferromagnets near the Curie temperature and the behaviour of a fluid in the vicinity of its critical point. Before doing so, we want briefly to summarise the way in which the structure factor $S(k)$ and the pair distribution function $g(r)$ of a fluid vary as the critical point is approached.

Some details are given in Appendix 8.3, but below we summarise the salient features of liquid structure as the critical point is approached.

8.3.2 Structure of fluid near the critical point

In equation (7.14), the long-wavelength limit $S(0)$ of the structure factor $S(k)$ is related to the isothermal compressibility κ_T. But we have seen above that κ_T diverges as $T \to T_c$ and hence, from equation (7.14), $S(0)$ diverges. But from the relation (7.13) between $S(k)$ and the radial distribution function $g(r)$, the divergence of $S(0)$ tells us that $g(r)$ must become so long-ranged as $T \to T_c$ that $\int [g(r)-1]\,\mathrm{d}\boldsymbol{r}$ diverges.

In fact the argument of Ornstein and Zernike, presented in Appendix 8.3, leads to the large-r form

$$g(r) - 1 \sim \text{constant} \times \mathrm{e}^{-\kappa r}/r. \tag{8.38}$$

This equation leads to an important characterisation of critical behaviour through a correlation length $\xi = \kappa^{-1}$, ξ becoming infinite as $T \to T_c$. This means from equation (8.38) that $g(r) - 1 \sim r^{-1}$ for large r as $T \to T_c$, which does indeed cause the integral of $g(r) - 1$ through the liquid volume to diverge. The correlation length ξ is characterised by a third critical exponent ν through

$$\xi \sim |T - T_c|^{-\nu}. \tag{8.39}$$

Actually, as with the van der Waals model above, the Ornstein–Zernike assumptions are oversimplified. In reality, at $T = T_c$, for large r,

$$g(r) - 1 \sim \text{constant}/r^{1+\eta} \tag{8.40}$$

where experimental evidence points to the fact that η is small (~ 0.1). If $g(r) \sim r^{-1}$ at large r then by Fourier transform $S(k) \sim \text{constant} \times k^{-2}$ for small k, showing how $S(k)$ diverges at T_c as k tends to zero. This latter result, of course, must be modified because in real fluids $\eta \neq 0$, the result being readily shown to take the form

$$S(k) \sim \text{constant}/k^{2-\eta} \qquad T = T_c,\ k \to 0. \tag{8.41}$$

Modern critical point theories lead to a relation between β, ν and η, namely

$$\beta/\nu = (1+\eta)/2. \tag{8.42}$$

With $\beta = \frac{1}{2}$, the van der Waals value, and $\eta = 0$ resulting from the Ornstein–Zernike treatment, we find from equation (8.42) that $\nu = 1$ which is again the Ornstein–Zernike value. In practice, $\beta \sim \frac{1}{3}$ for insulating liquids and hence, with η small, $\nu \simeq 2\beta \simeq \frac{2}{3}$. A little more will be done with critical exponents below.

8.4 Scaling laws

Widom (1965) and independently Kadanoff (1966) have put forward the idea that near T_c the range of the correlations is so great that local behaviour plays a negligible role and that the correlation length κ_c^{-1}, which diverges as $T \to T_c$, is the fundamental parameter, and the only one, entering the theory.

If this idea is correct, and it certainly seems to be very useful (though there may be departures from it occasionally), then there will be essentially a 'law of corresponding states' between systems having the same lattice structure. This can be expressed more precisely as follows. Suppose we fix attention on the magnetisation $M(H,T)$. Then, by integrating this with respect to H, we can obtain the free energy $F(H,T)$ and from it all thermodynamic properties. The existence of a scaling law implies that, if $T = T_c = \epsilon$, and we replace ϵ by $\lambda\epsilon$

$$\frac{M}{m(\lambda)} = f\left(\frac{H}{h(\lambda)}, \lambda\epsilon\right). \tag{8.43}$$

If $\lambda \to 0$, we expect $m(\lambda)$ and $h(\lambda)$ to increase and we therefore assume

$$\left.\begin{aligned} m(\lambda) &= \lambda^{-u} \\ h(\lambda) &= \lambda^{-v} \end{aligned}\right\} \quad u, v > 0. \tag{8.44}$$

It is easy to show that, as a consequence of the above assumptions, we can write, since f must be such that equation (8.43) is independent of λ,

$$\frac{M}{|\epsilon|^u} = f_\pm\left(\frac{H}{|\epsilon|^v}\right) \tag{8.45}$$

where f_+ must be used for $\epsilon > 0$ and f_- for $\epsilon < 0$. It is clear that, if this relation holds near the critical point, then a great simplification has been achieved by reducing $M = M(H,T)$ to an unknown function of a single variable only.

8.4.1 Identification of parameters with critical indices.

For $\epsilon > 0$, i.e. $T > T_c$, $M = 0$ and therefore $f_+(0) = 0$. Similarly $f_-(0)$ must be a finite constant if M is to decrease with $|\epsilon|$. We can thus draw up table 8.2. The

Table 8.2 The behaviour of relevant physical quantities as the critical point is approached. For details of specific heat c_H, see Jones and March (1973).

Physical quantity	In terms of u and v	In terms of critical indices	Consequences
Spontaneous magnetisation	$M = \|\epsilon\|^u f_\pm(0)$	$M \sim \|\epsilon\|^\beta$	$u = \beta$
Magnetisation, M as function of H on critical isotherm	$M = \|H\|^{u/v}$	$M \sim \|H\|^{1/\delta}$	$v = \beta\delta$
Initial susceptibility	$\chi \sim \|\epsilon\|^{\beta(1-\delta)} f'_\pm(0)$	$\chi \sim (T-T_c)^{-\gamma}$ or $(T_c-T)^{-\gamma'}$	$\beta(1-\delta) = -\gamma$ $= -\gamma'$
Specific heat c_H at $H = 0$	$c_H \sim \|\epsilon\|^{\beta+\beta\delta-2}$	$c_H \sim (T-T_c)^{-\alpha}$ or $(T_c-T)^{-\alpha'}$	$\alpha' = \alpha$ $\alpha' + \beta(1+\delta) = 2$

quantities of physical interest are expressed in terms of u and v in the first column and in terms of the critical indices in the second column. The consequences resulting from this are recorded in the third column.

Given the assumption (8.43), one gets back a theory which turns out to have only two critical exponents, which we might choose as β and γ say.

Various efforts have been made to support the ansatz (8.43), prominent work being that of Widom (1965) who assumed homogeneity of the free energy, cell scaling by Kadanoff (1966) and the droplet model proposed by Fisher (1967). No completely convincing argument from first principles has as yet been given (see, however, Wilson 1971).

8.4.2 Equation of state and its parametric representation

One of the consequences of the scaling law assumption is that the magnetic equation of state is of the form

$$\frac{M}{|\epsilon|^{\beta}} = f_{\pm}\left(\frac{H}{|\epsilon|^{\beta\delta}}\right). \tag{8.46}$$

Experimental evidence gives strong support to this form near to the critical point.

Schofield (1969) and Josephson (1969) have independently pointed out that it is valuable to use 'polar coordinates' in the H, T plane to represent this equation of state. The reason for this becomes clear if we refer to figure 8.5. One chooses a variable r related to the distance T_c in the H, T plane and θ, which varies continuously if we move from the coexistence line around the critical point and back to the coexistence line. Then it would appear that the

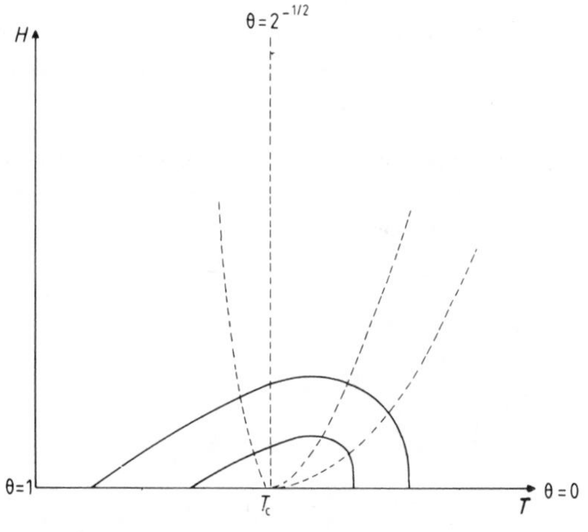

Figure 8.5 Schematic illustration of variables r and θ in parametric representation (8.49) of equation of state: ———, curves of constant r; – – – – –, curves of constant θ.

thermodynamic functions will be smooth functions of θ, but non-analytic functions of r as $r \to 0$.

One now chooses the parametrisation

$$\begin{aligned} H &= r^{\beta\delta} h(\theta) \\ \epsilon &= r t(\theta) \end{aligned} \tag{8.47}$$

which yields

$$\frac{H}{|\epsilon|^{\beta\delta}} = \frac{h(\theta)}{[t(\theta)]^{\beta\delta}} \tag{8.48}$$

and hence from equation (8.46)

$$M = r^{\beta} m(\theta) \tag{8.49}$$

where $m(\theta)$ is a regular, but unknown function.

8.4.3 *Use of molecular field theory*

The value of such a parametrisation is made clear by taking as an example molecular field theory. It can then be shown (see Kasteleyn 1971) that

$$\left(\frac{3\mu^2}{T_c}\right)^{1/2} \theta(1-\theta^2) = m(\theta)(1-2\theta^2) + \frac{T_c}{3\mu^2}[m(\theta)]^3 \tag{8.50}$$

which yields as a solution for $m(\theta)$ in equation (8.49) the form

$$m(\theta) = \left(\frac{3\mu^2}{T_c}\right)^{1/2} \theta. \tag{8.51}$$

Hence, in this case, the magnetisation varies linearly with θ. Experiments support the linear relation between $m(\theta)$ and θ as figure 8.6 shows. Thus the equation of state near the critical point can be represented as

$$\begin{aligned} H &= a r^{\beta\delta} \theta(1-\theta^2) \\ \epsilon &= r(1-2\theta^2) \\ M &= g r^{\beta} \theta, \end{aligned} \tag{8.52}$$

where a and g are constants. A fundamental explanation still seems to be lacking here however.

8.5 Phase diagrams and elementary excitations

As an example of the relationship between elementary excitations, in particular phonons, magnons and electronic single-particle excitations, and phase transitions, it will be helpful to make reference to the work of Bhatia *et al* (1976). These workers discuss the form of the phase boundary

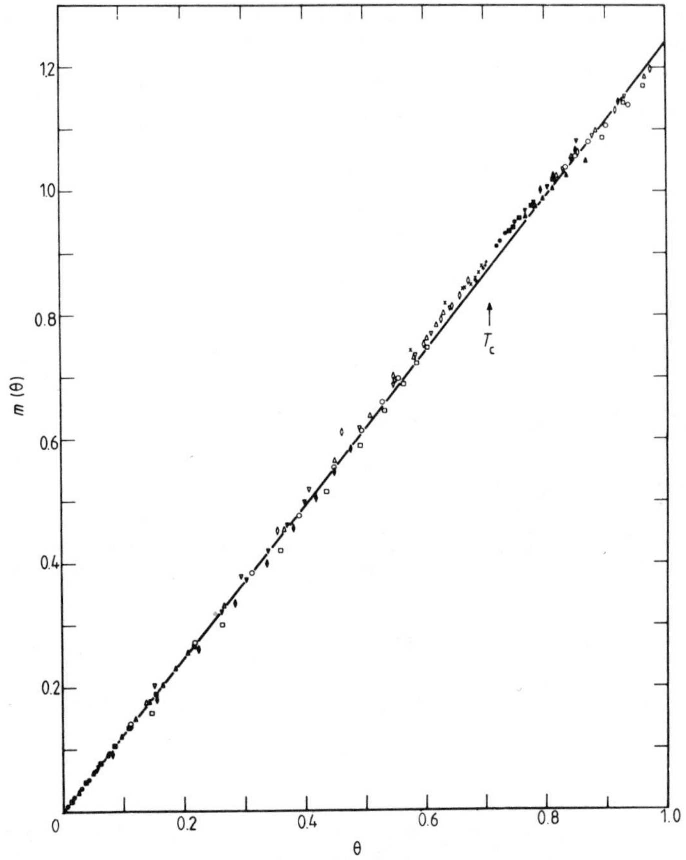

Figure 8.6 Experiments illustrating linearity of $m(\theta)$ with θ as in equation (8.51). Data shown are for $CrBr_3$ at different temperatures. The full line has equation $m(\theta) = 1.24\,\theta$. (From Ho J T and Litster J D 1969 *Phys. Rev. Lett.* **22** 603.)

separating two phases as $T \to 0$. In contrast to the second-order transitions discussed above, we are now concerned with first-order transitions.

The slope of such a phase boundary is determined by the Clausius–Clapeyron equation, namely

$$\frac{\mathrm{d}T}{\mathrm{d}p} = \frac{V_2 - V_1}{S_2 - S_1} \tag{8.53}$$

$V_2 - V_1$ and $S_2 - S_1$ being volume and entropy differences between the two phases. Suppose first we apply this equation to a class of transitions in which an insulating or semiconducting phase 1 is transformed under pressure to a metallic phase 2. The semiconducting materials InSb, Si and Ge are in this class, transforming to metallic phases under pressure.

Now the metal is the denser phase in such a transition and thus in equation (8.53) the volume change $V_2 - V_1$ is negative. But now, since the electronic contribution to the specific heat of the metal is proportional to T at low temperatures, and is associated with an entropy $S_2 \propto T$, whereas in phase 1 the phonons dominate the specific heat ($\propto T^3$) and hence $S_1 \propto T^3$ it follows that:

(a) $\mathrm{d}T/\mathrm{d}p$ is negative;

(b) $$\frac{\mathrm{d}T}{\mathrm{d}p} \sim \frac{\text{constant}}{T} \quad \text{as } T \to 0$$

or

$$\left.\frac{\mathrm{d}(T^2)}{\mathrm{d}p}\right|_{T=0} = \text{constant}, \tag{8.54}$$

the constant being a negative quantity. In fact, since c_V, the electronic specific heat of the metal, is determined by the density of electronic states, $N(E_F)$, at the Fermi level (cf §5.1), the constant in the above equation is determined by $N(E_F)$, together with the magnitude of the volume change. One qualitative consequence of (a) for the insulator–metal boundary is that the 'triple point' marked in figure 8.7 must lie at a lower pressure than that, p_0 say, required to induce the phase transition at $T = 0$.

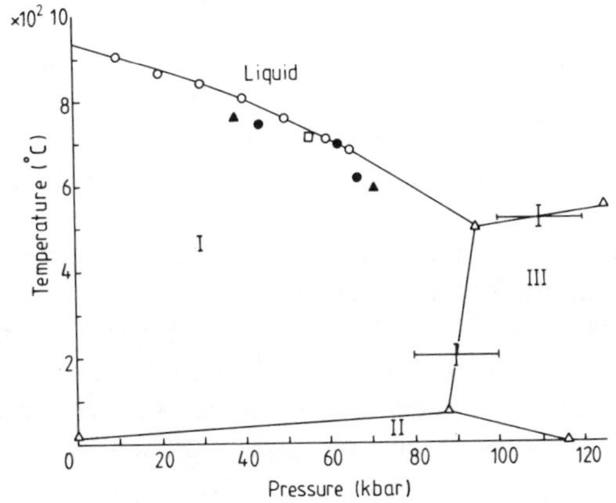

Figure 8.7 Phase diagram of germanium (from Cannon J F 1974 *J. Phys. Chem. Ref. Data* **3** No. 3). (The phase diagrams of InSb and Si are similar in character.)

As a second example, suppose that both crystalline phases involved are insulating. Then $S_2 - S_1 \propto T^3$ at low temperatures (for non-magnetic phases: see below for a discussion of the effect of cooperative magnetism). Thus the Clausius–Clapeyron equation yields

$$\left.\frac{\mathrm{d}(T^4)}{\mathrm{d}p}\right|_{T=0} = \text{constant}. \tag{8.55}$$

The difference between the results (8.54) and (8.55) clearly resides in the fact that in equation (8.54) the elementary excitations dominating the entropy difference are the single-particle electronic excitations (also the case in a metal–metal transition), whereas in the phase change between insulators the dominant low-lying excitations are phonon-like.

As a third and final example, let us consider the case of a transition between two crystalline insulating phases in which one or both is ferromagnetic at low temperatures. Then, as we saw in §6.6, spin wave (magnon) excitations contribute a term in the entropy proportional to $T^{3/2}$. Hence, provided only that the volume change at $T = 0$ is non-zero, we find

$$\left.\frac{\mathrm{d}(T^{5/2})}{\mathrm{d}p}\right|_{T=0} = \text{constant}. \tag{8.56}$$

The fact that, at $T = 0$, with non-zero volume change, there is an infinite (positive or negative) value of $\mathrm{d}T/\mathrm{d}p$ at $T = 0$, is of course a consequence of the third law of thermodynamics. The different powers of T required to make the pressure derivative finite reflect, as stressed above, the nature of the dominant elementary excitations.

8.6 Phase transition in one-component classical plasma

As a further example of a first-order phase transition, we shall discuss briefly the case of a classical one-component plasma. This, by definition, is an overall electrically neutral system. The electrons in such a plasma are correlated in their motions by the Coulombic repulsion but are assumed to move in a uniform background of positive charge.

It is a remarkable fact that such a system exhibits a phase transition, as demonstrated quantitatively by computer simulation (Brush *et al* 1966, Hansen 1976). It is not difficult to see, in this classical system, that its basic behaviour depends on whether or not the thermal energy $k_{\mathrm{B}}T$ dominates the Coulomb repulsion e^2/r_{s}, r_{s} being the average interelectronic distance. This classical plasma is therefore characterised by a quantity

$$\Gamma = \frac{e^2}{r_{\mathrm{s}} k_{\mathrm{B}} T}. \tag{8.57}$$

In the regime $\Gamma < 1$, the electrons behave like a classical gas, being negligibly correlated. The machine calculations show that as Γ is increased, the Coulomb repulsions lead to the formation of an electron liquid. The pair distribution function $g(r)$ is shown in figure 8.8 for $\Gamma = 120$, and features are seen which are quite similar to those found for the ions in a classical liquid metal (cf figure 7.2). Appreciable oscillations in $g(r)$ are what we imply when we use the term 'electron liquid'.

It is clear that, for small Γ, the electrons are delocalised. However, figure 8.8

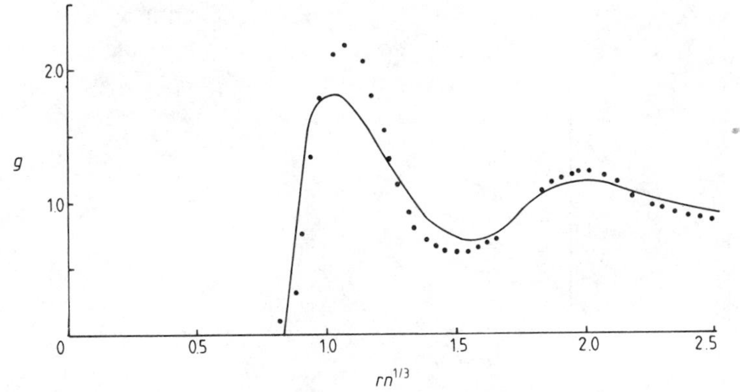

Figure 8.8 Form of pair distribution function in a classical one-component plasma corresponding to ratio Γ of 120. Points are from computer experiment; solid curve is from an approximate analytical treatment. n is the number density and so the unit of distance is the mean interparticle spacing. (From Gillan M J 1974 *J. Phys. C: Solid St. Phys.* **7** 11.)

shows the development of short-range order as the electron–electron interaction energy gets large compared with the thermal energy. As we might anticipate, the machine experiment shows that, on further increase in Γ (in fact to 155), electron crystallisation occurs, i.e. the electrons localise around the sites of a lattice, which is found to be body-centred cubic†.

8.6.1 Application to melting of metallic Na

No system in nature, to our knowledge, has so far been found which approximates to the above one-component electron plasma. However, there is no reason why we could not take the non-responsive uniform background to be electrons and the responsive component to be (point) positive ions.

But we know from extensive studies of metallic sodium that the conduction electrons form an almost uniform background, in which the ions sit on a BCC lattice. It is then important to note that, substituting into equation (8.57) for Γ the measured melting temperature of Na and the observed lattice spacing, Na melts when $\Gamma \sim 200$. Along the melting curve under pressure, which presumably reduces the residual non-uniformity in the conduction electron charge cloud, Γ reduces to around 155 (Ivanov *et al* 1974).

As Ferraz and March (1980) have pointed out, it is of interest to consider the transition from the liquid metal to the frozen solid by plotting the height S_{max} of the principal peak in the structure factor $S(k)$ as a function of the coupling strength Γ. In figure 8.9 S_{max} is seen to rise from its value of unity in the Debye–Hückel limit $\Gamma \rightarrow 0$ to a value of 2.71, the data plotted being taken from

† The reason is that the BCC lattice has the lowest Madelung energy (cf Wigner 1934, 1938 and also §8.8 below).

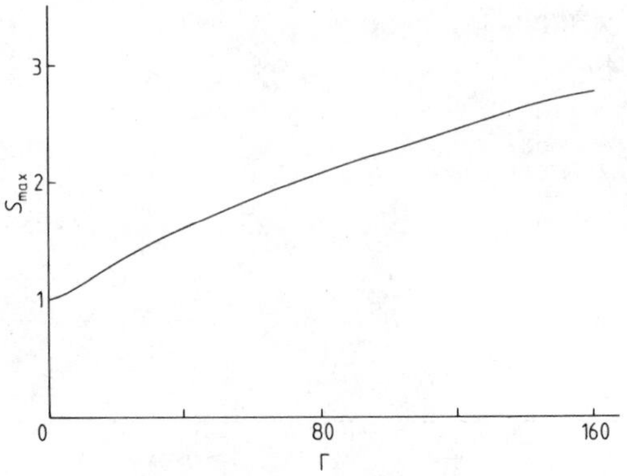

Figure 8.9 Structure factor $S(k)$ evaluated at principal peak, plotted as a function of dimensionless interaction strength Γ defined in equation (8.57) for the classical one-component plasma. Limit $\Gamma \to 0$ corresponds to the Debye–Hückel limit, whereas $\Gamma \to 155$ corresponds to the approach to the freezing transition. (After Ferraz A and March N H 1980 *Solid St. Commun.* **36** 977.)

Hansen (1976). It is of interest that the accurate x-ray measurements of $S(k)$ of Greenfield *et al* (1971) for liquid Na and K near their melting points lead to values of S_{max} of 2.80 and 2.73 respectively. Thus for these simple metals, with BCC structure and with weak electron–ion interaction, it is of considerable interest to observe that freezing occurs when the structure factor reaches a peak height S_{max} of about 2.7, in accord with the criterion obtained from the one-component plasma model. Of course, we emphasise again that the reason why this model can be used in these cases is that, in the metal, the ions can be treated classically at or near the melting point and, although the electrons are degenerate, this does not enter in lowest order for, with weak electron–ion interaction, they are genuinely providing a non-responsive uniform background in which the positive ions motions take place.

It would, of course, be folly to suppose that such a model, which may be useful for Na, had anything to do with the melting of a BCC metal like Fe, in which the itinerant electron distribution in which the ions move is vastly different from a uniform background. For Fe, and like materials, dislocation models of melting that have been proposed (see e.g. Edwards and Warner 1979 and other references given therein) may be more appropriate but their validity is not established in such three-dimensional situations at the present time. What has been proved is that, as with the one-component plasma model of the freezing of the alkali metals discussed above, the dislocation models lead correctly to a first-order transition.

Though we shall not go into more detail of the dislocation model in the three-dimensional situations referred to above, we shall now show how it can be employed successfully in a two-dimensional framework.

8.7 Dislocation model of melting in a two-dimensional assembly of interacting electrons

There are, by now, suitable experimental methods for studying two-dimensional systems of electrons. In particular, when electrons are trapped in a He liquid surface, their density can be so low that classical statistical mechanics is applicable.

This has prompted computer simulation studies of a two-dimensional classical one-component plasma, analogous to those reported in three dimensions in the previous section. These studies have shown that, also in two dimensions, the electron fluid undergoes crystallisation; often referred to as a Wigner phase transition. This happens, as in §8.6, as the density is lowered, the lattice structure for the two-dimensional assembly being triangular. The melting parameter corresponding to equation (8.57), which would be proportional to $n^{1/3}$ (n denoting the density), is in two dimensions now defined by

$$\Gamma = \frac{e^2 (\pi n)^{1/2}}{k_B T}. \tag{8.58}$$

At melting, the computer studies yield $\Gamma_m = 95 \pm 2$. In equation (8.58) the quantity n is now naturally the number of particles per unit area.

In order to provide a mechanism for this transition, and thereby to explain the results of the computer experiment, Thouless (1978) has studied the dislocation model already referred to, put forward much earlier for three-dimensional solids by Kuhlmann-Wilsdorf (1965) and considered in computer experiments by Cotterill and his co-workers (see also the work of Edwards and Warner 1979, referred to in §8.6). The application of the dislocation model to two-dimensional situations seems rather natural and convincing.

It is not possible to deal in any detail with dislocation theory here (see Read 1953, Nabarro 1970). These line defects in crystalline solids not only destroy the perfect crystalline order but also allow the two parts of the crystal to slip relative to one another. Thus it is plausible that if created in sufficient density, dislocations will drive the melting process.

The temperature at which melting will occur can be estimated by considering the free energy of a dislocation:

$$F = E - TS. \tag{8.59}$$

The energy E is due to the elastic distortion necessary to introduce in the crystal the additional semi-infinite line of atoms. We may estimate this elastic energy by dividing the region surrounding the edge of the dislocation into two parts. One of these is the region surrounding the edge of the dislocation, whose area is of the order of a^2 (a being the lattice parameter), where the crystal is very heavily deformed. From this region, a contribution E_0 to the total energy results, E_0 being large but finite. The second, outer region, is less strongly deformed and therefore one can apply elastic continuum theory. For the system we are treating, this then results in a further contribution to the total energy of the form

$$E \simeq ma^2 v_{st}^2\, n \ln (A/\pi a^2)/4\pi \tag{8.60}$$

where m is the mass of the particles, A the total area and v_{st} is the transverse sound velocity given by the shear elastic constant (cf Appendix 3.1). Clearly for $A \to \infty$, the contribution will dominate E_0 above. The dominant contribution to the entropy is also readily estimated. There are, in fact, about $A/\pi a^2$ ways of inserting a dislocation into the system. Therefore for $A \to \infty$

$$S = k_B \ln (A/\pi a^2) \tag{8.61}$$

and the total dislocation free energy is

$$F \simeq \left(\frac{mna^2 v_{st}^2}{4\pi} - k_B T\right) \ln\left(\frac{A}{\pi a^2}\right). \tag{8.62}$$

This becomes negative for

$$T \equiv T_m = \frac{nma^2 v_{st}^2}{4\pi k_B}. \tag{8.63}$$

From the arguments given above we may expect T_m to be the melting temperature. Using for v_{st} calculated values at $T = 0$, the resulting melting parameter Γ_m turns out to be $\Gamma_m \simeq 79$, in fair agreement with the computer experiment which gave 95. The reader interested in pursuing this area further is referred to the work of Nelson and Halperin (1979) and to other references given therein.

8.8 Metal–insulator phase transitions

The argument of §§8.6 and 8.7 for the classical one-component plasma has an interesting parallel in the opposite limit of the completely degenerate one-component plasma. This is simply the jellium model, in which we have interacting electrons moving in a fixed uniform background of neutralising positive charge. The parallel concerns the crystallisation discussed previously, when the ratio of potential energy to thermal energy exceeds a certain value ($\Gamma = (e^2/r_s)/k_B T > 155$ for the classical case). Wigner (1934, 1938), before the machine calculations on the classical one-component plasma, had argued that in the ground state of jellium, electron crystallisation would occur at sufficiently low density (i.e. large r_s, the mean interelectronic spacing). He predicted a BCC structure, in the ground state, the same lattice subsequently found in the machine experiments on the classical plasma. This prediction was based on the electrostatic energy of charges on different lattices in a uniform neutralising background. Though different structures have energies quite close together, it turns out that of the lattices so far examined the BCC lattice has the lowest energy. As Fuchs (1935) has shown, for this case the electrostatic

(Madelung) energy per electron (in rydbergs) is

$$E/N = -1.80/r_s \tag{8.64}$$

if r_s is in Hartree units ($a_0 = \hbar^2/m_e e^2 = 1$).

In contrast, the Slater determinant of plane waves, when used to calculate the expectation value of the jellium Hamiltonian in the limit $r_s \to \infty$ yields

$$E_{HF}/N = -0.916/r_s \text{ Ryd} \tag{8.65}$$

(where HF stands for Hartree–Fock), which is, of course, an extremely poor approximation compared with the crystal of electrons localised on the BCC Wigner lattice. Therefore, the delocalised picture of the free Fermi gas, where the mean energy per particle

$$E/N = \tfrac{3}{5}E_F = 2.2/r_s^2 \text{ Ryd}, \tag{8.66}$$

E_F being the Fermi energy, is obviously inappropriate as $r_s \to \infty$ and we must have a transition at $T = 0$ from the high-density metallic state, with a sharp Fermi surface at $|\boldsymbol{k}| = k_F$, to an insulating electron crystal at low density.

Parrinello and March (1976) have argued, by analogy with classical melting, that this Wigner transition is first order. This implies that a suitable order parameter must go to zero discontinuously at the transition†.

As March *et al* (1979) have argued, a suitable order parameter for metal–insulator transitions at $T = 0$ can be found by considering the momentum distribution of the electrons.

8.8.1 Order parameter for metal–insulator transitions at $T = 0$

Let us start with the Wigner first-order metal–insulator transition. Figure 8.10 shows the probability of occupation $n(k)$ of a state of wavevector $\boldsymbol{k}$ (momentum $\hbar\boldsymbol{k}$). As $r_s \to 0$, we have the Fermi sphere completely occupied by electrons for $|\boldsymbol{k}| < k_F$ and no electrons outside, in the ground state.

As the electron–electron interactions begin to play a role as r_s is increased from the limiting value of zero, we discussed in Chapter 2 how some electrons are promoted to states with $k > k_F$ leaving some unoccupied states (holes) inside the Fermi sphere. This is indicated by the two curves A in figure 8.10.

The important point to be made is that in figure 8.10 the discontinuity in $n(k)$ at $k = k_F$, which is evidently of magnitude unity as $r_s \to 0$, is reduced, but not removed. This conclusion is supported by the detailed calculations of Daniel and Vosko (1960), who used many-body perturbation theory in the high-density limit. No rigorous proof that the discontinuity remains has been given, but there is general agreement that it does. Experiment also supports the existence of remarkably sharp Fermi surfaces in pure metals at low temperature.

We now turn to the low-density side of the metal–insulator transition. Here

† In contrast to figure 8.3 for the spontaneous magnetisation which goes continuously to zero in a second-order transition.

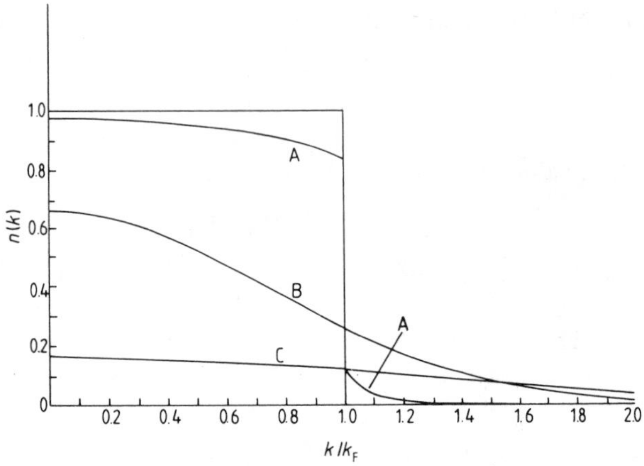

Figure 8.10 Schematic form of momentum distribution $n(k)$ in an electron gas. The square distribution is for a non-interacting gas, with all states inside the Fermi sphere completely full and all states outside empty. A, High-density theory, showing holes inside the Fermi surface and electrons outside; B, Low-density result, when in the insulating, electron crystal, phase; C, extremely low-density electron crystal.

one can argue that since the Wigner–Seitz cell of the BCC lattice, obtained by drawing planes perpendicularly bisecting lines joining a central atom to its neighbours, evidently has high symmetry, it can be well approximated by a sphere. Then the electron vibrating about the lattice site at the centre of this cell will feel mainly the uniform positive charge enclosed in a sphere of volume equal to that of the Wigner–Seitz cell. The other cells, being electrically neutral, will contribute only higher-order multiple terms to the potential in the central cell; these we shall neglect. The potential energy in which the electron moves, due to a uniform spherical distribution of total charge $|-e|$ is then

$$V(r) = \frac{e^2 r^2}{2r_s^3} + \text{constant} \tag{8.67}$$

and hence the ground-state wavefunction is that of an isotropic three-dimensional harmonic oscillator with force constant given by the above result. Explicitly the wavefunction is (cf March *et al* 1967)

$$\psi = \exp(-\alpha r^2), \qquad 2\alpha = 1/r_s^{3/2} \tag{8.68}$$

and one sees, by Fourier transform, that the momentum distribution is also Gaussian in form.

Such a distribution, valid in the low-density limit, is shown in figure 8.10. No semblance of a Fermi surface now remains, as curves B and C show.

Thus, the discontinuity, Δq say, in the momentum distribution $n(k)$ has gone to zero in the insulating phase. For the first-order Wigner transition, q goes to zero discontinuously at the metal–insulator transition.

For a first-order transition in jellium, one can in fact use the virial theorem (March 1958, see also Argyres 1967):

$$2K+V=-r_s\frac{\mathrm{d}E}{\mathrm{d}r_s}, \tag{8.69}$$

where K is the kinetic energy and V the potential energy, to obtain essentially the same results as those got by using q as an order parameter.

We shall therefore turn to the more fruitful consideration of the second-order metal–insulator transition which will occur for strong electron–electron correlations in narrow energy bands, when site interactions are emphasised rather than the long-range character.

8.8.2 Second-order transition for strong electron–electron correlations in a half-filled narrow energy band

The idea is to start out from a narrow energy band (or set of such bands). The electron–electron interactions are then introduced through an energy, C say, which it costs to bring two electrons with opposed spins on to the same site. We shall deal exclusively here with the case of one electron per atom, i.e. a half-filled band. The energy C is often called the Hubbard energy. Of course, the Pauli exclusion principle, or more precisely the Fermi hole discussed in §7.10, already keeps two electrons with parallel spins from coming close together.

March *et al* (1979) base their phenomenological treatment on the following assumptions:

(a) A suitable, generalised, order parameter, as discussed above, is the discontinuity, q say, at the Fermi surface in the metallic phase, of the single particle occupation number n_k.

(b) At a critical strength, C_0 say, of the Hubbard interaction C, a metal–insulator transition occurs, at which the discontinuity q is reduced to zero.

(c) The average number of doubly occupied sites, ν say, is a function of q, i.e. the interaction energy is $C\nu(q)$.

The ground-state energy $E(q)$ is now expanded about the singular point $q=0$ as

$$E(q)=E_0+E_1q+E_2q^2+\ldots. \tag{8.70}$$

The form of E_1 is assumed to be such that

$$E_1(C)=\alpha(C_0-C) \tag{8.71}$$

i.e. $E_1(C_0)=0$, $\alpha<0$, while $E_2(C_0)>0$.

The energy minimisation

$$\partial E/\partial q=0 \tag{8.72}$$

evidently yields

$$E_1+2E_2q+\ldots=0. \tag{8.73}$$

Near the metal–insulator transition, i.e. for very small q, we find

$$q = -\frac{E_1}{2E_2} = Q\left(1-\frac{C}{C_0}\right), \qquad Q = -\frac{\alpha C_0}{2E_2}. \tag{8.74}$$

Substituting q from equation (8.74) into (8.70), the minimum energy is found to be

$$E_{\min} = E_0 + (\alpha C_0 Q + E_2 Q^2)\left(1-\frac{C}{C_0}\right)^2. \tag{8.75}$$

Thus, the difference in energy between the metallic and the insulating state, ΔE say, is given by

$$\Delta E = (\alpha C_0 Q + E_2 Q^2)\left(1-\frac{C}{C_0}\right)^2. \tag{8.76}$$

The number of doubly occupied sites ν can be shown† in the Hubbard framework for describing the electron–electron interaction to be

$$\nu = \frac{\mathrm{d}(\Delta E)}{\mathrm{d}C} = -\frac{2}{C_0}(\alpha C_0 Q + E_2 Q^2)\left(1-\frac{C}{C_0}\right). \tag{8.77}$$

The phenomenological theory, with results exhibited in equations (8.74), (8.76) and (8.77), is equivalent to a variational calculation of Gutzwiller (1965, see also Brinkman and Rice 1970) as regards the dependences of q, ΔE and ν on $[1-(C/C_0)]$.

8.8.3 *Spin susceptibility near the metal–insulator transition*

As pointed out by Brinkman and Rice (1970) there is some experimental interest in connection with V_2O_3 in the way the electron spin susceptibility is enhanced by the interaction C in the metallic phase as the metal–insulator transition is approached.

The above phenomenological treatment can be generalised to allow for non-zero magnetisation m. If h is the applied magnetic field, we therefore write

$$E(m, q) = E_0(m) + am^2 - hm + E_1 q + E_2 q^2 + \dots + eqm^2. \tag{8.78}$$

Minimising with respect to m, the result is

$$2am + 2eqm = h \tag{8.79}$$

and the spin susceptibility is to be obtained from

$$m = \frac{1}{2[a(C)+eq]}h \equiv \chi h. \tag{8.80}$$

From equation (8.80), it can be seen that since $q \to 0$ as $C \to C_0$, the question of the enhancement of the susceptibility as the metal–insulator transition is approached rests on the behaviour of a as a function of interaction C. Provided

† Essentially from Feynman's theorem.

$a \to 0$ as $C \to C_0$ as $[1-(C/C_0)]$ or faster, then using equation (8.74) in (8.80) one obtains

$$\chi \propto \frac{1}{1-(C/C_0)} \tag{8.81}$$

which is the form given by Brinkman and Rice.

We conclude this discussion of the metal–insulator transition by mentioning two general references for the interested reader. The first is Mott (1974) and the second Friedman and Tunstall (1978), though this latter reference concentrates on the transition in disordered systems.

8.9 Structural phase transitions

The simplest theory of structural phase transitions is due to Landau (see Landau and Lifshitz 1969). It was applied in detail to ferroelectrics, independently, by Devonshire (1954). Briefly, in ferroelectric crystals one of the more interesting forms of phase transitions accompanied by a lattice distortion occurs. Here, the instability is in a $q = 0$ optic mode which has an electric dipole moment associated with it. The electric polarisation of such crystals increases dramatically as T_c is approached. $BaTiO_3$, having the so called perovskite crystal structure shown in figure 8.11, is a good example. In the distorted phase, the cations move relative to the anions. The Ti ion moves away from the centre of the tetrahedron of oxygen atoms. This distortion corresponds to a transverse optic (TO) mode. The phonon frequency can show

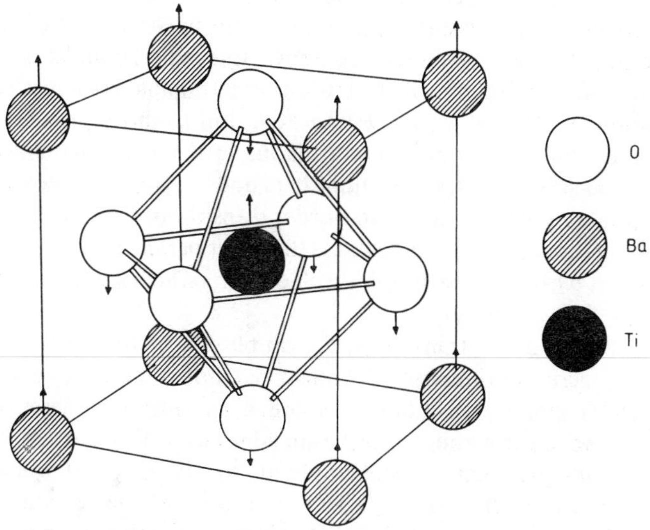

Figure 8.11 Shows perovskite structure possessed by $BaTiO_3$. Main distortion giving rise to ferroelectricity is also shown.

a temperature dependence of the form

$$\omega_{\mathrm{TO}}^2 = C(T - T_c) \tag{8.82}$$

and such a dependence is observed experimentally in $BaTiO_3$ by neutron scattering. We note that from the Lyddane–Sachs–Teller relation (4.28)

$$\epsilon(0) = \epsilon(\infty)\omega_{\mathrm{LO}}^2/\omega_{\mathrm{TO}}^2 \sim \Lambda/(T - T_c) \tag{8.83}$$

since only ω_{TO}^2 is expected to have a strong dependence on temperature. This dependence of $\epsilon(0)$ on T is typical of all ferroelectrics. Below a transition temperature T_c, there is spontaneous polarisation. Actually, most transitions are first order although $\epsilon(0)$ reaches a very high value and the discontinuity in the polarisation is small, as is the latent heat. However, not all such transitions can be discussed as above in terms of phonon modes for which a particular (soft) mode has a frequency which tends to zero at the critical temperature.

For example, in crystals like KH_2PO_4, the ferroelectricity is associated with the distribution of hydrogen atoms between anharmonic potential wells with double minima.

8.9.1 Precise form of order parameter

To return to the structural instabilities, let us consider a little more precisely the nature of the order parameter that one will require to describe structural phase transitions. The discussion below follows Cowley (1978). First, structural phase transitions occur when the symmetry of the crystal structure of a material changes, under the influence of a change in some external condition. For the example of $BaTiO_3$ discussed above, this material undergoes a phase transition from a cubic crystal structure to a tetragonal structure when it is cooled below 393 K. Below, we restrict attention to those phase transitions which are continuous (or nearly continuous in the sense discussed above). Thus, we treat second-order transitions here and the treatment below will not be appropriate to reconstructive first-order transitions, such as the BCC to FCC transition. In the continuous transitions the order parameter for the transition describes that pattern of the atomic displacements which is responsible for the lowering of the symmetry.

At continuous phase transitions, the amplitude of this order parameter decreases to zero continuously. At most, but not quite all (see Bruce and Cowley 1973) phase transitions, this order parameter is described by a particular wavevector within the Brillouin zone and will be written $P_j(\boldsymbol{q})$. The wavevector $\boldsymbol{q}$ in ferroelectrics, for example, as already remarked, will be $\boldsymbol{q} = \mathbf{O}$, while in the case of the crystal $SrTiO_3$ it turns out that $\boldsymbol{q}$ is at the zone boundary (111). The index $1 \leqslant j \leqslant n$ describes the number of components of the order parameter. In cubic ferroelectrics, n is 3, corresponding to polarisation along the x, y, or z crystal axes.

Working with the Landau theory, this takes the simplest possible case of the order parameter, $n = 1$ and $P_j(\boldsymbol{q}) = P$. It is assumed that the free energy of the crystal can be expanded in a power series in P:

$$F = AP^2 + BP^3 + DP^4. \tag{8.84}$$

Landau could then show that if the symmetry of the most symmetric phase is such that $B = 0$, a continuous phase transition will result at T_c if A is written as $a(T - T_c)$. For $T > T_c$, the average of the order parameter $\langle P \rangle = 0$ and the undistorted phase is the stable one. But for $T < T_c$, the minimum of the free energy corresponds to

$$\langle P \rangle^2 = (a/2D)(T_c - T) \tag{8.85}$$

and there is a distortion which increases as $(T_c - T)^{1/2}$.

As well as the distortion at T_c, the phase transition is accompanied by an anomaly in the susceptibility χ associated with the order parameter. Equation (8.85) leads to the result (cf equations (8.86) and (8.37))

$$\chi^{-1} = \begin{cases} a(T - T_c) & T > T_c \\ 2a(T_c - T) & T < T_c. \end{cases} \tag{8.86}$$

For ferroelectrics, where the order parameter is the dielectric polarisation, χ is the dielectric susceptibility which is frequently found to obey the Curie law given by Landau theory.

It should be added that the order parameter may also couple to other parameters, such as the macroscopic strain. The effect of these couplings can be incorporated into expansions of the type shown in equation (8.86).

8.9.2 Soft phonon modes

Cochran (1960) and also Anderson (1960) drew attention to the fact that if the leading term in equation (8.84) is compared with the expression of the free energy of a harmonic crystal, the implication is that an equation like (8.82), namely

$$[\omega(\boldsymbol{q})]^2 = K(T - T_c) \tag{8.87}$$

arises, where $\omega(\boldsymbol{q})$ denotes the frequency of one of the modes of vibration of the crystal with the wavevector corresponding to the distortion of the low-temperature phase.

This leads to the soft phonon mode referred to above. According to this concept, the structural phase transition arises from an instability of the crystal against a particular normal mode of vibration. In the low-temperature phase, the crystal distorts in order to stabilise this normal mode again.

Following the consequences of equation (8.87) being pointed out the prediction was confirmed in a number of different materials, one being the example of $SrTiO_3$ referred to briefly above, and this case is illustrated in figure 8.12.

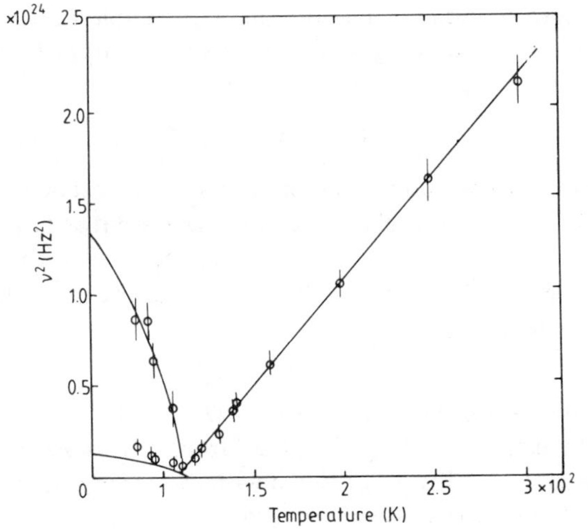

Figure 8.12 Showing soft mode in $SrTiO_3$. (From Cowley R A 1978 *Proc. Int. Conf. on Lattice Dynamics* (Paris: Flammarion Sciences) p. 625.) The square of the frequency above $T_c = 110$ K is approximately linear in temperature (cf equation (8.87)).

We note, with Cowley (1978) that equation (8.87) was obtained using harmonic or nearly harmonic lattice dynamics. This, of course, implies small vibrations about equilibrium and therefore equation (8.87) is applicable only to those crystals in which the amplitudes of motion of the atoms close to T_c are genuinely small.

8.9.3 Microscopic theory

Figure 8.12 can leave little doubt that the idea that structural phase transitions may arise from an instability against a normal mode of vibration is correct in essence. But it does not answer the question as to the physical mechanism for this temperature dependence.

It must be stressed that in different materials it can arise from very different mechanisms. In the one-dimensional conductors, to be discussed more fully in §8.12, it arises from the remarkable Kohn anomaly caused by the properties of the one-dimensional electron gas (cf §§2.7, 4.8). In the material Nb_3Sn, in contrast (see §8.11), it arises from the unusual electronic band structure, while in simple insulators the temperature dependence is due to the anharmonicity between the normal modes (cf Cowley 1965).

In these, and other cases, it is possible to develop theories which lead to equation (8.87) with varying degrees of sophistication. For the case of $SrTiO_3$ illustrated in figure 8.12, lowest-order anharmonic perturbation theory for the temperature dependence of the normal modes in a crystal (cf the machine computations shown in figures 3.14, 3.15) leads back to equation (8.87).

However, such an approach fails to explain why the particular zone boundary mode is unstable at 110 K, or why its temperature dependence is much larger and of opposite sign from that observed for the modes of the alkali halides.

It seems probable that the answers to these questions will not be found in qualitative treatments such as we have discussed above, but will require fully quantitative calculations on individual crystals.

8.10 Electron–phonon interaction and phase transitions

We turn now to discuss briefly the essentials of the theory of the electron–phonon interaction in the simple metals (i.e. with s and p conduction electrons) and then to go on to treat also the role of d electrons. This is, of course, with particular reference to phase transitions.

For the sp metals, the electron–phonon interaction can be usefully treated on the basis of two assumptions:

(a) The electron–ion interaction is weak. This can then be represented by some pseudopotential, which takes account of orthogonality of conduction band states to core states, and therefore describes weakened conduction electron–ion scattering.

(b) Nearly-free-electron screening, through, say the Lindhard dielectric function derived in Appendix 2.1, or better its refinements to include electron exchange and correlation (cf the reviews by Hedin and Lundqvist 1969 or March and Tosi 1976).

For transition metals and their compounds, the d electrons play a dominant role in the electron–phonon interaction. Here, one must emphasise the d electron screening by formulating the dielectric constant in terms of local orbitals; in lowest order from atomic orbitals for tightly bound electrons in solids.

There are several manifestations of the electron–phonon interaction in metals which are of considerable interest (cf Sham 1978). One, of long standing, is the temperature dependence of resistivity (cf Douglass 1976). The superconducting transition temperatures and the phonon frequency spectra are areas where the theory of the electron–phonon interaction can, of course, be subjected to quantitative tests. A dramatic consequence of the electron–phonon interaction is the occurrence of structural phase transitions. We shall consider below two examples:

(a) the so called A15 compounds (β tungsten structure), also important in superconductivity (see figure 8.13).
(b) the quasi-one dimensional conductors, the specific case considered being (cf the review by Sham 1978) potassium cyano-platinide (KCP).

The driving mechanisms in these transitions originate in the electron–phonon interaction; however impurities and defects can also play a substantial role.

8.11 Structural phase transition of A15 compounds

The class of intermetallic compounds, with chemical formula A_3B and in the crystallographic class A15, have a rich variety of properties, as surveyed by Testardi (1973) and also by Weger and Goldberg (1973). Here the specific examples considered are the microscopic theories of the structural transformations of Nb_3Sn and V_3Si from cubic to tetragonal structure on cooling. The A15 or β tungsten structure is shown in figure 8.13.

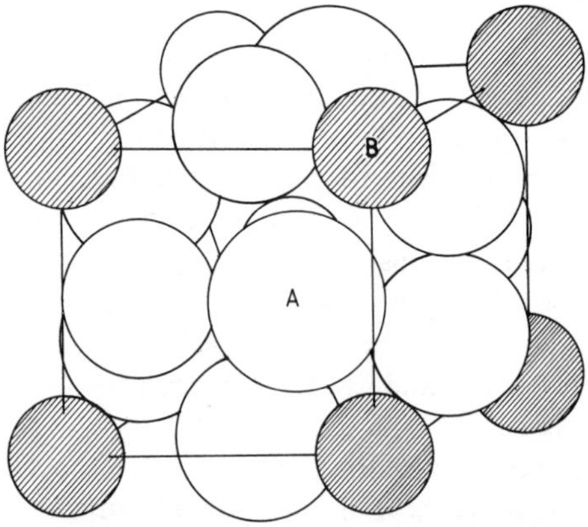

Figure 8.13 Shows A15 or β tungsten crystal structure for the compound A_3B. For the high T_c superconductors, A is a transition metal, usually V or Nb, and B is usually, but not always, a non-transitional metal (e.g. Sn).

The A15 compounds consist of mutually orthogonal and parallel linear chains of metal atoms in all three spatial directions. A given chain only comes close to the atoms of another chain at one point.

In the tetragonal phase of Nb_3Sn, there are relative displacements of atoms within a unit cell as shown in figure 8.14. The tetragonal distortion is caused by the coupling of the associated zero-wavevector optic mode to the long-wavelength elastic shear wave of the $c_{11}-c_{12}$ type (Sham 1971).

It is known that at temperatures above the structural transformation, the $c_{11}-c_{12}$ elastic shear modulus, or equivalently the transverse acoustic phonons with wavevector in the [110] direction, polarised in the $[1\bar{1}0]$ direction, soften. Labbé and Friedel (1966 a, b) proposed that this softening was driven by the electron–phonon interaction. They and Weger (1964) noted that in an A_3B compound with A15 structure, the nearest-neighbour A atoms form three sets of chains along the cubic crystal axes, as already remarked. The d electrons hopping along these chains will have the one-dimensional property (cf the phonon density of states in one dimension in problem 3.1) of high density of states near the band edges.

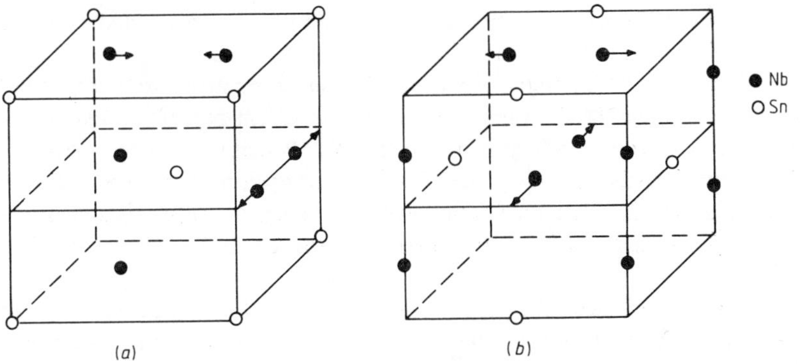

Figure 8.14 Illustrating the structural transformation in Nb_3Sn, by showing the Nb atom sublattice distortion. (a) Pm3n(O_h^3), $\Gamma_{12}(+)$; (b) P4_2/mmc (D_{4h}^9), origin shifted by ($\frac{1}{2}$, 0, 0).

Noolandi and Sham (1973), including the coupling between the relative ionic displacements and the elastic strain, and with a simple model for the d electrons, obtain:

(a) The result that from the equilibrium conditions, on cooling the cubic phase becomes tetragonal with a $c_{11}-c_{12}$ type strain and the type of displacements shown in figure 8.14(a), both of which have the same temperature dependence (cf Appendix 8.2 and problem 8.7 for quantitative details). This agrees with the experimental findings of Vieland *et al* (1971) and also of Shirane and Axe (1971).

(b) The elastic shear constant $c_{11}-c_{12}$ can be fitted over the whole measured temperature range in the cubic phase with values of the density of states and Fermi temperature quite close to those which fit the magnetic susceptibility (Rehwald *et al* 1972). The bare density of states is also consistent with that deduced from the measured electronic specific heat corrected for the electron–phonon contribution.

(c) From the parameters obtained by fitting the shear modulus, values of the frequencies of the relevant optical mode and for the deformation potential of the d band can be deduced.

These results show that the temperature dependence of the above properties can be explained quantitatively in terms of the one-dimensional model for the d electrons from the transition metal Nb in Nb_3Sn. The Fermi level at zero temperature is assumed to fall near the edge of a d band which has strong one-dimensional character and therefore a high density of states. The temperature dependence of the crystal properties comes from the temperature dependence of the electronic contribution from this d band which is no longer highly degenerate at finite temperatures. Some of these considerations are illustrated quantitatively in Appendix 8.3.

It has to be said here that because band structure calculations (Matthiess 1975) have not fully substantiated the one-dimensional character, some question must inevitably arise as to whether the Labbé–Friedel model is sufficiently realistic to describe the essence of the transition.

Gorkov (1973) has suggested that because of a particular Fermi surface property, a Peierls-like transition can occur. With regard to macroscopic properties, it appears difficult to distinguish between the Labbé–Friedel and Gorkov proposals. Both are based on the assumption of the one dimensionality of the bands and on the Fermi wavevector being close to a particular point (for the initiated, point X in the Gorkov treatment and Γ in the Labbé–Friedel scheme).

Work by Bhatt (see Sham 1978) has led to a possible way of reconciling the band structure of Matthiess (1975) with the ideas of Labbé and Friedel and of Gorkov. Bhatt found that in Matthiess' work, the x^2-y^2 orbitals make an important contribution to the density of states at the Fermi level. He therefore employed these orbitals to produce a set of three-dimensional energy bands. These, he showed, can cause phonon softening provided that the bandwidth due to interchain coupling is sufficiently small. He found that in his model both the Labbé–Friedel distortion and the Peierls distortion proposed by Gorkov occur. This means that in a three-dimensional narrow band, both the vicinities of Γ and X points contribute.

Although it is established beyond reasonable doubt that the driving mechanism for the lattice instabilities is the electron–phonon interaction, a complete theory of the structural phase transformation will eventually require:

(a) the inclusion of lattice anharmonicity (cf §3.8), which is known from experiment to be important (Testardi 1975);
(b) account of defects and impurities.

With regard to (b), Phillips *et al* (1976 and in Douglass 1976) have envisaged an important role for defects. The lattice instability driven by the electron–phonon interaction is, in their picture, released by strain around point defects and the structural phase transition consists of ordering of the strain fields around the defects.

It is worth adding here, not directly connected with the above discussion, that the topological transformation properties of the order parameters have important consequences in determining the stability of the defects that can be present in an ordered condensed phase. This promising area for future work is discussed by Toulouse (1977).

8.12 Structural phase transition of a quasi-one-dimensional conductor

A number of quasi-one-dimensional conductors show structural distortions at low temperatures. Tetrathiafulvalene–tetracyanoquinodimethane (TTF–TCNQ), for instance, undergoes a considerable variety of structural transformations. We follow Sham (1978) in focusing here on the example of KCP† in order to

† $K_2Pt(CN)_4X_{0.3}.3H_2O$ where X = Cl or Br.

illustrate a number of the more important ideas. (For a review of the physical properties of KCP, reference can be made to Renker and Comes (1975), Devreese *et al* (1979) and Berlinsky (1979).)

The bromine ions partially oxidise the platinum ions so that the d electrons along the platinum chains form a partially filled conduction band with the Fermi diameter $2k_F = 0.3\pi/c$ determined by the fraction of the oxidising agent, c being the Pt–Pt distance along the chain. A giant Kohn anomaly (cf §4.8) is observed (Renker *et al* 1973, Carneiro *et al* 1976) in the phonon spectrum at precisely the wavenumber $2k_F$, already at room temperature.

One-dimensional conductors are known to be unstable against distortions at low temperatures as discussed in §4.8. On cooling, KCP is known to show a tendency to develop a structural distortion. In elastic neutron scattering experiments (Renker *et al* 1974, Lynn *et al* 1975) a peak with a wavevector (π/a, π/a, $2k_F$), a denoting the lattice constant perpendicular to the chain direction, develops a high intensity on cooling to about 100 K. In mean field theory (see Appendix 8.1) there is a second-order phase transition at the temperature T_c where the phonon frequency at $2k_F$ vanishes and at low temperatures the excitation energy spectrum has a gap Δ of the same order as T_c. From infrared reflectivity data an estimate of Δ (Brüesch *et al* 1975) gives about 500 K. That T_c is much lower than 500 K indicates the suppression of T_c by the large fluctuation effects of the distortions along the individual chains (Scalapino *et al* 1972).

However, the distortions in KCP do not form truly one-dimensional systems, as indicated by the ordering at wavevector (π/a, π/a, $2k_F$), i.e. at the Brillouin zone edge, which can be understood in terms of the Coulomb interaction between chains, as proposed by Bjelis *et al* (1974). In each chain, the distortion creates a charge density wave† of electrons with wavenumber $2k_F$ along the chain direction. The Coulomb interactions between the charge density waves in nearest-neighbour chains are π out of phase. In the plane perpendicular to the chains, the charge distribution is like a two-dimensional NaCl structure.

The interchain interaction makes KCP a three-dimensional system and it is expected to undergo a structural phase transition (Dieterich 1974, Scalapino *et al* 1975) at a temperature lower than the mean-field temperature of about 500 K. If the transition occurred around 100 K, that would seem plausible. However, more detailed studies indicate that there is not a genuine phase transition (Renker *et al* 1974, Lynn *et al* 1975) around 100 K. The intensity of the peak at (π/a, π/a, $2k_F$), on cooling, increases drastically around 100 K but then saturates at low temperatures instead of approaching a critical behaviour. The coherence length in the direction perpendicular to the chains remains finite at the lowest temperatures examined (6 K) which means that no long-range order has developed. Thus the three-dimensional structural transition appears to be smeared out.

† For a review of the theory of charge density waves, see Berlinsky (1979).

Patton and Sham (1976) proposed that the disordered bromine ions (Williams *et al* 1974, Deiseroth and Schulz 1974) play a substantial role in accounting for the structural variation in KCP. They constructed a theory which incorporated:

(a) the electron–phonon interaction along each chain of the platinum ions, which produces the giant Kohn anomaly;
(b) the interchain Coulomb interaction, which suppresses the fluctuations in the chains; and
(c) the disordered impurities.

In a chain where the Kohn phonon is soft, an impurity can also induce a strong static Friedel oscillation (cf §2.7) of the electron density with wavenumber $2k_F$, which in turn induces a static lattice distortion of the same wavenumber where the lattice is soft. Averaging over the random distribution of impurities leads to a finite mean-square displacement of the atoms, which accounts for the presence of the central peak found experimentally even at room temperature (Renker *et al* 1974, Lynn *et al* 1975). The charge density oscillations induced by the impurity are sufficiently pronounced to destroy long-range coherence even at the lowest temperatures. The theory of Sham and Patton (1976) gives a fairly satisfactory account of the measured temperature dependence of the peak at $(\pi/a, \pi/a, 2k_F)$ and of the transverse coherence length.

8.13 Solitons as quasi-particles in non-linear treatments

In almost the whole of this book, the description of collective modes has been reduced to a linearised framework.

An area which is obviously going to grow in importance is that of non-linear equations, especially those which possess 'particle'- or 'pulse'-like running wave solutions. It is now recognised that there are types of non-linear equations which can be treated without linearisation and that the non-linear solutions can possess highly distinctive physical properties (see Bishop 1978). These are, of course, the more interesting when they cannot be regained by any finite-order perturbation approaches.

Non-linearity is an important aspect of critical phenomena at many levels. Below we can only indicate briefly (cf Bishop 1978) the physical role of solutions to non-linear equations of motion governing the order parameter in a small selection of problems. We emphasised critical exponents in §8.4; here we conclude by giving some attention to real-space configurations. Models now exist where strongly non-linear effects are important in energetic and dynamic considerations and can play an essential role in static and dynamic critical behaviour.

8.13.1 Example of sine–Gordon equation

To take an example which combines non-linearity and dispersion, but which nevertheless can support a spatially limited travelling wave that propagates without change of shape (a so called solitary wave), consider the so called sine–Gordon equation for a field $\psi(x, t)$:

$$\frac{\partial^2\psi}{\partial t^2} - c_0^2 \frac{\partial^2\psi}{\partial x^2} + \omega_0^2 \sin\psi = 0. \tag{8.88}$$

Clearly if the $\sin\psi$ term were absent, this would be a one-dimensional classical wave equation with characteristic velocity c_0. Evidently ω_0 is a characteristic frequency. This equation could, for instance, represent coupled pendulums with large amplitude.

A large-amplitude 'kink' or solitary wave solution of velocity V may be shown to be

$$\psi_{\pm}^{V}(x, t) = 4\tan^{-1}\{\exp[\pm(x - Vt)/d(V)]\} \tag{8.89}$$

where

$$d(V) = d\left(1 - \frac{V^2}{c_0^2}\right)^{1/2} \qquad d = \frac{c_0}{\omega_0}. \tag{8.90}$$

The width of this solitary wave solution is of the order of $2d(V)$.

The excess energy $E(V)$ associated with this solitary wave solution can also be obtained as (cf Berlinsky 1979, p 1275)

$$E(V) = E(0)\left(1 - \frac{V^2}{c_0^2}\right)^{-1/2} \qquad E(0) = 8Ac_0\omega_0 \tag{8.91}$$

where A determines the energy scale of the problem. Such large-amplitude modes are quite different from the approximate, small-amplitude solutions of the linearised equation obtained by replacing $\sin\psi$ by ψ, namely

$$\psi(x, t) \propto \exp[\mathrm{i}(kx - \omega_k t)] \tag{8.92}$$

where

$$\omega_k^2 = \omega_0^2 + c_0^2 k^2.$$

The linear normal modes of the above form, to be likened to the phonons, magnons etc. discussed in detail in earlier chapters, are approximately non-interacting, i.e. independent in $\boldsymbol{k}$-space whereas the 'kinks' are much more nearly independent in real configuration space.

The solitary waves $\psi_\pm$ which are solutions of the sine–Gordon equation are known to mathematicians as solitons (+ sign) and antisolitons (− sign).

Some very special Hamiltonians, including the sine–Gordon example above, are exactly soluble and the resultant non-linear equations of motion have solitons as solutions. The allowed excitations, notwithstanding the non-linearity, have infinite lifetime. In the sine–Gordon case there are modes corresponding to bound soliton–antisoliton pairs and these are of potential interest in a number of scientific applications.

The importance of all this is that there are in nature situations where it is sometimes essential to transcend linear normal modes. These, and perturbation expansions based on them, prove sufficient to treat, say, weak anharmonic effects in crystal vibrations at temperatures that are not too high. But they can fail completely to account for distinctive non-linear excitations. One practical example which can be cited is the demonstration of solitons in polyacetylene by Su *et al* (1980).

It is increasingly recognised that kinks and other non-linear excitations can represent quite different aspects of a problem from linear modes, and should be considered as 'quasi-particles' in their own right. This has led to attempts to construct, for example, a theory of type-II superconductors, incorporating all solutions, including non-linear ones, which in this example are vortices.

Another example is the large-area Josephson junction (see e.g. Bullough 1974). The sine–Gordon equation applies here, the two Josephson relations corresponding to equations (5.46) and (5.44) having the form

$$\partial\sigma/\partial t = 2eV/\hbar, \qquad j_z = j_{z0}\sin\sigma. \tag{8.93}$$

One can show that σ, which can be identified with the phase difference across the junction, satisfies a form of the sine–Gordon equation.

8.14 Dynamics in critical phenomena

In the previous section on solitons, we referred to the relevance to static and dynamic critical phenomena. In this connection we discussed static scaling in §8.4 and we conclude the book by briefly considering the dynamical generalisation.

The oldest theory of critical dynamics goes back to Van Hove (1954). Consider some variable, $Q(\boldsymbol{r}, t)$ say, which relaxes to its equilibrium value via a single, non-propagating mode. If Q is chosen such that it is conserved by the equations of motion of the system (that is $\int Q(\boldsymbol{r}, t)\,d\boldsymbol{r}$ is independent of time) then at any temperature $T \neq T_c$, the relaxation of Q in the long-wavelength limit $k \to 0$ is given by a diffusion equation. The characteristic frequency associated with the relaxation rate of Q can be written as

$$\bar{\omega} = \lambda_Q k^2/\chi_Q \tag{8.94}$$

where λ_Q is the transport coefficient for the variable Q while χ_Q is the appropriate susceptibility describing the response of the system to an applied field. In the case of entropy fluctuations, for example, λ_E is the thermal conductivity while χ_E is the heat capacity per unit volume at constant pressure, ρc_p, and hence

$$\bar{\omega}_E = \lambda_E k^2/\rho c_p. \tag{8.95}$$

If the variable Q, on the other hand, is not conserved, then the relaxation to equilibrium should take place at a finite rate even at $k = 0$ and hence

$$\bar{\omega} = \Lambda_Q/\chi_Q \tag{8.96}$$

where Λ_Q is the 'kinetic coefficient' for Q.

In Van Hove's theory, the assumption made is that the transport coefficients λ_Q and also the kinetic coefficients Λ_Q are finite and non-zero at the critical point.

If the variable Q is, in fact, the order parameter (say ψ) of the system, then χ_Q diverges near T_c as the inverse coherence length κ, to the power $2-\eta$:

$$\chi_Q \propto \kappa^{2-\eta} \tag{8.97}$$

where η is a small positive quantity. If the relaxation of ψ is described by a single non-propagating mode, then the Van Hove theory predicts a 'critical slowing down' of the order parameter fluctuations as $k \to 0$, near T_c with

$$\bar{\omega} \propto \begin{cases} \kappa^{2-\eta}k^2 & \text{if } \psi \text{ is conserved} \\ \kappa^{2-\eta} & \text{if } \psi \text{ is not conserved.} \end{cases} \tag{8.98}$$

Unfortunately, there is a variety of situations in which both first principles theory, and experiment, are not in agreement with the predictions of this simple theory.

If we focus on, say, the liquid–gas critical point, the order parameter is conserved (also true for, say, critical point for phase separation) and is dominated by a non-propagating diffusive mode at long wavelength, both above and below the transition. For instance, (compare §8.3) in the liquid–gas transition the order parameter may be taken as the density difference $\rho-\rho_c$ and long-wavelength density fluctuations near T_c are dominated by entropy fluctuations at constant pressure, whose relaxation is described by the thermal conductivity formula (8.95). Nevertheless, a mode–mode analysis (see e.g. Kawasaki 1970, Stanley 1971) shows that the conventional theory is not correct. It is also of interest to mention here the work of Swift and Kadanoff (1968) on the transport coefficients of ^{4}He near the λ point discussed in the previous chapter.

8.14.1 Universality

Let us briefly motivate the ensuing discussion by considering again the notion of universality. As with static critical phenomena, universality does not mean that all systems have the same critical exponents, but that there exist large classes of systems with the same critical behaviour. As an example, all Ising models (cf problem 6.3) of a given lattice dimensionality d, with short-range forces, are supposed to have the same static critical behaviour, this being independent of the details of the underlying lattice. In the language of the renormalisation group referred to very briefly in Appendix 8.3, details of the lattice are irrelevant variables, which disappear from the problem near T_c.

Such ideas applied now to dynamic critical phenomena clearly imply that most of the details of the dynamics are also irrelevant near T_c and that large classes of systems will show the same dynamic critical behaviour. It is known from various examples, however, that a class of systems showing the same static critical behaviour may be divided into several classes of dynamic behaviour. One known example is the classical isotropic Heisenberg ferromagnets and antiferromagnets (cf §6.3); they have identical equilibrium properties but very different time dependences, both in the hydrodynamic and in the critical regimes. It should be said, however, that the dynamical characteristics of the liquid–vapour critical phenomena may have somewhat complex features. For a discussion of these difficulties the reader may refer to Kadanoff (1970).

We come then, finally, to dynamic scaling. Following Halperin and Hohenberg (1967, see also Halperin 1973) we shall say that the correlation function $Q(\boldsymbol{k}, \omega)$ obeys dynamic scaling if:

(a) At $T = T_c$, for sufficiently small $\boldsymbol{k}$,

$$Q(\boldsymbol{k}, \omega) \sim 2\pi\bar{\omega}_k^{-1} Q(\boldsymbol{k}) f_Q(\omega/\bar{\omega}_{\boldsymbol{k}}) \tag{8.99}$$

$$\bar{\omega}_k = \text{constant} \times k^z$$

where $Q(\boldsymbol{k})$ is the Fourier transform of the time $t = 0$ form of Q, z is an as yet unspecified exponent and f_Q is a function normalised to unity and depending only on $\omega/\bar{\omega}_{\boldsymbol{k}}$. The dynamic scaling hypothesis supposes also that at T_c, f_Q is a well behaved function of its argument, with a characteristic width of order unity. Thus the spectrum of $Q(\boldsymbol{k}, \omega)$, at fixed $\boldsymbol{k}$, has a single relaxation frequency $\bar{\omega}_{\boldsymbol{k}}$.

(b) For temperatures near T_c, the dynamic scaling hypothesis requires that the function $Q(\boldsymbol{k}, \omega)$ has a form similar to equation (8.99) but now the functions depend in an essential way on the ratio k/κ, where the inverse coherence length κ goes to zero at T_c. (Along the coexistence curve or the 'critical isochore', κ varies as $(T - T_c)^\nu$ (cf equation (8.39)).) For example, away from the critical point, $\bar{\omega}_{\boldsymbol{k}}$ is assumed to take the form

$$\bar{\omega}_{\boldsymbol{k}} \propto k^z \Omega(k/\kappa) \tag{8.100}$$

where Ω depends on k only through the ratio k/κ. In the limit $k/\kappa \to 0$, the characteristic frequency will have the form

$$\bar{\omega} \sim k^z (k/\kappa)^x \tag{8.101}$$

where x is determined by macroscopic or hydrodynamic considerations. Usually x will be 0, 1 or 2, depending on the variable Q under consideration.

As Halperin (1973) discusses in detail, the exponent z depends not only on whether or not the order parameter ψ is conserved, but also on whether or not the energy is conserved. When ψ is not conserved, conventional theory would yield $z = 2 - \eta$, whereas when ψ is conserved $z = 4 - \eta$.

When neither ψ nor E are conserved, the renormalisation methods mentioned in Appendix 8.4 have been applied to some simple models. The exponent z in this case is found to be greater than the conventional results.

When ψ is conserved but not E, evidence points to the applicability of conventional theory. When ψ is not conserved but E is, the conventional theory again appears to fail, as discussed more fully by Halperin *et al* (1972, see also Halperin 1973).

Problems

8.1 Show that the relation between the liquid–gas critical values of p, V and T given by Dieterici's equation of state

$$p(V-b) = RT\exp(-a/RTV) \tag{P8.1}$$

is

$$\frac{p_c V_c}{RT_c} = 2e^{-2} = 0.271. \tag{P8.2}$$

(Note that the experimental values found for Ne, Ar, Kr and Xe are 0.30, 0.29, 0.29 and 0.28 respectively.)

8.2 In quantum theory, the Langevin theory must be transcended by considering a spin S as having energy levels which are Zeeman split in a magnetic field according to the Hamiltonian

$$H_Z = -g\mu_B \boldsymbol{S}\cdot\boldsymbol{H} = -g\mu_B H S_z \tag{P8.3}$$

when the field is taken in the z-direction. S_z can take the $(2S+1)$ values from S to $-S$.

Show that:

(a) The partition function Z_1 is given by

$$Z_1 = \frac{\sinh[(S+\frac{1}{2})x]}{\sinh(\frac{1}{2}x)} \qquad x = \frac{g\mu_B H}{k_B T}. \tag{P8.4}$$

(b) For N independent spins the partition function $Z_N = (Z_1)^N$.

Use the relation

$$M = -\left(\frac{\partial F}{\partial H}\right)_T \tag{P8.5}$$

between magnetic moment M and Helmholtz free energy F to show that

$$M = Ng\mu_B B_S(y) \qquad y = \frac{g\mu_B H}{2k_B T} \tag{P8.6}$$

where the Brillouin function $B_S(y)$ is given by

$$B_S(y) = (S+\tfrac{1}{2})\coth[(2S+1)y] - \tfrac{1}{2}\coth y. \qquad \text{(P8.7)}$$

Contrast the phase transition behaviour with the Weiss theory.

8.3 For a Coulomb system, the virial theorem is

$$2K + V = 3p\mathscr{V} \qquad \text{(P8.8)}$$

where K is the kinetic energy, V the potential energy, p the pressure and $\mathscr{V}$ the volume of the assembly.

The equilibrium condition for the fluid–crystal transition is

$$G_1 = G_2 \qquad p, T \text{ constant}$$

where G is the Gibbs free energy.

Show from this condition plus the virial theorem that the changes ΔK, ΔV and ΔS in K, V and entropy S across the transition are related by

$$T\Delta S = \tfrac{5}{3}\Delta K + \tfrac{4}{3}\Delta V. \qquad \text{(P8.9)}$$

For the classical case, K is simply $\tfrac{3}{2}k_B T$ per particle, and therefore $\Delta S = \tfrac{4}{3}\Delta V/T$. Use the virial theorem to show that

$$\Delta\mathscr{V} = \tfrac{1}{3}\Delta V/p. \qquad \text{(P8.10)}$$

Use the Clausius–Clapeyron equation to show that the melting curve in this classical regime is

$$p = AT^4. \qquad \text{(P8.11)}$$

8.4 Rework problem 8.3, under conditions of constant volume and constant T, in which case the condition of equilibrium between the fluid and crystal phases is that the Helmholtz free energies are equal, to show that the melting curve is

$$\mathscr{V} = \text{constant} \times T^{-3}. \qquad \text{(P8.12)}$$

Express this in terms of the density ρ, and use the result for the critical value of Γ found from the machine calculations reported in §8.6 to show that

$$T = \chi\rho^{1/3} \qquad \chi = (Ze)^2(\tfrac{4}{3}\pi)^{1/3}/158k_B, \qquad \text{(P8.13)}$$

Ze being the charge on the point ions.

8.5 For the same Coulomb system considered in problems 8.3 and 8.4, but in d dimensions, show that the virial theorem is

$$2K + (d-2)V = dp\mathscr{V}. \qquad \text{(P8.14)}$$

8.6 Connect the soft mode result in equation (8.87) with the free energy expansion of a harmonic crystal.

8.7 In the A15 (β tungsten) structure, discussed in §8.11 and in Appendix 8.2

show that the distortion d and the strain ϵ satisfy

$$\frac{\epsilon(T)}{\epsilon(0)} = \frac{d(T)}{d(0)} = \frac{\Delta(T)}{\Delta(0)} = \frac{v_1(T) - v_2(T)}{v_1(0) - v_2(0)}. \qquad \text{(P8.15)}$$

Hence show that d and ϵ have the same temperature dependence.

Use the theory in Appendix 8.2 to show that for the cubic phase, in which $v_\lambda = v$, the elastic constant difference $c_{11} - c_{12}$ is given by

$$c_{11} - c_{12} = V_0^{-1}\left[8A - \frac{\partial v}{\partial \mu} U^2 \left(1 - 6\frac{\sigma^2}{\Omega^2}\frac{\partial v}{\partial \mu}\right)^{-1}\right], \qquad \text{(P8.16)}$$

where V_0 is the volume of the unit cell.

8.8 Write the expansion of the equation of state of a fluid around its critical point in equation (8.30) in terms of volume V rather than density ρ. For suitable values of pressure p, assume the resulting equation in V has three real roots, $V_1 < V_2 < V_3$.

Derive the pressure, p_0 say, for any temperature $T < T_c$ at which condensation and evaporation take place at this temperature, by using the 'rule of equal areas' in the form

$$\int_{V_1}^{V_3} p \, dV = p_0 (V_3 - V_1) \qquad \text{(P8.17)}$$

the roots V_1 and V_3 being then the volumes V_l and V_g of the liquid and vapour phases, respectively, at this temperature. Hence show that if one uses the lowest-order approximation

$$p_0 = p_c + \text{constant } (T_c - T) \qquad \text{(P8.18)}$$

then the roots for V are

$$\begin{aligned} V_1 &= V_c - \text{constant } (T_c - T)^{1/2} \\ V_2 &= V_c \\ V_3 &= \bar{V}_c + \text{constant } (T_c - T)^{1/2} \end{aligned} \qquad \text{(P8.19)}$$

where the two constants in equation (P8.19) are identical. Derive these (equal) constants using Dieterici's equation of state (P8.1).

Appendix 2.1

SCREENING AND DIELECTRIC FUNCTION OF A DEGENERATE UNIFORM ELECTRON GAS

Our purpose here is to give a simple treatment of the screening of a positive test charge, Ze say, introduced into a degenerate electron gas which initially has a uniform density n_0.

First, consider the gas in its initial uniform state. Electrons occupy states out to a maximum momentum p_F. If the volume of the gas is Ω, then the volume of occupied phase space is the product of $\frac{4}{3}\pi p_F^3$, the volume of occupied momentum space, and Ω. But, as a consequence of the Heisenberg uncertainty principle, a cell in phase space of volume h^3 can contain a maximum of two electrons, provided their spins are opposed. At absolute zero therefore we can write, for N electrons in volume Ω

$$N = 2\frac{\frac{4}{3}\pi p_F^3 \Omega}{h^3}, \tag{A2.1.1}$$

the factor of two coming from doubly occupied states by electrons with opposite spins. Hence $n_0 = N/\Omega$ is related to the maximum momentum $p_F = m_e v_F$, where v_F is the Fermi velocity:

$$n_0 = \frac{8\pi}{3h^3} p_F^3. \tag{A2.1.2}$$

A.2.1.1 Semiclassical theory of screening

When we introduce the test charge Ze at the origin of coordinates, the uniform density n_0 changes to $n(\boldsymbol{r})$ and a self-consistent field is produced, described by a potential energy $V(\boldsymbol{r})$. Then, provided we assume that $V(\boldsymbol{r})$ is slowly varying in space, we can apply the free-electron relation (A2.1.2) locally to obtain

$$n(\boldsymbol{r}) = \frac{8\pi}{3h^3} p_F^3(\boldsymbol{r}) \tag{A2.1.3}$$

where clearly $p_F(\boldsymbol{r})$ is the maximum momentum of electrons around position $\boldsymbol{r}$. But now we can write the classical energy equation for the fastest electron as

$$E_F = \frac{p_F^2(\boldsymbol{r})}{2m_e} + V(\boldsymbol{r}) \tag{A2.1.4}$$

where the condition of equilibrium is that the maximum or Fermi energy E_F

must not depend on $\boldsymbol{r}$, for otherwise electrons could redistribute in space to lower the energy. Substituting for $p_F(\boldsymbol{r})$ from equation (A2.1.4) into equation (A2.1.3) yields

$$n(\boldsymbol{r}) = \frac{8\pi}{3h^3}(2m_e)^{3/2}[E_F - V(\boldsymbol{r})]^{3/2} \tag{A2.1.5}$$

provided $E_F - V(\boldsymbol{r}) \geqslant 0$, or otherwise $n(\boldsymbol{r}) = 0$, which means that electrons are not allowed in the classically forbidden region. This result (A2.1.5) is the Thomas–Fermi density–potential relation. If we assume that Ze is a test charge, and linearise equation (A2.1.5) we obtain

$$-e^2(n(\boldsymbol{r}) - n_0) = \frac{q^2 V}{4\pi} \qquad q^2 = \frac{4k_F}{\pi a_0} \tag{A2.1.6}$$

where $\hbar k_F = p_F$, i.e. k_F is the Fermi wavenumber, while a_0 is the Bohr radius $\hbar^2/m_e e^2$. Combining the expression (A2.1.6) for the charge displaced by the test charge with the requirement of self-consistency embodied in Poisson's equation, yields

$$\nabla^2 V = q^2 V, \tag{A2.1.7}$$

which was first given by Mott (1936). Obviously the solution of equation (A2.1.7), subject to the conditions that

$$V(\boldsymbol{r}) \to \begin{cases} -Ze^2/r & \text{as } r \to 0 \\ 0 \text{ faster than } r^{-1} & \text{as } r \to \infty \end{cases} \tag{A2.1.8}$$

yields

$$V(\boldsymbol{r}) = -\frac{Ze^2}{r}\exp(-qr). \tag{A2.1.9}$$

For a good metal such as Cu, we can readily estimate the screening length q^{-1} from equation (A2.1.6) and we obtain the value of about 1 Å referred to in the main text. Equation (A2.1.9) is evidently the basis for writing the screened electron–electron interaction in the approximate form (2.13), with $l = q^{-1}$.

A2.1.2 Wave theory of screening

The above argument neglects the wave character of the electrons. But the Friedel oscillations discussed in §2.7 have wavelength π/k_F and it is essential to incorporate into the theory the de Broglie wavelength for an electron at the Fermi surface. Thus, diffraction of the electron waves by the test charge must be treated. One solves the Schrödinger equation for the perturbed wavefunction $\psi_{\boldsymbol{k}}(\boldsymbol{r})$ derived from the unperturbed state $\exp(\mathrm{i}\boldsymbol{k}\cdot\boldsymbol{r})$ by switching on the (weak) potential energy $V(\boldsymbol{r})$. Then a straightforward calculation by first-order perturbation theory (Born approximation: cf March 1968) yields ($\hbar = m_e = 1$)

$$n(\boldsymbol{r}) = \sum_{|\boldsymbol{k}|<k_F} \psi_{\boldsymbol{k}}^* \psi_{\boldsymbol{k}} = n_0 - \frac{k_F^2}{2\pi^3} \int \frac{j_1(2k_F|\boldsymbol{r}-\boldsymbol{r}'|)}{|\boldsymbol{r}-\boldsymbol{r}'|^2} V(\boldsymbol{r}')\,\mathrm{d}\boldsymbol{r}', \quad \text{(A2.1.10)}$$

where the spherical Bessel function $j_1(x) = (\sin x - x\cos x)/x^2$ is a typical wave factor representing the diffraction process. Combining equation (A2.1.10) with Poisson's equation yields the desired wave formulation transcending equation (A2.1.7) as (March and Murray 1960)

$$\nabla^2 V = \frac{2k_F^2}{\pi^2} \int \frac{j_1(2k_F|\boldsymbol{r}-\boldsymbol{r}'|)}{|\boldsymbol{r}-\boldsymbol{r}'|^2} V(\boldsymbol{r}')\,\mathrm{d}\boldsymbol{r}'. \quad \text{(A2.1.11)}$$

While $n - n_0$ in equation (A2.1.10) would reduce to equation (A2.1.6) if $V(\boldsymbol{r}')$ was sufficiently slowly varying to be replaced by $V(\boldsymbol{r})$, the solution of equation (A2.1.11) subject to the boundary conditions (A2.1.8) shows that, in fact, both $V(\boldsymbol{r})$ and $n - n_0 \sim \cos(2k_F r)/r^3$ at large r. This then is the origin of the Friedel oscillations.

This asymptotic form for the density would, in fact, be recovered from (A2.1.10) if we took $V(\boldsymbol{r})$ as the zero-range delta function potential $\lambda\delta(\boldsymbol{r})$.

A2.1.3 Dielectric function

If we take the Fourier transform $V(\boldsymbol{k})$ of the screened potential $V(\boldsymbol{r})$ of the semiclassical theory we obtain from equation (A2.1.9) the result (cf equation (2.16))

$$V(\boldsymbol{k})_{\text{semiclassical}} = -\frac{4\pi Z e^2}{k^2 + q^2}. \quad \text{(A2.1.12)}$$

Introducing the $\boldsymbol{k}$-dependent dielectric function $\epsilon(\boldsymbol{k})$ through (cf equation (2.15))

$$V(\boldsymbol{k}) = -\frac{4\pi Z e^2}{k^2 \epsilon(\boldsymbol{k})}, \quad \text{(A2.1.13)}$$

which is a natural way to screen the bare Coulomb potential $-4\pi Ze^2/k^2$ in $\boldsymbol{k}$-space, we see immediately from equation (A2.1.12) that $\epsilon(\boldsymbol{k})$ is given by equation (2.17). This is a correct result in the long-wavelength limit $k \to 0$ but fails for larger k. One must then solve the first-order equation (A2.1.11) of the wave theory of screening when one finds

$$\epsilon(\boldsymbol{k}) = \left[k^2 + \frac{k_F}{\pi} g\left(\frac{k}{2k_F}\right)\right] k^{-2} \quad \text{(A2.1.14)}$$

where

$$g(x) = 2 + \frac{x^2 - 1}{x} \ln\left|\frac{1-x}{1+x}\right|.$$

This result was obtained by Lindhard (1954).

Appendix 2.2

HYDRODYNAMIC MODEL OF BULK AND SURFACE PLASMA MODES

This appendix is, fundamentally, based on the work of Bloch (1933, 1934) who essentially generalised the static density theory of Thomas (1926) and Fermi (1928) to apply to the dynamics of a continuous electron fluid. However, the detailed presentation given below follows closely that of Barton (1979) rather than that in the original papers of Bloch.

The so called hydrodynamic model of Bloch leads naturally to the bulk and surface plasmons discussed in the main text, since the normal modes of the model subdivide into bulk modes, whose amplitudes extend throughout the whole electron fluid and surface modes, with amplitudes which decay away exponentially with distance from the surface. In this picture, the hydrodynamic pressure of the fluid is responsible for the spatial dispersion; i.e. for the variation of the mode frequencies with wavenumber.

A2.2.1 Formulation of hydrodynamic model

It is assumed that the fluid has mass density $m_e(n+\Delta n)$, charge density $e(n+\Delta n)$ plus an immobile uniform overall neutralising background of charge density $-en$, both extending through the half-space $z \leqslant 0$. All equations will be made linear in the deviation Δn from the equilibrium value n, or equivalently in the displacement $\xi(\boldsymbol{r})$ of the fluid from equilibrium (cf equation (2.7) of the main text):

$$\Delta n = -n \operatorname{div} \xi \qquad \text{(A2.2.1)}$$

The deviation Δp of the hydrodynamic pressure from its equilibrium value is

$$\Delta p = -n m_e \beta^2 \operatorname{div} \xi \qquad \text{(A2.2.2)}$$

where β would be the sound velocity if the medium were neutral. It is treated here as a parameter of the model, responsible for dispersion (see §2.3 of main text).

As in the discussion of the main text, we neglect retardation. The electric field inside the plasma is governed by Poisson's equation

$$\mathscr{E} = -\operatorname{grad} \Phi, \qquad \nabla^2 \Phi = 4\pi n e \operatorname{div} \xi. \qquad \text{(A2.2.3)}$$

The force per unit volume is $(ne\mathscr{E} - \operatorname{grad} \Delta p)$ and for normal modes having time dependence $\exp(-i\Omega t)$ the equation of motion is

$$-\Omega^2 m_e \xi = -\operatorname{grad}(e\Phi - m_e \beta^2 \operatorname{div} \xi). \tag{A2.2.4}$$

Whereas in Bloch's theory it was traditional to work with a velocity potential (see e.g. March 1975), Barton prefers to use an equivalent procedure in terms of a displacement potential Ψ defined by

$$\xi = -\operatorname{grad} \Psi. \tag{A2.2.5}$$

We follow him here in choosing to work with this potential Ψ as the basic variable. Substituting equation (A2.2.5) into equation (A2.2.4) one can immediately integrate once (the integration constant having no physical significance) to obtain

$$\Phi = -(m_e/e)(\Omega^2 + \beta^2 \nabla^2)\Psi \qquad z \leqslant 0. \tag{A2.2.6}$$

Operating on both sides of this equation with ∇^2 and using equation (A2.2.3) one finds the basic differential equation for the plasma normal modes:

$$\nabla^2(\Omega^2 - \omega_p^2 + \beta^2 \nabla^2)\Psi = 0 \tag{A2.2.7}$$

where ω_p is the plasma frequency given by equation (2.12). Outside the plasma (i.e. for $z > 0$), Φ obeys Laplace's equation (2.39).

A2.2.2 Specification of boundary conditions

At infinity, we must insist that any exponential increase is ruled out. At the surface of the plasma, the only condition which is natural in the hydrodynamic model is that the normal component of the displacement vanishes there, and thus we take

$$\xi_z = 0 = \partial\Psi/\partial z \qquad \text{at } z = 0^-. \tag{A2.2.8}$$

This is readily shown from equations (A2.2.3) and (A2.2.4) to be equivalent to

$$\frac{\partial}{\partial z}(e\Phi + m_e \beta^2 \nabla^2 \Psi) = 0 \qquad \text{at } z = 0^-. \tag{A2.2.9}$$

These conditions allow the volume charge density to remain non-zero at the surface, but because of the finite hydrodynamic pressure no true (singular) surface charge density can arise (in contrast to the more primitive model used to derive the surface plasma frequency in §2.6.2 of the main text). Hence the conditions on Φ are that Φ and its derivative $\partial\Phi/\partial z$ are continuous across the plane $z = 0$.

It should be stressed that although equation (A2.2.7) is a local differential equation for Ψ alone, it cannot be solved directly as such, for lack of sufficient local boundary conditions involving only Ψ and a finite number of its derivatives. But enough non-local boundary conditions on Ψ do result if $\Phi(\boldsymbol{r})$ in equation (A2.2.9) is expressed by Coulomb's law as $\int \mathrm{d}\boldsymbol{r}' |\boldsymbol{r}-\boldsymbol{r}'|^{-1} ne\nabla'^2 \Psi(\boldsymbol{r}')$.

A2.2.3 Translationally invariant solutions for displacement potential

Writing $\boldsymbol{r} = (\boldsymbol{X}, z)$ with $\boldsymbol{X}$ a vector parallel to the surface as in §2.6.1 (and $\boldsymbol{k}$ likewise), the condition of translational invariance parallel to the surface (cf equation (2.40)) allows the individual normal modes to be written

$$\Psi = \exp(\mathrm{i}\boldsymbol{k}\cdot\boldsymbol{X})\psi(z), \qquad \Phi = \exp(\mathrm{i}\boldsymbol{k}\cdot\boldsymbol{X})\phi(z). \tag{A2.2.10}$$

Inside the plasma, equations (A2.2.6) and (A2.2.7) then become respectively

$$(-e/m_\mathrm{e})\phi = (\Omega^2 - \beta^2 k^2 + \beta^2 \mathrm{d}^2/\mathrm{d}z^2)\psi \qquad z \leqslant 0 \tag{A2.2.11}$$

and

$$(-k^2 + \mathrm{d}^2/\mathrm{d}z^2)(\Omega^2 - \omega_\mathrm{p}^2 + \beta^2 k^2 - \beta^2 \mathrm{d}^2/\mathrm{d}z^2)\psi = 0 \qquad z \leqslant 0. \tag{A2.2.12}$$

Outside the plasma, we have from equation (2.39)

$$\phi = \text{constant} \times \exp(-kz) \qquad z > 0. \tag{A2.2.13}$$

A2.2.4 Energy density in the hydrodynamic model

Following Bloch, the energy H is the sum of the kinetic energy, with density such that

$$\text{kinetic energy density} = \tfrac{1}{2} m_\mathrm{e} n \dot{\xi}^2, \tag{A2.2.14}$$

hydrodynamic compressional energy, with density

$$\text{compressional energy density} = \tfrac{1}{2} m_\mathrm{e} n \beta^2 (\operatorname{div} \xi)^2 \tag{A2.2.15}$$

and electrical energy, expressible through a density given by

$$\text{electrical energy density} = -\tfrac{1}{2} ne\Phi \operatorname{div} \xi. \tag{A2.2.16}$$

Thus we have

$$H = \int_{-\infty}^{\infty} \mathrm{d}^2 X \int_{-\infty}^{0} \mathrm{d}z \left[\tfrac{1}{2} m_\mathrm{e} n (\nabla\dot{\Psi})^2 + \tfrac{1}{2} m_\mathrm{e} n \beta^2 (\nabla^2 \Psi)^2 + \tfrac{1}{2} ne\Phi \nabla^2 \Psi\right]. \tag{A2.2.17}$$

If we wish to express the energy H in terms of the displacement potential alone, then we readily obtain the non-local form

$$H = \int d\boldsymbol{r}[\tfrac{1}{2}nm_e(\nabla\Psi)^2 + \tfrac{1}{2}m_e n\beta^2(\nabla^2\Psi)^2]$$

$$+\tfrac{1}{2}\int\int d\boldsymbol{r}\, d\boldsymbol{r}' \frac{n^2e^2}{|\boldsymbol{r}-\boldsymbol{r}'|}\nabla^2\Psi(\boldsymbol{r})\nabla'^2\Psi(\boldsymbol{r}'). \qquad \text{(A2.2.18)}$$

where we have not indicated integration limits explicitly, and volume integrals over the interior, like those in equation (A2.2.17) are denoted by $\int d\boldsymbol{r}$.

A2.2.5 Dispersion relation for surface modes

As an application of Bloch's hydrodynamic model, we shall derive the dispersion relation for surface plasmon modes referred to in §2.6.6 of the main text. Ruling out from the boundary conditions stated above an exponential increase as $z \to \infty$, the only solution of equations (A2.2.12) and (A2.2.8) when $\Omega^2 < (\omega_p^2 + \beta^2k^2)$ is

$$\psi_k(z) = N_k[p_s \exp(kz) - k\exp(p_s z)] \qquad \text{(A2.2.19)}$$

where N_k is a normalisation constant. After setting $\Omega^2 = \Omega_s^2(k)$, p_s is defined by

$$\Omega_s^2(k) = \omega_p^2 + \beta^2k^2 - \beta^2p_s^2. \qquad \text{(A2.2.20)}$$

The eigenvalue Ω_s is determined, via p_s, by using equation (A2.2.19) together with continuity equations for Φ and $d\Phi/dz$ at $z = 0$, (A2.2.11) and (A2.2.13) for ϕ. One then obtains in a straightforward manner

$$2\Omega_s^2 p_s - \omega_p^2(p_s + k) = 0 \qquad \text{(A2.2.21)}$$

and eventually

$$\Omega_s^2(k) = \tfrac{1}{2}[\omega_p^2 + \beta^2k^2 + \beta k(2\omega_p^2 + \beta^2k^2)^{1/2}] \qquad \text{(A2.2.22)}$$

$$\beta p_s = \tfrac{1}{2}[-\beta k + (2\omega_p^2 + \beta^2k^2)^{1/2}] \qquad \text{(A2.2.23)}$$

and

$$\phi_k = -N_k(m_e/e)\exp(-kz)(\Omega_s^2 p_s - \omega_p^2 k) \qquad z \geqslant 0. \qquad \text{(A2.2.24)}$$

For $z \leqslant 0$, ϕ_k is given by equation (A2.2.11). Notice that:

(a)
$$\Omega_s(k=0) = \omega_p/\sqrt{2}. \qquad \text{(A2.2.25)}$$

(b) Ω_s has a term linear in k for small k, in contrast to equation (2.21) for the bulk mode. (The present treatment can in fact be used in the bulk case to estimate α in equation (2.21), with the result, $\frac{3}{5}v_F^2$, quoted in equation (2.22), as discussed, for example, by March and Tosi (1972).)

(c) Ω_s increases monotonically with k.

The dispersion relation (A2.2.22) for surface plasmons is implicit in the discussion by Ritchie (1963) for a finite slab (see also Sturm 1968, Ritchie and Wilems 1969).

The quantum mechanical generalisation of the above semiclassical theory due to Bloch has been given by March and Tosi (1972) (see in addition Mukhopadhyay and Lundqvist (1975), who also derive α in equation (2.21) as $\frac{3}{5}v_F^2$ as an example of the use of this density oscillation theory).

Appendix 2.3

ENERGY LOSS OF FAST ELECTRONS BY PLASMON EXCITATION

In order to calculate the energy dissipated by a fast charge traversing a metallic medium, we shall have recourse to the following formula of classical electrodynamics:

$$P = \frac{1}{4\pi}\mathscr{E}\cdot\frac{\partial \boldsymbol{D}}{\partial t} \tag{A2.3.1}$$

which relates the power dissipated in unit volume to the electric and displacement fields. In the present case, $\boldsymbol{D}$ is the field due to the external charge e and is given by

$$\boldsymbol{D} = -\nabla\frac{e}{|\boldsymbol{r}-\boldsymbol{v}t|}. \tag{A2.3.2}$$

Here $\boldsymbol{v}$ is the velocity vector of the charge, and this is assumed to be so large that all recoil effects can be neglected. The next step is to Fourier analyse $\boldsymbol{D}$ in the form

$$\boldsymbol{D}(\boldsymbol{r}, t) = \int\frac{\mathrm{d}\boldsymbol{k}\,\mathrm{d}\omega}{(2\pi)^4}\exp(\mathrm{i}\boldsymbol{k}\cdot\boldsymbol{r}-\omega t)\boldsymbol{D}(\boldsymbol{k}, \omega) \tag{A2.3.3}$$

from which, using equation (A2.3.2), one obtains the explicit result

$$\boldsymbol{D}(\boldsymbol{k}, \omega) = -\mathrm{i}\boldsymbol{k}\frac{4\pi e}{\boldsymbol{k}^2}\delta(\omega-\boldsymbol{k}\cdot\boldsymbol{v}). \tag{A2.3.4}$$

Each Fourier component is related to the corresponding component of the internal field by

$$\mathscr{E}(\boldsymbol{k}, \omega) = \boldsymbol{D}(\boldsymbol{k}, \omega)/\epsilon(\boldsymbol{k}, \omega) \tag{A2.3.5}$$

where $\epsilon(\boldsymbol{k}, \omega)$ is the frequency- and wavenumber-dependent dielectric function of the metallic medium, which is assumed to be uniform and isotropic.

The corresponding contribution to the power dissipation can now be calculated using equation (A2.3.1) as

$$\begin{aligned} P(\boldsymbol{k}, \omega) &= \frac{1}{4\pi}\{\mathrm{Re}[\mathscr{E}(\boldsymbol{k}, \omega)\mathrm{e}^{-\mathrm{i}\omega t}]\,\mathrm{Re}[-\mathrm{i}\omega\boldsymbol{D}(\boldsymbol{k}, \omega)\mathrm{e}^{-\mathrm{i}\omega t}]\} \\ &= \frac{1}{4\pi}\boldsymbol{D}(\boldsymbol{k}, \omega)\left[\mathrm{Re}\left(\frac{1}{\epsilon}\cos(\omega t)\right)+\mathrm{Im}\left(\frac{1}{\epsilon}\sin(\omega t)\right)\right] \\ &\quad\times\boldsymbol{D}(\boldsymbol{k}, \omega)[-\omega\sin(\omega t)] \end{aligned} \tag{A2.3.6}$$

and averaging over a cycle one finds

$$P(\boldsymbol{k}, \omega) = -\frac{\omega}{8\pi}\operatorname{Im}\left(\frac{1}{\epsilon(\boldsymbol{k}, \omega)}\right)[D(\boldsymbol{k}, \omega)]^2 \tag{A2.3.7}$$

which is the desired result. Clearly there will be a peak in the power dissipation whenever collective modes are excited which is equivalent to saying that $\epsilon(\boldsymbol{k}, \omega) = 0$. This last equation gives the condition for plasmon excitation, since it follows from equation (A2.3.5) that this latter equation can be satisfied then for $\boldsymbol{D}(\boldsymbol{k}, \omega) = 0$, i.e. for no external charges. It follows that one can then have $\mathscr{E}(\boldsymbol{k}, \omega) \neq 0$ even in the absence of external charges. This condition is indeed realised when a plasma oscillation is set up in the metallic medium.

For an $\boldsymbol{r}$-space formulation of the energy loss problem (discussed above in $\boldsymbol{k}$-space) generalised to include a weak lattice potential, the reader is referred to the work of March and Tosi (1973).

Appendix 3.1

WAVES IN AN ELASTIC CONTINUUM

For wavelengths more than an order of magnitude greater than interatomic spacings, a solid behaves as an elastic continuum. It is therefore of interest as a limiting case of microscopic phonon theory to discuss wave propagation in an anisotropic elastic medium.

We define the six independent strain components e_{xx}, e_{yy}, e_{zz}, e_{xy}, e_{xz} and e_{yz} for Cartesian axes x, y, z. Take a rectangular block of the solid; e_{xx} is the fractional change in the length of the block in the x-direction due to stress; e_{yy} and e_{zz} being similarly defined. e_{xy} is the shear angle in the (x, y) plane, and similarly for e_{xz} and e_{yz}. If the strains vary from point to point the block must be taken as infinitesimal in size.

Next, define the six independent stress components X_x, Y_y, Z_z, Y_z, Z_x and X_y. These quantities are forces per unit area, the capital letters denote their directions and the subscripts indicate the normals to the faces of the block at which they are applied. To maintain equilibrium, it is assumed that equal forces act on opposite sides of the block. It will be noted that we have not included Z_y, X_z and Y_x; this is because we are not interested in forces producing pure rotations of the body. Therefore only forces for which the total torque is zero are considered and then (see figure A3.1) it is clear that $X_y = Y_x$, etc.

As we are treating an elastic medium the strains must be linear functions of the stresses so that, for instance

$$e_{xx} = s_{11}X_x + s_{12}Y_y + s_{13}Z_z + s_{14}Y_z + s_{15}Z_x + s_{16}X_y \qquad \text{(A3.1.1)}$$

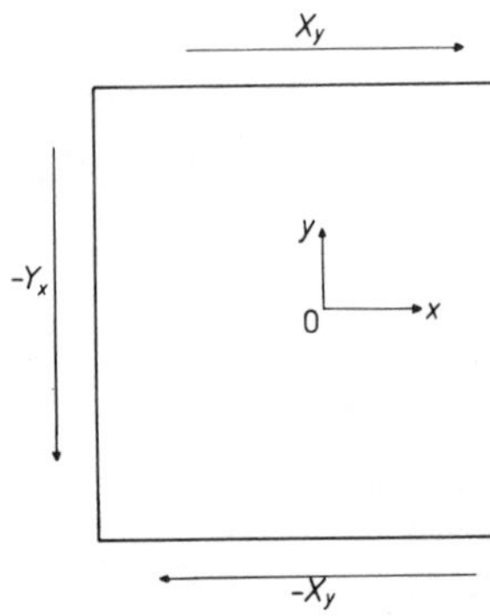

Figure A3.1 Demonstrates that forces X_y and Y_x must be equal for no rotation of the block to occur.

where the s's are constants. The matrix equation

$$\begin{pmatrix} e_{xx} \\ e_{yy} \\ e_{zz} \\ e_{yz} \\ e_{zx} \\ e_{xy} \end{pmatrix} = (s_{ij}) \begin{pmatrix} X_x \\ Y_y \\ Z_z \\ Y_z \\ Z_x \\ X_y \end{pmatrix} \tag{A3.1.2}$$

summarises these relations, the s_{ij}'s being the elastic compliance constants. If the inverse matrix of (s_{ij}) is denoted by (c_{ij}) then

$$\begin{pmatrix} X_x \\ Y_y \\ Z_z \\ Y_z \\ Z_x \\ X_y \end{pmatrix} = (c_{ij}) \begin{pmatrix} e_{xx} \\ e_{yy} \\ e_{zz} \\ e_{yz} \\ e_{zx} \\ e_{xy} \end{pmatrix}. \tag{A3.1.3}$$

The 36 elastic moduli c_{ij} can be reduced by considering the energy change δU per unit volume under infinitesimal strains of the block of material, which will, for convenience, now be taken as a cube of side L. Consider the work done under infinitesimal shear δe_{xy}. Although other forces might be required to produce this shear, it is clear that only the force $X_y L^2$ contributes to the first-order energy change. In figure A3.2 it can be seen that the upper (x, y) face of the block moves through a horizontal distance $L\delta\theta$, with energy change $X_y L^3\,\delta\theta$. There is also an energy change $X_y L^3\,\delta\phi$. Since $\delta e_{xy} = \delta\theta + \delta\phi$, it follows that the energy change per unit volume is $X_y \delta e_{xy}$. The energies involved in other infinitesimal deformations can be calculated in a similar manner and one finds for the energy change per unit volume

$$\delta U = X_x \delta e_{xx} + Y_y \delta e_{yy} + Z_z \delta e_{zz} + Y_z \delta e_{yz} + Z_x \delta e_{xz} + X_y \delta e_{yx}. \tag{A3.1.4}$$

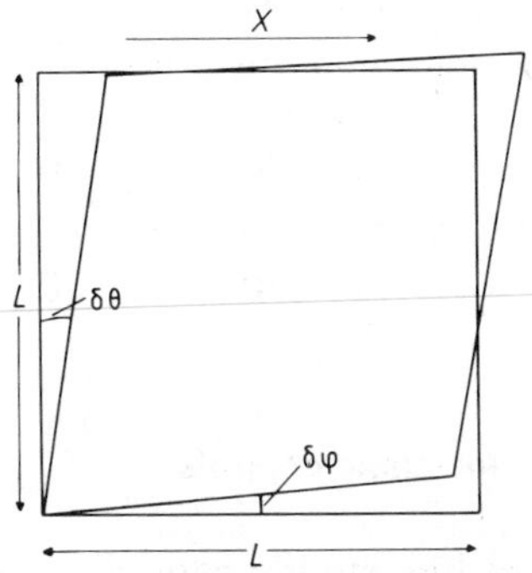

Figure A3.2 Shows shear of elastic block. From this the energy change follows as $X_y L^3(\delta\theta + \delta\phi)$.

Examples of results which now follow are:

$$\partial U/\partial e_{xx} = X_x, \qquad \partial U/\partial e_{yy} = Y_y. \tag{A3.1.5}$$

On further differentiation one finds

$$\partial X_x/\partial e_{yy} = \partial Y_y/\partial e_{xx}. \tag{A3.1.6}$$

Now from equation (A3.1.3), it can be seen that $\partial X_x/\partial e_{yy} = c_{12}$, and $\partial Y_y/\partial e_{xx} = c_{21}$. Thus $c_{12} = c_{21}$ and in general $c_{ij} = c_{ji}$ which means that the matrix (c_{ij}) is thus Hermitian. There are six diagonal elements and 15 off-diagonal components, or 21 elastic moduli in all.

A3.1.1 Further reduction for cubic crystals

Fortunately, there is a great reduction in this number, in fact to three, for cubic crystals. Application of a stress X_x cannot produce a shear for this would imply asymmetric behaviour of the crystal in the direction y or z. This argument must also apply to stresses Y_y and Z_z. It can also be seen that such asymmetric behaviour would occur if the stresses Z_x or X_y produced a shear in the (x, y) plane. Thus e_{yz} can only depend on Y_z and

$$Y_z = c_{44} e_{yz}. \tag{A3.1.7}$$

Equivalence of axes means that $c_{66} = c_{55} = c_{44}$ and also that $c_{ij} = 0$ if $i \neq j$ and either i or j (or both) is greater than 3, where use has been made of the property that (c_{ij}) is Hermitian. Hence X_x cannot depend on e_{xy}, e_{xz} or e_{yz}. Since its dependence on e_{yy} and e_{zz} must be the same we can write

$$X_x = c_{11} e_{xx} + c_{12} e_{yy} + c_{12} e_{zz}. \tag{A3.1.8}$$

Thus, for cubic crystals we can write the explicit form for the matrix of the elastic moduli:

$$(c_{ij}) = \begin{pmatrix} c_{11} & c_{12} & c_{12} & 0 & 0 & 0 \\ c_{12} & c_{11} & c_{12} & 0 & 0 & 0 \\ c_{12} & c_{12} & c_{11} & 0 & 0 & 0 \\ 0 & 0 & 0 & c_{44} & 0 & 0 \\ 0 & 0 & 0 & 0 & c_{44} & 0 \\ 0 & 0 & 0 & 0 & 0 & c_{44} \end{pmatrix}. \tag{A3.1.9}$$

A3.1.2 Wave equations for cubic crystals

In propagation of waves, the strains naturally vary from place to place. Although below we shall only write down the wave equations for cubic crystals, for the appropriate definitions of the strain components we can go back to the general case. We write the position of any element of the elastic medium as

$$\boldsymbol{r} = u\hat{\boldsymbol{x}} + v\hat{\boldsymbol{y}} + w\hat{\boldsymbol{z}} \tag{A3.1.10}$$

where $\hat{\boldsymbol{x}}$, etc denote unit vectors. When stress is applied, the element goes to

$$\boldsymbol{r}' = (u + \delta u)\hat{\boldsymbol{x}} + (v + \delta v)\hat{\boldsymbol{y}} + (w + \delta w)\hat{\boldsymbol{z}}. \tag{A3.1.11}$$

Considering now an infinitesimal block, with sides δx, δy, δz, one can immediately write

$$e_{xx} = \partial u/\partial x, \qquad e_{yy} = \partial v/\partial y, \qquad e_{zz} = \partial w/\partial z. \tag{A3.1.12}$$

To express the other strain components in a similar form, let us return to figure A3.1.2 where the rectangular block is now sheared through the finite angle $e_{xy} = \theta + \phi$. From this figure, with a block of sides δx and δy rather than L as before, we can see that

$$\theta = \delta u/\delta y, \qquad \phi = \delta v/\delta x,$$

$$e_{xy} = \frac{\partial v}{\partial x} + \frac{\partial u}{\partial y}, \qquad e_{xz} = \frac{\partial w}{\partial x} + \frac{\partial u}{\partial z} \tag{A3.1.13}$$

$$e_{yz} = \frac{\partial w}{\partial y} + \frac{\partial v}{\partial z}.$$

The equations for elastic waves can now be found. By considering a block with sides δx, δy, δz and density ρ, the equation of motion in the x-direction takes the form

$$\rho\frac{\partial^2 u}{\partial t^2} = \frac{\partial X_x}{\partial x} + \frac{\partial X_y}{\partial y} + \frac{\partial X_z}{\partial z}. \tag{A3.1.14}$$

For a cubic crystal use of equations (A3.1.3) and (A3.1.9) yields

$$\rho\frac{\partial^2 u}{\partial t^2} = c_{11}\frac{\partial e_{xx}}{\partial x} + c_{12}\left(\frac{\partial e_{yy}}{\partial y} + \frac{\partial e_{zz}}{\partial z}\right) + c_{44}\left(\frac{\partial e_{xy}}{\partial y} + \frac{\partial e_{xz}}{\partial z}\right). \tag{A3.1.15}$$

Using the definitions (A3.1.12) and (A3.1.13) of the strain components, this equation reads

$$\rho\frac{\partial^2 u}{\partial t^2} = c_{11}\frac{\partial^2 u}{\partial x^2} + c_{44}\left(\frac{\partial^2 u}{\partial y^2} + \frac{\partial^2 u}{\partial z^2}\right) + (c_{12} + c_{44})\left(\frac{\partial^2 v}{\partial x \partial y} + \frac{\partial^2 w}{\partial x \partial z}\right). \tag{A3.1.16}$$

Equations for $\partial^2 v/\partial t^2$ and $\partial^2 w/\partial t^2$ can readily be obtained from this equation by symmetry.

It should be noted that, although these equations lead to no dispersion, the velocity of sound will vary with direction.

A3.1.3 Isotropic medium

The elastic medium propagates shear and compression waves without dispersion. Thus there is a linear relation between frequency and wavenumber and the travelling wave solutions take the form

$$u_{\boldsymbol{k}}(\omega, \boldsymbol{r}) = Q_{\boldsymbol{k}} \epsilon_{\boldsymbol{k}} \exp[\mathrm{i}(\omega t - \boldsymbol{k} \cdot \boldsymbol{r})] \tag{A3.1.17}$$

with

$$\omega = v_{\mathrm{s}} k, \tag{A3.1.18}$$

v_{s} being the velocity of sound. This takes the value

$$v_{\mathrm{st}} = (n/\rho)^{1/2} \tag{A3.1.19}$$

for shear waves, where n is the shear modulus and ρ the density, whereas for compression waves

$$v_{\mathrm{sl}} = \left(\frac{K + 4n/3}{\rho} \right)^{1/2} \tag{A3.1.20}$$

where K is the bulk modulus.

In equation (A3.1.17), $\epsilon_{\boldsymbol{k}}$ is the unit polarisation vector, which is parallel to $\boldsymbol{k}$ for longitudinal (compression) waves and perpendicular to $\boldsymbol{k}$ for transverse (shear) waves. Hence, for each value of $\boldsymbol{k}$ there are three independent solutions of the wave equations and two possible values of ω.

Appendix 3.2

PHONONS IN A FACE-CENTRED CUBIC LATTICE WITH CENTRAL, SHORT-RANGE FORCES

As an example of phonon theory, we shall consider here a particular model in which central forces are assumed, but in addition the force range is taken to be so short that only nearest-neighbour interactions need be taken into account.

This calculation, for an FCC lattice, is not purely academic, for it has some relevance to the lattice vibrational spectrum of crystalline argon, where the attractive forces are of van der Waals type and therefore relatively short range.

If we had a simple cubic rather than an FCC structure, we should have found that no transverse waves are propagated when only nearest-neighbour central interactions are included, for the lattice then offers no resistance to shear. This point therefore draws attention to the likely limitations of such a short-range model.

In the main text we had that, for central forces, described by a pair potential $\Phi(\boldsymbol{r})$, the dynamical matrix takes the form, with m the atomic mass and $\boldsymbol{l}$ the direct lattice vectors,

$$D_{\alpha\beta}(\boldsymbol{k}) = m^{-1} \sum_{\boldsymbol{l}}{}' [1 - \exp(-i\boldsymbol{k}\cdot\boldsymbol{l})] \left(\frac{\partial^2 \Phi}{\partial x_\alpha \partial x_\beta} \right)_{\boldsymbol{r}=\boldsymbol{l}} \tag{A3.2.1}$$

where the prime on the summation shows that the term $\boldsymbol{l} = \boldsymbol{0}$ is to be omitted.

Let us write the derivatives at the nearest neighbours as†

$$\xi = \left(\frac{\partial^2 \Phi}{\partial r^2} - \frac{1}{r}\frac{\partial \Phi}{\partial r} \right) \qquad \eta = \frac{1}{r}\frac{\partial \Phi}{\partial r}. \tag{A3.2.2}$$

The dynamical matrix can then be put into the form

$$\mathbf{D}(\boldsymbol{k}) = \begin{pmatrix} d_{11} & d_{12} & d_{13} \\ d_{21} & d_{22} & d_{23} \\ d_{31} & d_{32} & d_{33} \end{pmatrix} \tag{A3.2.3}$$

where the diagonal elements are given by

$$d_{11}(\boldsymbol{k}) = (2\xi + 4\eta)[2 - \cos(k_x a)\cos(k_y a) - \cos(k_x a)\cos(k_z a)] \tag{A3.2.4}$$

$$d_{22}(\boldsymbol{k}) = (2\xi + 4\eta)[2 - \cos(k_x a)\cos(k_y a) - \cos(k_y a)\cos(k_z a)] \tag{A3.2.5}$$

† The first derivative $\partial\Phi/\partial r$ is zero for equilibrium under ordinary pair forces. But in metals for instance this is not so, an additional pressure term being needed to maintain equilibrium.

$$d_{33}(\boldsymbol{k}) = (2\xi+4\eta)[2-\cos(k_x a)\cos(k_z a)-\cos(k_y a)\cos(k_z a)]. \tag{A3.2.6}$$

The off-diagonal elements are

$$d_{21}(\boldsymbol{k}) = d_{12}(\boldsymbol{k}) = -2\xi\sin(k_x a)\sin(k_y a) \tag{A3.2.7}$$

$$d_{31}(\boldsymbol{k}) = d_{13}(\boldsymbol{k}) = -2\xi\sin(k_x a)\sin(k_z a) \tag{A3.2.8}$$

$$d_{32}(\boldsymbol{k}) = d_{23}(\boldsymbol{k}) = -2\xi\sin(k_y a)\sin(k_z a). \tag{A3.2.9}$$

Let us now consider the various symmetry directions, starting with [1 0 0].

(a) [1 0 0]. The matrix reduces to diagonal form

$$\begin{pmatrix} (2\xi+4\eta)[2-2\cos(k_x a)] & 0 & 0 \\ 0 & (2\xi+4\eta)[1-\cos(k_x a)] & 0 \\ 0 & 0 & (2\xi+4\eta)[1-\cos(k_x a)] \end{pmatrix}$$

and the eigenvalues are immediately evident.

(b) [1 1 1]. $\mathbf{D}(\boldsymbol{k})$ now takes the form

$$\begin{pmatrix} A & B & B \\ B & A & B \\ B & B & A \end{pmatrix}$$

where $A = (2\xi+4\eta)[2-2\cos^2(k_x a)]$ and $B = -2\xi\sin^2(k_x a)$.

The longitudinal (L) mode has polarisation vector

$$\begin{pmatrix} 1 \\ 1 \\ 1 \end{pmatrix}$$

and so

$$\begin{pmatrix} A & B & B \\ B & A & B \\ B & B & A \end{pmatrix}\begin{pmatrix} 1 \\ 1 \\ 1 \end{pmatrix} = (A+2B)\begin{pmatrix} 1 \\ 1 \\ 1 \end{pmatrix}. \tag{A3.2.10}$$

To obtain the degenerate transverse (T) eigenvalue, we have

$$\begin{pmatrix} A & B & B \\ B & A & B \\ B & B & A \end{pmatrix}\begin{pmatrix} 1 \\ 1 \\ -2 \end{pmatrix} = (A-B)\begin{pmatrix} 1 \\ 1 \\ -2 \end{pmatrix}. \tag{A3.2.11}$$

Thus we are led to

$$\omega_{\mathrm{L}}^2(\boldsymbol{k}) = (2\xi+4\eta)[2-2\cos^2(k_x a)]-4\xi\sin^2(k_x a) \tag{A3.2.12}$$

$$\omega_{\mathrm{T}}^2(\boldsymbol{k}) = (2\xi+4\eta)[2-2\cos^2(k_x a)]+2\xi\sin^2(k_x a). \tag{A3.2.13}$$

(c) [1 1 0]. The matrix becomes

$$\begin{pmatrix} A & C & 0 \\ C & A & 0 \\ 0 & 0 & B \end{pmatrix}$$

with now

$$A = (2\xi + 4\eta)[2 - \cos^2(k_x a) - \cos(k_x a)]$$
$$B = (2\xi + 4\eta)[2 - 2\cos(k_x a)]$$
$$C = -2\xi \sin^2(k_x a).$$

One eigenvalue is immediately evident. The polarisation vectors are (1, 1, 0), (1, −1, 0) and (0, 0, 1), the first being for the longitudinal mode and the other two exactly transverse. For the frequencies we find

$$\omega_L^2(k_x, k_x, 0) = A + C \qquad \text{(A3.2.14)}$$
$$\omega_{T_1}^2(k_x, k_x, 0) = A - C \qquad \text{(A3.2.15)}$$
$$\omega_{T_2}^2(k_x, k_x, 0) = B. \qquad \text{(A3.2.16)}$$

Thus, in each of these three symmetry directions we have obtained quite explicit expressions for the dispersion relations $\omega(\boldsymbol{k})$ in terms of the lattice constant and the derivatives of the pair potential representing the central force.

Appendix 3.3

NEUTRON SCATTERING AND DEBYE–WALLER FACTOR

Following Marshall and Lovesey (1971) we outline first the steps in the calculation of the cross sections for the elastic coherent and incoherent scattering of neutrons by nuclei undergoing harmonic vibrations. Secondly, we consider anharmonic corrections and finally inelastic processes.

A3.3.1 Elastic coherent cross section

The form of this cross section is given by

$$\left(\frac{\mathrm{d}\sigma}{\mathrm{d}\Omega}\right)_{\substack{\text{elastic}\\ \text{coherent}}} = \frac{\sigma_{\mathrm{c}}}{4\pi}\left|\int \mathrm{d}\boldsymbol{r}\exp(\mathrm{i}\boldsymbol{\kappa}\cdot\boldsymbol{r})\sum_l \langle \delta(\boldsymbol{r}-\boldsymbol{R}_l(0))\rangle\right|^2. \qquad \text{(A3.3.1)}$$

Here σ_{c} is the single (bound) nucleus coherent cross section, $\boldsymbol{\kappa}$ is the scattering vector $\boldsymbol{k}-\boldsymbol{k}'$, $\boldsymbol{k}$ being the wavevector of the incident neutron and $\boldsymbol{k}'$ referring to the scattered neutron, while $\boldsymbol{R}_l(0)$ are the instantaneous positions of the nuclei.

We can also write

$$\sum_l \langle \delta(\boldsymbol{r}-\boldsymbol{R}_l(0))\rangle = \frac{1}{(2\pi)^3}\int \mathrm{d}\boldsymbol{k}\sum_l \exp[\mathrm{i}\boldsymbol{k}\cdot(\boldsymbol{r}-\boldsymbol{l})]\,\langle \exp[-\mathrm{i}\boldsymbol{k}\cdot\boldsymbol{u}_l(0)]\rangle \qquad \text{(A3.3.2)}$$

in terms of the dispacements $\boldsymbol{u}_l(0)$ from equilibrium. Equation (A3.3.2) can be rewritten, when we employ the identity

$$\langle \exp[-\mathrm{i}\boldsymbol{k}\cdot\boldsymbol{u}_l(0)]\rangle = \exp\{-\tfrac{1}{2}\langle[\boldsymbol{k}\cdot\boldsymbol{u}_l(0)]^2\rangle\} \qquad \text{(A3.3.3)}$$

and use the phonon Bose operators (see Marshall and Lovesey 1971), as

$$\langle \exp[-\mathrm{i}\boldsymbol{k}\cdot\boldsymbol{u}_l(0)]\rangle = \exp\left(-\frac{\hbar}{4NM}\sum_{j,\boldsymbol{q}}\frac{|\boldsymbol{k}\cdot\epsilon^j(\boldsymbol{q})|^2}{\omega_j(\boldsymbol{q})}[2n_j(\boldsymbol{q})+1]\right) \qquad \text{(A3.3.4)}$$

where $\omega_j(\boldsymbol{q})$ are the phonon frequencies, ϵ the polarisation vectors and $n_j(\boldsymbol{q})$ represents the boson occupation number:

$$n_j(\boldsymbol{q}) = \{\exp[\beta\hbar\,\omega_j(\boldsymbol{q})]-1\}^{-1} \qquad \text{(A3.3.5)}$$

where $\beta = 1/k_{\mathrm{B}}T$. The right-hand side of equation (A3.3.4) is the Debye–Waller factor and is written simply as $\exp[-W(\boldsymbol{k})]$.

To evaluate the cross section from equation (A3.3.1), we have

$$\int d\boldsymbol{r} \exp(i\boldsymbol{\kappa}\cdot\boldsymbol{r}) \sum_l \langle \delta(\boldsymbol{r}-\boldsymbol{R}_l(0)) \rangle$$

$$= \sum_l \frac{1}{(2\pi)^3} \int d\boldsymbol{k} \int d\boldsymbol{r} \exp[i\boldsymbol{r}\cdot(\boldsymbol{\kappa}+\boldsymbol{k}) - i\boldsymbol{k}\cdot\boldsymbol{l}] \exp[-W(\boldsymbol{k})]$$

$$= \sum_l \exp(i\boldsymbol{\kappa}\cdot\boldsymbol{l}) \exp[-W(\boldsymbol{\kappa})]. \qquad \text{(A3.3.6)}$$

Thus the cross section for coherent elastic scattering is

$$\left(\frac{d\sigma}{d\Omega}\right)_{\substack{\text{elastic}\\ \text{coherent}}} = \frac{N\sigma_c}{4\pi}\frac{(2\pi)^3}{V_0}\sum_{\boldsymbol{\tau}} \delta(\boldsymbol{\kappa}-\boldsymbol{\tau}) \exp[-2W(\boldsymbol{\kappa})] \qquad \text{(A3.3.7)}$$

which is simply the result for scattering by a static lattice with reciprocal lattice vectors $\boldsymbol{\tau}$, multiplied by the factor $\exp[-2W(\boldsymbol{\kappa})]$. By definition

$$2W(\boldsymbol{\kappa}) = \langle [\boldsymbol{\kappa}\cdot\boldsymbol{u}_l(0)]^2 \rangle \qquad \text{(A3.3.8)}$$

so that $2W$ is the mean square displacement of a nucleus multiplied by κ^2. The factor $\exp[-2W(\boldsymbol{\kappa})]$ clearly implies a decrease of the intensity of the Bragg peaks with increasing $|\boldsymbol{\kappa}|$. The quantity V_0 in equation (A3.3.7) is the volume of the unit cell.

The Debye–Waller factor simplifies if we use the normalised density of states $g(\omega)$ given by (cf §3.5.1)

$$g(\omega) = \frac{1}{3N}\sum_{j,\boldsymbol{q}} \delta(\omega-\omega_j(\boldsymbol{q})) \qquad \text{(A3.3.9)}$$

when we find, if we denote the maximum phonon frequency by $\omega_{\max}$,

$$W(\boldsymbol{\kappa}) = 3N\int_0^{\omega_{\max}} d\omega \frac{g(\omega)\hbar}{4\omega NM}[2n(\omega)+1]\,(|\boldsymbol{\kappa}\cdot\boldsymbol{\epsilon}|^2)_{\text{av}}. \qquad \text{(A3.3.10)}$$

where the subscript av denotes an average over a surface with a given ω and $n(\omega)$ is the Bose–Einstein distribution function. In cubic symmetry, the average is simply $\frac{1}{3}\kappa^2$ and using this we find

$$W(\boldsymbol{\kappa}) = \frac{\hbar\kappa^2}{4M}\int_0^{\omega_{\max}} d\omega \frac{g(\omega)}{\omega}\coth(\tfrac{1}{2}\hbar\omega\beta). \qquad \text{(A3.3.11)}$$

A3.3.2 Temperature dependence for Debye spectrum

To estimate the temperature dependence of the Debye–Waller factor, we insert the Debye spectrum (cf equation (3.5)), namely

$$g(\omega) \to \frac{3\omega^2}{\omega_D^3} \qquad \hbar\omega_D = k_B\Theta_D \qquad \text{(A3.3.12)}$$

into equation (A3.3.11). If the temperature T is very much greater than the Debye temperature Θ_D then in equation (A3.3.11) we can use the approximation

$$\coth\left(\tfrac{1}{2}\hbar\omega\beta\right) \simeq 2/\hbar\omega\beta \qquad \text{(A3.3.13)}$$

and hence we find

$$W(\kappa) \sim 3\left(\frac{\hbar^2\kappa^2}{2M}\right)\left(\frac{1}{\hbar\omega_D}\right)\left(\frac{T}{\Theta_D}\right) \qquad T \gg \Theta_D. \qquad \text{(A3.3.14)}$$

It should be noted that $\hbar^2\kappa^2/2M$ is the energy of recoil of a nucleus of mass M, initially at rest.

In the opposite limit of low temperatures, $W(\kappa)$ reflects the zero-point motion of the nuclei and is given by

$$W(\kappa) = \frac{3}{4}\left(\frac{\hbar^2\kappa^2}{2M}\right)\left(\frac{1}{\hbar\omega_D}\right) \qquad T \to 0. \qquad \text{(A3.3.15)}$$

The incoherent elastic cross section can be calculated similarly and is again found to be the static lattice result times the factor $\exp[-2W(\kappa)]$.

A3.3.3 Anharmonic corrections to Debye–Waller factor

At this point we shall discuss briefly the way in which the Debye–Waller factor is influenced by anharmonic effects. The most important point is that $g(\omega)$, the frequency spectrum, then depends on temperature as can be seen in figure 3.15.

But in addition there are explicit corrections to add because equation (A3.3.3) is not true in the presence of anharmonicity. The way to handle the anharmonicity (cf Marshall and Lovesey 1971) is to expand in a power series in the scattering vector $\boldsymbol{\kappa}$, whence one finds

$$W(\kappa) = \tfrac{1}{2}\langle(\boldsymbol{\kappa}\cdot\boldsymbol{u})^2\rangle - \tfrac{1}{6}\mathrm{i}\langle(\boldsymbol{\kappa}\cdot\boldsymbol{u})^3\rangle - \tfrac{1}{24}[\langle(\boldsymbol{\kappa}\cdot\boldsymbol{u})^4\rangle - 3\langle(\boldsymbol{\kappa}\cdot\boldsymbol{u})^2\rangle^2] + \dots \qquad \text{(A3.3.16)}$$

The leading term in this expression is the harmonic term. The second term vanishes if every atom is at a centre of symmetry, but if this is not true it can give contributions to the coherent cross section. In the presence of anharmonicity, the next term does not vanish. According to the calculations of Ambegaokar and Maradudin (1964), these explicit anharmonic terms are less important than the temperature dependence of the frequency spectrum $g(\omega)$.

A3.3.4 Inelastic one-phonon scattering

As the final topic of this appendix we shall discuss one-phonon cross sections (for Bravais lattices). Following Marshall and Lovesey (1971), the inelastic partial coherent differential cross section can be written in the form, with E' the final energy of the neutron,

$$\left(\frac{d^2\sigma}{d\Omega dE'}\right)_{\substack{\text{inelastic}\\\text{coherent}}} = \frac{N\sigma_c}{4\pi}\frac{k'}{k}\frac{1}{2\pi\hbar}\int_{-\infty}^{\infty} dt \exp(-i\omega t) \times \int d\boldsymbol{r} \exp(i\boldsymbol{\kappa}\cdot\boldsymbol{r})G'(\boldsymbol{r}, t) \tag{A3.3.17}$$

where

$$G'(\boldsymbol{r}, t) = G(\boldsymbol{r}, t) - G(\boldsymbol{r}, \infty).$$

The pair function $G(\boldsymbol{r}, t)$ is given by

$$G(\boldsymbol{r}, t) = \frac{1}{(2\pi)^3 N}\sum_{l,l'}\int d\boldsymbol{k} \exp(-i\boldsymbol{k}\cdot\boldsymbol{r}) \times \langle \exp(-i\boldsymbol{k}\cdot\boldsymbol{R}_l(0) \exp(i\boldsymbol{k}\cdot\boldsymbol{R}_{l'}(t))\rangle. \tag{A3.3.18}$$

The correlation function in $G(\boldsymbol{r}, t)$ can be written after some manipulation (see Marshall and Lovesey 1971) as

$$\exp[-2W(\boldsymbol{k})] \exp[\langle \boldsymbol{k}\cdot\boldsymbol{u}_l(0)\, \boldsymbol{k}\cdot\boldsymbol{u}_{l'}(t)\rangle]$$

where the $\boldsymbol{u}$ as usual are the lattice displacements. The cross section (A3.3.17) can then be rewritten as

$$\left(\frac{d^2\sigma}{d\Omega\, dE'}\right)_{\substack{\text{inelastic}\\\text{coherent}}} = \frac{\sigma_c}{4\pi}\frac{k'}{k}\frac{1}{2\pi\hbar}\int_{-\infty}^{\infty} dt \exp(-i\omega t)\exp[-2W(\boldsymbol{\kappa})] \times \sum_{l,l'} \exp[-i\boldsymbol{\kappa}\cdot(\boldsymbol{l}-\boldsymbol{l}')]\{\exp[\langle\boldsymbol{\kappa}\cdot\boldsymbol{u}_l(0)\, \boldsymbol{\kappa}\cdot\boldsymbol{u}_{l'}(t)\rangle]-1\}. \tag{A3.3.19}$$

Similarly the inelastic incoherent cross section is

$$\left(\frac{d^2\sigma}{d\Omega\, dE'}\right)_{\substack{\text{inelastic}\\\text{incoherent}}} = \frac{\sigma_i}{4\pi}\frac{k'}{k}\frac{1}{2\pi\hbar}\int_{-\infty}^{\infty} dt \exp(-i\omega t)\exp[-2W(\boldsymbol{\kappa})] \times \sum_{l}\{\exp[\langle\boldsymbol{\kappa}\cdot\boldsymbol{u}_l(0)\boldsymbol{\kappa}\cdot\boldsymbol{u}_l(t)\rangle]-1\}. \tag{A3.3.20}$$

Equations (A3.3.19) and (A3.3.20) are exact within the harmonic approximation.

If we now expand the exponentials in equations (A3.3.19) and (A3.3.20) which contain the displacement correlation functions, the first non-zero contributions give the one-phonon cross sections, the second terms the two-phonon cross sections, and so on.

A3.3.5 One-phonon cross sections

Developing the correlation function $\langle \boldsymbol{\kappa}\cdot\boldsymbol{u}_l(0), \boldsymbol{\kappa}\cdot\boldsymbol{u}_{l'}(t)\rangle$ following Marshall and Lovesey (1971) yields

$$\left(\frac{\mathrm{d}^2\sigma}{\mathrm{d}\Omega\,\mathrm{d}E'}\right)_{\substack{\text{inelastic}\\ \text{incoherent}}} = \frac{\sigma_c}{4\pi}\frac{k'}{k}\frac{(2\pi)^3}{V_0}\frac{1}{2M}\sum_{\boldsymbol{r}}\exp[-2W(\boldsymbol{\kappa})]$$

$$\times\sum_{j,\boldsymbol{q}}\frac{|\boldsymbol{\kappa}\cdot\boldsymbol{\epsilon}^j(\boldsymbol{q})|^2}{\omega_j(\boldsymbol{q})}\{n_j(\boldsymbol{q})\,\delta(\omega+\omega_j(\boldsymbol{q}))\,\delta(\boldsymbol{\kappa}+\boldsymbol{q}-\boldsymbol{\tau})$$

$$+[n_j(\boldsymbol{q})+1]\delta(\omega-\omega_j(\boldsymbol{q}))\delta(\boldsymbol{\kappa}-\boldsymbol{q}-\boldsymbol{\tau})\}. \qquad \text{(A3.3.21)}$$

The cross section in equation (A3.3.21) is seen to be the sum of two terms. The first, which contains $\delta(\omega+\omega_j(\boldsymbol{q}))\delta(\boldsymbol{\kappa}+\boldsymbol{q}-\boldsymbol{\tau})$ represents a scattering process in which one phonon is annihilated. The second term contains $\delta(\omega-\omega_j(\boldsymbol{q}))\delta(\boldsymbol{\kappa}-\boldsymbol{q}-\boldsymbol{\tau})$ and represents a process in which one phonon is created.

In the limit $T\to 0$, only the second process can occur, since there are no phonons to be annihilated at absolute zero.

A3.3.6 Structure of one-phonon annihilation cross section and determination of phonon dispersion relations

The two delta functions associated with this scattering process represent conservation of energy and momentum and show that the scattering process satisfies the conditions

$$E' = E + \hbar\omega_j(\boldsymbol{q}) \qquad \text{(A3.3.22)}$$

and

$$\boldsymbol{k}' = \boldsymbol{k} + \boldsymbol{q} - \boldsymbol{\tau}. \qquad \text{(A3.3.23)}$$

These conditions are so restrictive that for a given scattering angle only phonons of a particular $\boldsymbol{q}$ and $\omega_j(\boldsymbol{q})$ can give scattering. This fact can be utilised to determine the phonon spectrum $\omega_j(\boldsymbol{q})$ as a function of $\boldsymbol{q}$.

Let us suppose that we have a monochromatic beam incident on a crystal and that we measure the energy of the neutrons scattered through a given angle. From the conditions (A3.3.22) and (A3.3.23) these scattered neutrons will in general have one of three energies corresponding to the three choices of j. If a choice is made of one of these energies, then knowing the scattered energy means that $\boldsymbol{k}'$ is known. From equation (A3.3.23), since $\boldsymbol{q}$ must lie in the first Brillouin zone, we have both $\boldsymbol{q}$ and $\boldsymbol{\tau}$. Hence from this experiment we have both $\boldsymbol{q}$ and $\omega_j(\boldsymbol{q})$ for a phonon; i.e. a point on the dispersion curve.

Appendix 3.4

COULOMB TERMS AND PHONON DISPERSION IN THE RIGID-ION MODEL OF IONIC CRYSTALS

We shall divide this appendix into two parts. In the first, Ewald's method of handling lattice sums involving long-range Coulomb potentials will be outlined in general terms. In the second part of the appendix, both the Coulomb contribution $D^{\mathrm{C}}_{\alpha\beta}$ to the dynamical matrix $D_{\alpha\beta}$ in equation (3.50) and the repulsive contribution $D^{\mathrm{r}}_{\alpha\beta}$ will be discussed for the specific case of ionic crystals with the NaCl structure.

A3.4.1 Coulomb coefficients and Ewald's transformation

To deal with the Coulomb terms we follow the treatment of Kellerman (1940) for NaCl. To see how to handle the long-range of the Coulomb interaction, let us first consider a lattice sum of the form

$$F_{\boldsymbol{k}}(\boldsymbol{r}) = \sum_{l} f(\boldsymbol{r}-\boldsymbol{R}_l)\exp(\mathrm{i}\boldsymbol{k}\cdot\boldsymbol{R}_l) \qquad \text{(A3.4.1)}$$

where the $\boldsymbol{R}_l$ denote direct lattice vectors. The function f will later be corrected with the Coulomb interaction. This sum can be rewritten as follows:

$$\begin{aligned} F_{\boldsymbol{k}}(\boldsymbol{r}) &= \exp(\mathrm{i}\boldsymbol{k}\cdot\boldsymbol{r})\sum_{l} f(\boldsymbol{r}-\boldsymbol{R}_l)\exp[\mathrm{i}\boldsymbol{k}\cdot(\boldsymbol{R}_l-\boldsymbol{r})] \\ &= \exp(\mathrm{i}\boldsymbol{k}\cdot\boldsymbol{r})\sum_{h} F^{h}_{\boldsymbol{k}}\exp(\mathrm{i}\boldsymbol{K}_h\cdot\boldsymbol{r}). \end{aligned} \qquad \text{(A3.4.2)}$$

Here use has been made of the Fourier expansion of a periodic function in the direct lattice in terms of reciprocal lattice vectors $\boldsymbol{K}_h$. The Fourier coefficients $F^{h}_{\boldsymbol{k}}$ can evidently be found from equation (A3.4.2) as

$$F^{h}_{\boldsymbol{k}} = \frac{1}{\Omega}\int f(\boldsymbol{r})\exp[-\mathrm{i}(\boldsymbol{K}_h+\boldsymbol{k})\cdot\boldsymbol{r}]\,\mathrm{d}\boldsymbol{r} \equiv \frac{1}{\Omega}\phi(\boldsymbol{K}_h+\boldsymbol{k}) \qquad \text{(A3.4.3)}$$

where ϕ is a function of $(\boldsymbol{K}_h+\boldsymbol{k})$ only, as can be seen immediately from the integral form, while Ω is the volume per unit cell.

If we now choose

$$f(r) = \frac{2}{\pi^{1/2}}\exp(-\epsilon^2 r^2) \qquad \text{(A3.4.4)}$$

then

$$\int_0^\infty f(r)\mathrm{d}\epsilon = 1/r. \tag{A3.4.5}$$

Determining ϕ in equation (A3.4.3) by Fourier transforming equation (A3.4.4) we find

$$\begin{aligned} F_k(\boldsymbol{r}) &= \frac{2}{\pi^{1/2}} \sum_l \exp[-\epsilon^2(\boldsymbol{r}-\boldsymbol{R}_l)^2 + \mathrm{i}\boldsymbol{k}\cdot\boldsymbol{R}_l] \\ &= \frac{2\pi}{\Omega} \sum_h \frac{1}{\epsilon^3} \exp\left(-\frac{1}{4\epsilon^2}(\boldsymbol{K}_h+\boldsymbol{k})^2 + \mathrm{i}(\boldsymbol{K}_h+\boldsymbol{k})\cdot\boldsymbol{r}\right) \end{aligned} \tag{A3.4.6}$$

which is Ewald's transformation. The sum over l is rapidly convergent if ϵ is large, while that over h is favourable for evaluation if ϵ is small. In practice, the trick is to divide the sum into two parts: one sum over l and the other over h. The resulting form of $F_k(\boldsymbol{r})$ involves a compromise between the number of terms in each part of the sum.

A3.4.2 Structure of NaCl and symmetry properties of dynamical matrix

This is the point at which we return to the dynamical matrix in the form of equation (3.50). The argument follows that of Reissland (1973). The NaCl structure is such that if we write the displacements in the form

$$\boldsymbol{R}^l_{\kappa\kappa'} = \boldsymbol{R}^l + \boldsymbol{R}_{\kappa\kappa'} \tag{A3.4.7}$$

then

$$\exp(\mathrm{i}\boldsymbol{k}\cdot\boldsymbol{R}^l_{\kappa\kappa'}) \rightarrow \cos(\boldsymbol{k}\cdot\boldsymbol{R}^l_{\kappa\kappa'}). \tag{A3.4.8}$$

Further, from lattice symmetry properties it follows that

$$D_{\alpha\beta}(\kappa, \kappa') = D_{\alpha\beta}(\kappa', \kappa) = D_{\beta\alpha}(\kappa, \kappa'). \tag{A3.4.9}$$

To be quite explicit about the structure, NaCl is such that the sodium ions and the chlorine ions form an FCC structure: the basis vectors of one of these cells being

$$\boldsymbol{a}_1 = \tfrac{1}{2}a(0, 1, 1) \qquad \boldsymbol{a}_2 = \tfrac{1}{2}a(1, 0, 1) \qquad \boldsymbol{a}_3 = \tfrac{1}{2}a(1, 1, 0) \tag{A3.4.10}$$

and the vector joining the ions $\kappa = 1$ and $\kappa = 2$ is

$$\boldsymbol{R}_{12} = \tfrac{1}{2}a(1, 1, 1) \tag{A3.4.11}$$

the volume of a unit cell being then $V = a^3/4$. Thus the lattice vectors are

$$\boldsymbol{R}^l = \tfrac{1}{2}a(l_1, l_2, l_3) \tag{A3.4.12}$$

and the position vectors of the other atom in the unit cell are

$$\boldsymbol{R}^l_{12} = \tfrac{1}{2}a(l_1+1, l_2+1, l_3+1) \equiv \tfrac{1}{2}a(m_1, m_2, m_3). \tag{A3.4.13}$$

The allowed values for an FCC structure are such that

$$\begin{aligned} l_1+l_2+l_3 &= \text{even integer} \qquad & l^2 &= l_1^2+l_2^2+l_3^2 \\ m_1+m_2+m_3 &= \text{odd integer} \qquad & m^2 &= m_1^2+m_2^2+m_3^2. \end{aligned} \tag{A3.4.14}$$

The reciprocal lattice vectors are

$$\boldsymbol{K}_h = \frac{2\pi}{a}\boldsymbol{h} \qquad \boldsymbol{h} = (h_x, h_y, h_z) \tag{A3.4.15}$$

h_x, h_y and h_z being all even or all odd integers. A wavevector $\boldsymbol{k}$ in the Brillouin zone will be measured in reduced units such that

$$\boldsymbol{k} = \frac{2\pi}{a}\boldsymbol{q} \qquad \boldsymbol{q} = (q_x, q_y, q_z). \tag{A3.4.16}$$

A3.4.3 Coulomb part of dynamical matrix

Using the Ewald method outlined above, the Coulomb part, D^{C}, of the dynamical matrix

$$D_{\alpha\beta}(\kappa,\kappa') = (m_\kappa m_{\kappa'})^{-1/2}\sum_l \phi^l_{\alpha\beta}(\kappa,\kappa')\exp(\mathrm{i}\boldsymbol{k}\cdot\boldsymbol{R}^l_{\kappa\kappa'}) \tag{A3.4.17}$$

for $\kappa = \kappa'$ can be written in the following form:.

$$D^{\mathrm{C}}_{\alpha\beta}(\kappa,\kappa) = \frac{e^2}{4\pi\epsilon_0 V}\left(\sum_h G^{\kappa\kappa}_{\alpha\beta}(h) + \sum_l H_{\alpha\beta}(l) + \frac{8\epsilon^3}{3\pi^{1/2}}\delta_{\alpha\beta}\right) \tag{A3.4.18}$$

and for $\kappa \neq \kappa'$ as

$$D^{\mathrm{C}}_{\alpha\beta}(\kappa,\kappa') = \frac{-e^2}{4\pi\epsilon_0 V}\left(\sum_h G^{\kappa\kappa'}_{\alpha\beta}(h) + \sum_m H_{\alpha\beta}(m)\right). \tag{A3.4.19}$$

Here the forms of G and H can be shown to be

$$G^{\kappa\kappa}_{\alpha\beta}(h) = G^{11}_{\alpha\beta}(h) = -\frac{4\pi(h_\alpha+q_\alpha)(h_\beta+q_\beta)}{|\boldsymbol{h}+\boldsymbol{q}|^2}\exp\left(\frac{-\pi^2}{4\epsilon^2}(\boldsymbol{h}+\boldsymbol{q})^2\right) \tag{A3.4.20}$$

$$G^{\kappa\kappa'}_{\alpha\beta}(h) = G^{12}_{\alpha\beta}(h) = G^{11}_{\alpha\beta}\cos[\pi(h_x+h_y+h_z)] \tag{A3.4.21}$$

$$H_{\alpha\beta}(l) = 2\left(\frac{l_\alpha l_\beta}{l^2}g(l) - f(l)\delta_{\alpha\beta}\right)\cos(\pi\boldsymbol{q}\cdot\boldsymbol{l}) \tag{A3.4.22}$$

where

$$f(l) = \frac{2\epsilon}{\pi^{1/2}l^2}\exp(-\epsilon^2 l^2) + \frac{1}{l^3}[1 - G(\epsilon l)] \tag{A3.4.23}$$

$$g(l) = \frac{4}{\pi^{1/2}}\epsilon^3\exp(-\epsilon^2 l^2) + 3f(l) \tag{A3.4.24}$$

$G(\epsilon l)$ being defined through

$$G(y) = \frac{2}{\pi^{1/2}} \int_0^y \exp(-x^2)\mathrm{d}x. \tag{A3.4.25}$$

A3.4.4 Repulsive contribution to dynamical matrix

As in discussing cohesive energy and compressibility of ionic crystals we write the ionic potential energy in the form

$$U(r) = \frac{A}{r} + \sum_{i,j} \phi(r_{ij}) \tag{A3.4.26}$$

where A is the Madelung constant, and $\phi(r)$ represents the short-range repulsive interactions, then one finds that the repulsive contribution D^{r} is of the diagonal form

$$D^{\mathrm{r}}_{\alpha\beta}(\kappa, \kappa') \propto \delta_{\alpha\beta}. \tag{A3.4.27}$$

For $\kappa = \kappa'$

$$D^{\mathrm{r}}_{\alpha\beta}(\kappa, \kappa) = -\frac{e^2}{4\pi\epsilon_0 V}(A + 2B) \tag{A3.4.28}$$

where $B = \frac{1}{2}\partial^2\phi/\partial r^2$ and V is the volume. For $\kappa \neq \kappa'$:

$$D^{\mathrm{r}}_{xx}(\kappa, \kappa') = A\cos(\pi q_x) + B[\cos(\pi q_y) + \cos(\pi q_z)]. \tag{A3.4.29}$$

Corresponding expressions for D^{r}_{yy} and D^{r}_{zz} can be obtained from equation (A3.4.29) by cyclic permutation of the suffixes.

The six eigenvalues for a given $\boldsymbol{k}$, $\omega^2(\boldsymbol{k}, j)$ say, can be found by adding D^{C} and D^{r} to obtain the full dynamical matrix. One must then diagonalise this. This diagonalisation can be carried out for any value of $\boldsymbol{k}$ (except $\boldsymbol{k} = \boldsymbol{0}$, which presents special problems as some terms of the Coulomb part D^{C} diverge there) and hence the entire phonon spectrum may be found in this rigid-ion model. For detailed calculations the reader is referred to Kellerman (1940) for NaCl and to Karo (1959, 1960) who has applied the rigid-ion method to the structurally similar halides of Li, Na, K, Rb and Cs.

Finally, ways of transcending the rigid-ion model are briefly summarised in §4.6.

Appendix 3.5

TWO-PHONON PROCESSES IN SEMICONDUCTORS

A photon can be absorbed by a process involving two phonons. Lax and Burstein (1955, see also the summary by Reissland 1973) have proposed two possible mechanisms for this coupling, namely:

(a) The photon creates a transverse optic phonon which decays into two phonons. The intermediate state need not conserve energy. Momentum must be conserved at all stages.

(b) The photon couples directly with the two phonons. In germanium, for example, this can occur by a displaced atom distorting the charge distribution of neighbouring atoms. When they, in their turn, are displaced by another phonon, a dipole moment is created.

Mechanism (a) can apply only in polar crystals, as a linear electric moment is required. In mechanism (b), one phonon induces a charge and another phonon displaces it to create a second-order electric moment term. This can occur in covalent or in polar crystals. In cases where both processes are possible, their relative importance depends on the relative magnitude of the second-order electric moment and the anharmonic coupling between the transverse optic mode and the two phonons involved.

It can be shown that the absorption has the following form:

summation (s) band (two phonons created)

$$K_{\mathrm{s}}(\omega) = \text{constant} \times \sum_{\boldsymbol{k},j,j'} \frac{H'(\boldsymbol{k},j,j')}{\omega(\boldsymbol{k},j)\omega(\boldsymbol{k},j')}[1+n(\boldsymbol{k},j)n(\boldsymbol{k},j')]$$
$$\times\, \delta(\omega-\omega(\boldsymbol{k},j)-\omega(\boldsymbol{k},j'))$$

difference (d) band (phonon $\boldsymbol{k},j$ created, phonon $\boldsymbol{k},j'$ destroyed)

$$K_{\mathrm{d}}(\omega) = \text{constant} \times \sum_{\boldsymbol{k},j,j'} \frac{H'(\boldsymbol{k},j,j')}{\omega(\boldsymbol{k},j)\omega(\boldsymbol{k},j')}[n(\boldsymbol{k},j)-n(\boldsymbol{k},j')]$$
$$\times\, \delta(\omega-\omega(\boldsymbol{k},j)-\omega(\boldsymbol{k},j')).$$

Here $H'(\boldsymbol{k},j,j')$ denotes the coupling coefficient for the photon–two-phonon process.

Though these are complex equations as they stand, Johnson (1965) has made a comparison between these expressions and measured absorption curves.

Appendix 3.6

EFFECT OF DISORDER ON ATOMIC DYNAMICS OF A LINEAR CHAIN

We have dealt at some length with phonons, the quanta of the vibrational modes, in a one-dimensional periodic chain in the main text. As also discussed briefly there, a problem of considerable importance is the way in which the atomic dynamics is affected by disordering such a chain. This is relevant to the properties of glasses and amorphous semiconductors.

We introduced Dyson's (1953) study of the problem of a disordered chain in §3.9. Our purpose here is to give a little more detail than that presented in the body of the text for the type (called (1) in §3.9) of disordered chain in which each of the force constants λ_j in equations (3.66) and (3.67) is an independent random variable with probability distribution function $G(\lambda)$. Specifically, Dyson considered a chain, C_n say, in which each λ_j is an independent random variable with the probability distribution

$$G_n(\lambda) = [n^n/(n-1)!]\,\lambda^{n-1}\exp(-n\lambda). \qquad \text{(A3.6.1)}$$

The integer n takes values 1, 2, 3, . . . , the distribution G_n has mean value unity and standard deviation $n^{-1/2}$. Thus C_1 represents a highly disordered chain, C_2 has less disorder while, in the limit $n \to \infty, C_n$ becomes the uniform chain with all force constants $\lambda_j = 1$.

The function $M_n(\mu)$ in equation (3.68), the proportion of the eigenfrequencies ω_j for which $\omega_j^2 \leqslant \mu$, can be obtained analytically with the choice of $G_n(\lambda)$ in equation (A3.6.1); but as the expressions are complicated, we merely note first that as $n \to \infty$ the perfectly periodic linear chain spectrum in equation (3.69) is recovered. In figure 3.14, Dyson's numerical results for the very disordered chain $n = 1$ (curve B) are compared with the result of equation (3.69) (curve A).

The limiting behaviour of M_n for large μ can be obtained as

$$M_n(\mu) \sim 1 - 2[\ln(n\mu) - s_{n-1} + \gamma]\exp(-n\mu) \times (n\mu)^{2n-1}[(n-1)!]^{-2} \qquad \text{(A3.6.2)}$$

where $s_j = \sum_{l=1}^{j} l^{-1}$ and γ is Euler's constant.

The singularity (referred to as a Van Hove singularity) at $\mu = 4$ in the result for $M_\infty(\mu)$ in equation (3.69) is blurred out in the very disordered chain corresponding to $n = 1$.

This general area of vibrational modes in disordered systems remains a problem of considerable interest on which a great deal of work still needs to be done.

Appendix 4.1

POLARON HAMILTONIAN IN CONTINUUM MODEL

We are primarily interested here in a single electron with Hamiltonian (neglecting the periodic potential)

$$H_{\mathrm{e}} = \frac{p^2}{2m} + c\Delta(\boldsymbol{r}) \qquad \text{(A4.1.1)}$$

where $c\Delta(\boldsymbol{r})$ is the deformation potential due to dilatation $\Delta(\boldsymbol{r})$. (More precisely, if we have a long wavelength disturbance in the lattice producing a local dilatation $\Delta(\boldsymbol{r})$, the crystal potential can be considered changed by the deformation potential. In general c will be an integral operator.) Without going into detail, we note that a uniform dilatation Δ will modify the one-electron energies, if isotropy can be assumed, to yield

$$E(\boldsymbol{k}) \simeq E_0(\boldsymbol{k}) + c\Delta, \qquad \text{(A4.1.2)}$$

where $E_0(\boldsymbol{k})$ is the eigenvalue in the unstrained crystal. For a long-wavelength disturbance, equation (A4.1.2) is generalised to equation (A4.1.1).

To see the basic significance of the constant c in equation (A4.1.1) we note, following Fröhlich (1954a), that the interaction between an excess electron and a dielectric continuum can be written as

$$\begin{aligned} H_{\mathrm{int}} &= -\int \boldsymbol{D}(\boldsymbol{r}-\boldsymbol{r}_{\mathrm{e}})\cdot \boldsymbol{P}(\boldsymbol{r})\mathrm{d}\boldsymbol{r} \\ &= e\int \nabla_{\boldsymbol{r}}(|\boldsymbol{r}-\boldsymbol{r}_{\mathrm{e}}|^{-1})\boldsymbol{P}(\boldsymbol{r})\mathrm{d}\boldsymbol{r}. \end{aligned} \qquad \text{(A4.1.3)}$$

This corresponds to the classical interaction between an electron with electric displacement $\boldsymbol{D}(\boldsymbol{r}-\boldsymbol{r}_{\mathrm{e}}) = -e\nabla_{\boldsymbol{r}}(|\boldsymbol{r}-\boldsymbol{r}_{\mathrm{e}}|^{-1})$ and the induced longitudinal polarisation field $\boldsymbol{P}(\boldsymbol{r})$. The interaction vanishes for transverse fields. The total longitudinal polarisation of the actual ionic lattice, caused by a slow conduction electron, can be separated into two parts:

$$\boldsymbol{P}_{\mathrm{tot}}(\boldsymbol{r}) = \boldsymbol{P}_{\mathrm{ir}}(\boldsymbol{r}) + \boldsymbol{P}_{\mathrm{opt}}(\boldsymbol{r}). \qquad \text{(A4.1.4)}$$

The first part, describing the displacement polarisation of the ionic lattice, is the infrared component, while the second contribution, taking into account the polarisation of each ion, is the optical component. The interaction of the second term $\boldsymbol{P}_{\mathrm{opt}}(\boldsymbol{r})$ with the excess electron is of no interest to us here, since the core electrons follow adiabatically the motion of the slow electron. Thus the optical polarisation is always excited, independent of the electron velocity. This interaction therefore contributes a term periodic in the lattice to the Hamiltonian and can be accounted for by choosing the mass appropriately, within the effective mass description. The term of interest for present purposes is the displacement polarisation $\boldsymbol{P}_{\mathrm{ir}}(\boldsymbol{r})$ which we write as $\boldsymbol{P}(\boldsymbol{r})$. In order to relate $\boldsymbol{P}$ and $\boldsymbol{D}$ in terms of an effective dielectric constant, it is useful (cf Appel 1968) to consider: (a) a test charge at rest, when one can write

$$\boldsymbol{D} = \epsilon_{\mathrm{s}} \mathscr{E} = \mathscr{E} + 4\pi \boldsymbol{P}_{\mathrm{tot}}; \tag{A4.1.5}$$

and (b) a test charge oscillating in a rigid lattice (no displacement) with a frequency higher than the optical frequencies, when

$$\boldsymbol{D} = \epsilon_{\infty} \mathscr{E} = \mathscr{E} + 4\pi \boldsymbol{P}_{\mathrm{opt}}. \tag{A4.1.6}$$

Taking the difference between $\boldsymbol{P}_{\mathrm{tot}}$ and $\boldsymbol{P}_{\mathrm{opt}}$ from equations (A4.1.5) and (A4.1.6) one finds

$$\boldsymbol{P}(\boldsymbol{r}) = \frac{\boldsymbol{D}}{4\pi}\left(\frac{1}{\epsilon_{\infty}} - \frac{1}{\epsilon_{\mathrm{s}}}\right) \tag{A4.1.7}$$

in terms of the static and high-frequency dielectric constants ϵ_{s} and ϵ_{∞}.

Then the interaction (A4.1.3) can readily be transformed to relate to the Hamiltonian (A4.1.1) if one relates Δ to a displacement polarisation. In this way one can find explicitly the value of c in equation (A4.1.1) in terms of the coupling constant α of the main text.

Appendix 5.1

FLUX QUANTISATION IN A SUPERCONDUCTOR

We discuss here, briefly, the phenomenon of flux quantisation in a superconductor. To do so, let us consider a cylindrical piece of superconductor, with a hole in the centre, as illustrated in figure A5.1 (cf Appendix 7.2).

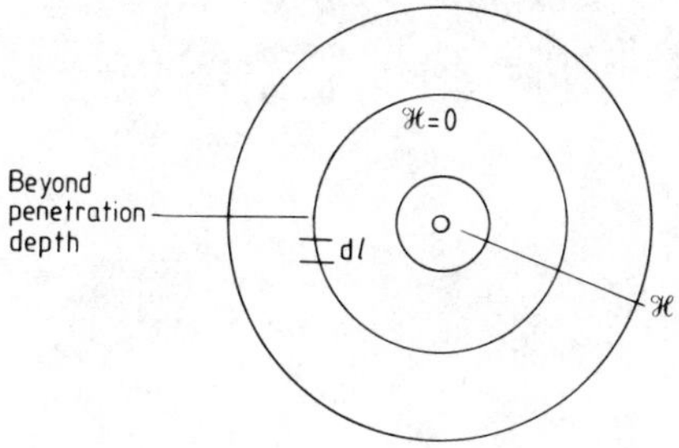

Figure A5.1 Hollow piece of superconductor, illustrating configuration used in discussing flux quantisation. Field $\mathscr{H}$ exists in hole.

We write the order parameter $\Delta(\boldsymbol{r})$ in terms of a phase $\chi(\boldsymbol{r})$ through

$$\Delta(\boldsymbol{r}) \propto \exp[\mathrm{i}\chi(\boldsymbol{r})]. \qquad \text{(A5.1.1)}$$

Since χ is a phase factor, the change in χ in going round a ring must equal $2\pi\nu$, where ν is an integer. Suppose now that a field $\mathscr{H}$ exists in the hole. There will be penetration of the field to a distance Λ say. Within the cylinder, beyond the penetration depth, $\boldsymbol{j}=0$ and hence from the equation

$$\boldsymbol{j} = 2en_s\boldsymbol{P}_s/m_{\text{eff}} \qquad \text{(A5.1.2)}$$

we have that $\boldsymbol{P}_s = \boldsymbol{0}$. But $\boldsymbol{P}_s$ is related to χ through

$$\boldsymbol{P}_s = \hbar\nabla\chi - \frac{2e}{c}\mathscr{A}, \qquad \text{(A5.1.3)}$$

where $\mathscr{A}$ is the vector potential, and hence when $\boldsymbol{P}_s = 0$

$$\hbar\nabla\chi = 2e\mathscr{A}/c. \qquad \text{(A5.1.4)}$$

Thus we have that

$$\oint \mathscr{A} \cdot \mathrm{d}\boldsymbol{l} = \frac{c\hbar}{2e} \times \text{ (change in } \chi \text{ in going round loop)}$$

$$= \frac{ch}{2e} \nu. \tag{A5.1.5}$$

But $\oint \mathscr{A} \cdot \mathrm{d}\boldsymbol{l}$ is simply the flux enclosed, and hence the flux is quantised in units of $ch/2e(=2 \times 10^{-7}\ \mathrm{G\ cm^2})$.

Appendix 5.2

PHENOMENOLOGICAL TREATMENT OF GINZBURG–LANDAU EQUATIONS

One of the most fruitful approaches to superconductivity has developed from the work of Ginzburg and Landau (1950), referred to already in §5.7. Before their work, Landau had developed the theory of second-order phase transitions which we shall refer to Chapter 8. This theory was based on the idea that a phase transition could be characterised by some order parameter and a simple assumption as to the form of the free energy as a function of the order parameter. Though we shall see in Chapter 8 that the form postulated by Landau does not always apply, it is in fact valid for superconductors.

In the Ginzburg–Landau theory, an essential point is that for a superconductor the order parameter must be identified with the macroscopic wavefunction Ψ†. Two important characteristics are, from the previous discussion: (a) the order parameter is complex; and (b) in general the order parameter varies in space.

Once the free energy has been expanded as a function of ψ and the vector potential $\boldsymbol{A}$, minimisation gives an equation of motion for ψ and an equation for the supercurrent in terms of $\boldsymbol{A}$. The latter equation has the form of the London equation so that the Ginzburg–Landau theory represents a generalisation of the London theory to situations in which ψ is spatially varying (cf §5.7).

Since the Ginzburg–Landau equations are much simpler than the microscopic theory, they are generally used when they are known to be valid. In addition the Ginzburg–Landau equations give valuable insight into the general qualitative behaviour of superconductors. They are particularly helpful in giving an understanding of the relationship between the various lengths (penetration depths, coherence lengths) involved in superconductivity.

As in the main text (cf equation (5.31)), the starting point is to define a function $\Psi(\boldsymbol{r})$ such that $|\Psi(\boldsymbol{r})|^2$ is the density of superconducting electrons at position $\boldsymbol{r}$. Thus Ψ is a kind of effective wavefunction which in the Gorkov (1959) derivation, for instance, turned out to be proportional to the local energy gap parameter $\Delta(\boldsymbol{r})$. This is a very reasonable result since the ability to create superconducting electrons is directly related to the size of Δ.

In terms of Ψ, the total free energy of the superconducting state in the presence of a magnetic field $H(\boldsymbol{r})$ is written plausibly in the form

† In the Appendix we reserve ψ for the ratio of Ψ to its zero field value Ψ_0 (see equation (A5.2.3)).

$$F = \int \mathrm{d}\boldsymbol{r} F_H^{\mathrm{s}} = \int \mathrm{d}\boldsymbol{r}\left(F_0^{\mathrm{s}}(\boldsymbol{r}) + \frac{1}{2m_{\mathrm{e}}}\left|-\mathrm{i}\hbar\nabla\Psi - \frac{e^*}{c}\boldsymbol{A}(\boldsymbol{r})\Psi\right|^2 + \frac{H^2(\boldsymbol{r})}{8\pi}\right). \tag{A5.2.1}$$

Here, as usual, e^* is the charge of a Cooper pair, i.e. $e^* = 2e$. The term F_0^{s} is written

$$F_0^{\mathrm{s}} = \frac{H_{\mathrm{c}}^2(T)}{8\pi}\left(1 - 2\left|\frac{\Psi}{\Psi_0}\right|^2 + \left|\frac{\Psi}{\Psi_0}\right|^4\right) \tag{A5.2.2}$$

where Ψ_0 denotes the zero field value of Ψ. The right-hand side of equation (A5.2.2) may be viewed as a truncated power series in Ψ and is thus most appropriate near the transition point. The coefficients are chosen to lead to an expression with the properties: (a) $F_0^{\mathrm{s}} = H_{\mathrm{c}}^2(T)/8\pi$ when $\Psi = 0$; and (b) $\partial F_0^{\mathrm{s}}/\partial\Psi = 0 = F_0^{\mathrm{s}}$ when $\Psi = \Psi_0$.

At this point, it is convenient to follow Abrikosov (1957) by using reduced variables. Thus, the following dimensionless quantities are introduced:

$$\frac{\Psi}{\Psi_0} = \psi, \quad \frac{A}{2^{1/2}H_{\mathrm{c}}(T)} = \boldsymbol{a}, \quad \frac{H}{2^{1/2}H_{\mathrm{c}}(T)} = \boldsymbol{h}, \quad K = \frac{2^{1/2}e^*}{\hbar c}H_{\mathrm{c}}(T)\lambda_{\mathrm{L}}^2. \tag{A5.2.3}$$

In the definition of K, the fundamental length λ_{L} given by (cf equation (5.39))

$$\lambda_{\mathrm{L}}^2 = \frac{m_{\mathrm{e}}c^2}{4\pi e^{*2}|\Psi_0|^2} \tag{A5.2.4}$$

is conveniently used as the unit of length in rewriting the free energy as

$$\frac{4\pi F}{H_{\mathrm{c}}^2(T)} = \int \mathrm{d}\boldsymbol{r}[\tfrac{1}{2} - |\psi|^2 + \tfrac{1}{2}|\psi|^4 + h^2(\boldsymbol{r}) + |\mathrm{i}K^{-1}\nabla\psi + \boldsymbol{a}\psi|^2]. \tag{A5.2.5}$$

The next step is to vary ψ and $\boldsymbol{a}$ in equation (A5.2.5). The respective Euler equations are

$$[(\mathrm{i}K^{-1})\nabla + \boldsymbol{a}]^2\psi = \psi - |\psi|^2\psi \tag{A5.2.6}$$

and, since $\boldsymbol{h} = \operatorname{curl}\boldsymbol{a}$,

$$-\operatorname{curl}\operatorname{curl}\boldsymbol{a} = |\psi^2|\boldsymbol{a} + (\mathrm{i}/2K)(\psi^*\nabla\psi - \psi\nabla\psi^*). \tag{A5.2.7}$$

Equations (A5.2.6) and (A5.2.7) are the Ginzburg–Landau equations. For solutions ψ of these equations, (A5.2.5) becomes

$$\frac{4\pi F}{H_{\mathrm{c}}^2(T)} = \int \mathrm{d}\boldsymbol{r}[\tfrac{1}{2} - \tfrac{1}{2}|\psi|^4 + h^2(\boldsymbol{r})]. \tag{A5.2.8}$$

The quantities $H_{\mathrm{c}}(T)$, λ_{L} and K are the three essential parameters of the theory. $H_{\mathrm{c}}(T)$, as defined above, is the thermodynamic bulk critical field and is given in terms of the magnetisation M by

$$H_c^2(T)/8\pi = -\int_0^\infty M\,\mathrm{d}H. \tag{A5.2.9}$$

For type-I superconductors, equation (A5.2.9) is hardly necessary for $H_c(T)$ is then the usual critical field, above which $M = 0$ and below which we have the property $M = -H/4\pi$ of a perfect diamagnet. But in type-II materials, the $M(H)$ relationship is more complicated, as illustrated in figure A5.2. Here the observed magnetisation M versus field H curves for typical type-I and -II superconductors are shown. Whereas in the first case there is the usual single transition field H_c already referred to, in type-II materials it will be seen that there are two transitions, one at H_{c1}, corresponding to an initial penetration of flux, and the other at H_{c2} when the penetration is complete.

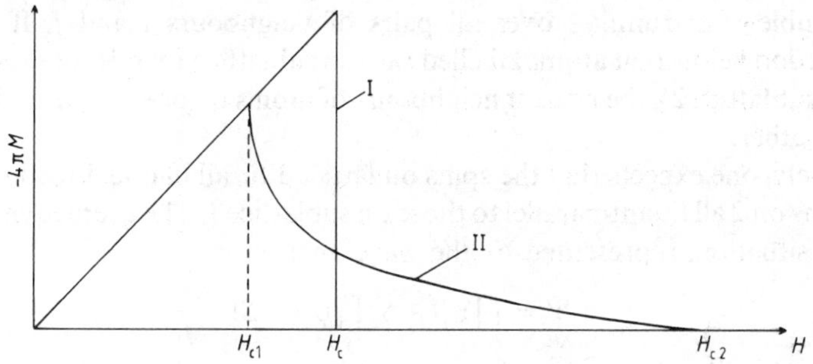

Figure A5.2 Observed forms of magnetisation M versus field H curves for typical type-I and type-II superconductors. Note that in type-I materials there is a single critical field H_c. In type-II materials in contrast there are two transitions, one at H_{c1} corresponding to initial penetration of flux, and the second at H_{c2} when the penetration is complete.

Actually (Saint-James and de Gennes 1963) it has been found that because of a very persistent superconducting sheath near the surface, an extremely weak diamagnetic effect can obtain up to some maximum value H_{c3}, but we shall not go into this here. If the two curves in figure A5.2 enclose the same area, then because of equation (A5.2.9), H_c will be the thermodynamic critical field for the type-II case.

Appendix 6.1

BOUNDS ON THE ENERGY OF THE ANTIFERROMAGNETIC GROUND STATE

The energy of the ground state of a simple antiferromagnetic lattice is not known exactly, except in the simplest case of atoms with $S = \frac{1}{2}$ on a linear chain (Hulthén, 1938).

For such a lattice, the Heisenberg Hamiltonian reads

$$H = J\sum_i\sum_j \boldsymbol{S}_i \cdot \boldsymbol{S}_j \tag{A6.1.1}$$

the double sum running over all pairs of neighbours i and j. It will be understood below that atoms labelled i are on sublattice 1 while those labelled j are on sublattice 2†, the nearest neighbours of atoms on one sublattice lying all on the other.

Naively, one expects that the spins on lattice 1 lie all in one direction, while the spins on 2 all lie antiparallel to those on sublattice 1. The average energy for such a situation, represented by the wavefunction

$$\Psi = \prod_i \psi_i(+S)\prod_j \psi_j(-S) \tag{A6.1.2}$$

is rather readily obtained as

$$\langle \Psi H \Psi \rangle = -\tfrac{1}{2}NZJS^2 \tag{A6.1.3}$$

where N is the number of atoms and Z is the number of nearest neighbours. Unfortunately, Ψ above is not an eigenstate of H, and all that we can therefore conclude is that the ground state energy E_g satisfies

$$E_g < -\tfrac{1}{2}NZJS^2 \tag{A6.1.4}$$

from the variational principle (Anderson 1951).

A lower bound may also be obtained by the following argument, also due to Anderson (1951). Divide H into a sum of terms:

$$H = \sum_i H_i \qquad H_i = J\sum_j \boldsymbol{S}_i \cdot \boldsymbol{S}_j. \tag{A6.1.5}$$

The sum over j in H_i now runs over all atoms j of sublattice 2 which are neighbours to the atom i of sublattice 1. But the diagonal energies of the H_i can be obtained (see, e.g. Weiss 1948). The lowest energy is

$$E_i = -JS(ZS+1) \tag{A6.1.6}$$

† It is here assumed that the lattice is sufficiently simple (e.g. BCC) so that it can be divided into two sublattices (simple cubic for BCC case).

which is found when the cluster of neighbour spins S_j is given its maximum angular momentum $\Sigma_j S_j = ZS$ and S_i is then made as nearly antiparallel to the surrounding cluster as it can be.

Now the lowest eigenvalue of the total Hamiltonian must be greater than the sum of the least eigenvalues of its parts:

$$E_g > \sum_i [-JS(ZS+1)]$$
$$E_g > -\tfrac{1}{2}NZJS^2(1+1/ZS). \qquad \text{(A6.1.7)}$$

This follows from the fact that the ground state energy must be the sum of the diagonal elements of H_i. But the variational principle says that all possible diagonal elements of any matrix are greater than or equal to the least element. Hence we can write the inequalities:

$$-\tfrac{1}{2}NJZS^2 > E_g > -\tfrac{1}{2}NJZS^2(1+1/ZS).$$

In an example such as MnF_2, $Z = 8$ and $S = \frac{5}{2}$ and the fractional variation allowed by the inequalities for E_g is only 1/20. This is an indication that the wavefunction of the ground state is not very different from our naive picture based on the two-sublattice model.

However, in Hulthén's soluble case, he found

$$E_g(S = \tfrac{1}{2}, Z = 2) = -1.773\ (\tfrac{1}{2}NZ)JS^2. \qquad \text{(A6.1.8)}$$

The above limits are correct, though somewhat wide, in this unfavourable case. Anderson (1952) also gives a semiclassical argument leading to the approximate dispersion relation

$$\omega = 2JS \sin(kR) \qquad \text{(A6.1.9)}$$

for antiferromagnetic spin waves†. This argument parallels that given a little earlier by Klein and Smith (1951) for ferromagnetic spin waves. The above dispersion relation shows that $\omega \propto k$ in the long-wavelength limit, in contrast to the quadratic dispersion relation for long-wavelength ferromagnons. For a discussion of phenomenology of antiferromagnetism, the reader may consult Nagamiya *et al* (1955).

† Mattheiss' work discussed in Appendix 6.2 confirms the validity of equation (A6.1.9).

Appendix 6.2

ANTIFERROMAGNETIC LINEAR CHAIN

Mattheiss (1961) has carried out many-electron-configuration interaction calculations on a system of six hydrogen atoms arranged in a regular hexagonal array with a variable lattice spacing.

The approximate wavefunctions for this system have been expressed as linear combinations of the $(2 \times 6)!/(6!) = 924$ determinantal functions which can be formed from atomic 1s functions.

In this manner, the effects of ionic configurations containing as many as three pairs of doubly filled orbitals have been introduced into the calculations. All three- and four-centre integrals have been taken into account. Instead of working with non-orthogonal hydrogenic 1s orbitals localised on the different atomic sites, Mattheiss works with orthonormal Wannier functions (see e.g. Jones and March 1973).

His main result is that the effects of configuration interaction (correlation) can be represented quite accurately at large internuclear separations in terms of a parameter J' (analogous to a near-neighbour exchange integral) which assumes a *negative* value in a non-ferromagnetic system like H_6. This provides a first principles justification for the use of the Heisenberg exchange operator

$$-2J'\left(\sum_i \boldsymbol{S}_i \cdot \boldsymbol{S}_{i+1} - \tfrac{1}{4}\right)$$

to describe the magnetic interaction at large separations in this system.

In addition, Mattheiss's work shows:

(a) H_6 is bound with respect to six separated H atoms;
(b) H_6 is unstable with respect to three molecules;
(c) The ground-state is a singlet at all internuclear separations.

The general form of the energy curves as a function of internuclear separation turns out to be extremely similar to that obtained for the hydrogen molecule.

It should be clear to the reader that this system of six H atoms arranged in a regular hexagonal array is qualitatively similar to that of a linear chain of six hydrogen atoms with periodic boundary conditions†.

The fundamental interest in Mattheiss's work is that, at large separations, a first principles justification is thereby provided for the use of a Heisenberg Hamiltonian. Also, Anderson's semiclassical result that the energy (frequency) of spin waves, $\omega \propto \sin(kR)$ (see Appendix 6.1), is given a basic justification.

† To the reader initiated in group theory, this statement can be sharpened up by saying that the group of the Hamiltonian (C_{6v}) is the same for both systems.

Appendix 6.3

SPIN WAVES IN THE ONE-DIMENSIONAL ANTIFERROMAGNET $CsCoCl_3$

An example of a system which, to a first approximation, can be described as an antiferromagnetic chain of N spins with $S = \frac{1}{2}$ is $CsCoCl_3$ (see Tellenbach and Arend 1977 and other references given there).

The Hamiltonian may be written

$$H = 2J \sum_i [S_i^z S_{i+1}^z + \beta(S_i^x S_{i+1}^x + S_i^y S_{i+1}^y)]. \tag{A6.3.1}$$

The spin-wave spectrum of the isotropic ($\beta = 1$) antiferromagnetic chain has been calculated exactly by des Cloiseaux and Pearson (1962).

Tellenbach and Arend report the extension to the anisotropic case $\beta \neq 1$, and they obtain for the dispersion relations

$$\hbar\omega(k) = 2J\,\epsilon\,(\phi_k) \qquad 0 \leqslant k \leqslant \pi \tag{A6.3.2}$$

where

$$\epsilon(\phi_k) = \frac{a^2-1}{a^2+1}\left(\frac{a-1}{a+1} + 2\sum_{m=1}^{\infty} \frac{1}{a^m + a^{-m}} \cos(m\phi_k) - 2\sum_{m=1}^{\infty} \frac{(-1)^m}{(a^{2m}+1)a^m}\right). \tag{A6.3.3}$$

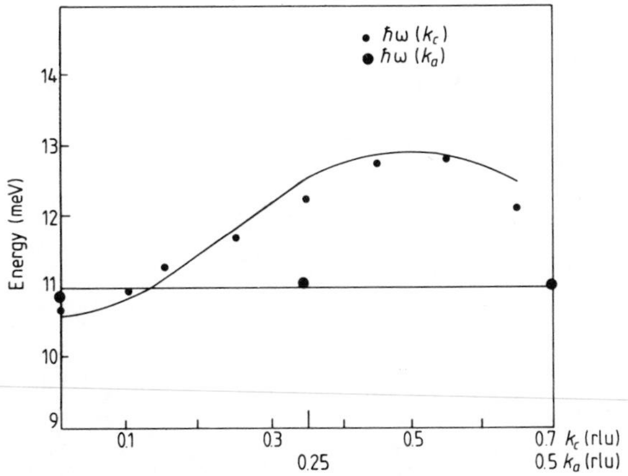

Figure A6.1 Shows measured points for spin wave dispersion parallel (k_c) and perpendicular (k_a) to the magnetic chains measured relative to (001) in $CsCoCl_3$ at 23 K. Full curve is theoretical result fitted to experiment. Upper scale on horizontal axis is for k_c, lower for k_a, and both are measured in reciprocal lattice units (rlu). (From Tellenbach U and Arend H 1977 *J. Phys. C: Solid St. Phys.* **10** 1311.) One-dimensionality is reflected in no dispersion along k_a perpendicular to the chain.

The angle ϕ_k and the quantity a are to be found from the equations

$$k = \tfrac{1}{2}(\phi_k + \pi) + 2 \sum_{m=1}^{\infty} \frac{\sin(m\phi_k)}{m(a^m + a^{-m})} \tag{A6.3.4}$$

and

$$\beta = 2a/(1 + a^2). \tag{A6.3.5}$$

Figure A6.1 shows the measured points for the spin wave dispersion parallel (k_c) and perpendicular (k_a) to the magnetic chains.

Tellenbach and Arend used the above theory to extract the values

$$J = (6.44 \pm 0.06)\text{meV}$$
$$\beta = 0.094 \pm 0.007$$

by least squares fitting of the theoretical spin wave frequencies to the experiments.

Appendix 7.1

ELECTRICAL CONDUCTIVITY AND VAN HOVE FUNCTION $S(K, \omega)$

As emphasised by Baym (1964), the inelastic scattering of a slow neutron from a metal is a process which is quite similar to the inelastic scattering of a conduction electron from the lattice vibrations in a metal. In each case, the coupling is not to individual phonons, but to the ion density. Below, we follow Baym in assuming that in each case the Born approximation may be employed.

The Bloch electrons, in a metal with rigid-ion cores, are scattered via a screened potential by the fluctuations of ion density. The scattering probability is thus proportional to the number $S'(q, \omega)$ of available states for the density fluctuations. This number is just the Van Hove correlation function $S(q, \omega)$ with the Bragg peaks subtracted out.

One now employs the Boltzmann equation as the basis of the transport theory. In a collision with the lattice, an electron in a Bloch state $\boldsymbol{k}$ with energy $E_{\boldsymbol{k}}$ is scattered into a state $\boldsymbol{p}$ with energy $E_{\boldsymbol{p}}$ by creating a density fluctuation with momentum $\boldsymbol{K}$ and energy ω, putting $\hbar = 1$. Let us suppose that the matrix element of the screened potential U is $\langle \boldsymbol{p} U(q, \omega) \boldsymbol{k} \rangle$ for this process. The scattering rate is then given by

$$2\pi\delta(E_{\boldsymbol{p}} - E_{\boldsymbol{k}} - \omega) f(\boldsymbol{k})[1 - f(\boldsymbol{p})]\rho_i S'(q, \omega) |\langle \boldsymbol{p} U \boldsymbol{k} \rangle|^2 \qquad \text{(A7.1.1)}$$

where $f(\boldsymbol{k})$ is the density of electrons in the Bloch state $\boldsymbol{k}$, $1 - f(\boldsymbol{p})$ is the density of available final states $\boldsymbol{p}$, while ρ_i is the ionic density. The collision term may then be written

$$\left(\frac{\partial f(\boldsymbol{k})}{\partial t}\right)_{\text{coll}} = \sum_{\boldsymbol{p}, \boldsymbol{q}} \int_{-\infty}^{\infty} \mathrm{d}\omega\, \delta(E_{\boldsymbol{p}} - E_{\boldsymbol{k}} - \omega)\, \rho_i |\langle \boldsymbol{p} U(q, \omega) \boldsymbol{k} \rangle|^2$$
$$\times \{f(\boldsymbol{p})[1 - f(\boldsymbol{k})]S'(-q, -\omega) - f(\boldsymbol{k})[1 - f(\boldsymbol{p})]S'(q, \omega)\}. \qquad \text{(A7.1.2)}$$

The identification of S, minus the Bragg elastic peaks, with the experimentally observed equilibrium density of states for lattice fluctuations assumes that the phonon rate of approach to equilibrium is much more rapid than electron–phonon scattering rates. One must add the condition of detailed balance:

$$S'(-q, -\omega) = \exp(-\beta\omega) S'(q, \omega) \qquad \text{(A7.1.3)}$$

where $\beta = 1/k_{\mathrm{B}}T$.

The customary variational calculation with the Boltzmann equation (see e.g. Ziman 1960), in which we linearise the collision term, assumes free electrons with effective mass m_e, and takes the matrix element to depend only on q; i.e.

$$|\langle \boldsymbol{p}U(q,\omega)\boldsymbol{k}\rangle|^2 \rightarrow |U(q)|^2. \tag{A7.1.4}$$

This then leads to the electrical conductivity as the usual form, with ρ_0 the electron density,

$$\sigma = \frac{\rho_0 e^2}{m_e}\tau \tag{A7.1.5}$$

where the inverse relaxation time is found to be

$$\frac{1}{\tau} = \frac{m_e}{12\pi^3 Z}\int_0^{2k_F} \mathrm{d}q q^3 |U(q)|^2 \int_{-\infty}^{\infty} \frac{\mathrm{d}\omega}{2\pi} \frac{S'(q,\omega)\beta\omega}{e^{\beta\omega}-1} \tag{A7.1.6}$$

with Z written for the valence.

This equation was also derived by Mannari (1961) for the special case of a liquid metal. This formula goes over into the usual weak-scattering formula for a liquid metal (Ziman 1961, see also Krishnan and Bhatia 1945) since $\beta\omega/(e^{\beta\omega}-1)$ can then be replaced by unity and the integral over ω obviously gives the static structure factor $S(q)$ from equation (7.18).

Appendix 7.2

RELATION BETWEEN SUPERFLUIDITY AND SUPERCONDUCTIVITY

In this appendix, we first make some general observations, of a qualitative kind, about superfluids. Secondly, in He II we discuss the quantisation of circulation (see equation (A7.2.4) below for its definition) and compare it with flux quantisation already discussed in Appendix 5.1. Finally, we add a short argument relating to the form of the free energy used in equation (5.44) and the preceding equation.

In §7.8.1 we treated Bose–Einstein condensation in an ideal gas. The physical origin of the long-range order characteristic of superfluids (liquid ^{4}He and superconductors, which we can view as a neutral and a charged fluid respectively) can be traced to such Bose–Einstein condensation (cf Geballe and White 1980, for solids). This is, now in the interacting fluid, the occupation of a single quantum state by a macroscopic number of particles (in Bose liquids) or particle pairs (in Fermi liquids; and in particular Cooper electron pairs in superconductors). One can think of the condensed units as held together by exchange forces and as a result the whole condensate has negligible fluctuations and moves as a whole. This leads then to superfluid behaviour, persistent currents and to other macroscopic manifestations.

A7.2.1 Quantisation of circulation

Following these general remarks on the relation of superfluidity and superconductivity, let us turn to a specific topic: the quantisation of circulation in superfluid He II. We shall stress the relation to flux quantisation in a superconductor.

The description of a superfluid in terms of a wavefunction of the form $|\psi| \exp(\mathrm{i}\chi)$ allows one to write the superfluid currents as

$$j_s = \hbar|\psi|^2\nabla\chi \tag{A7.2.1}$$

and (cf equation (5.34))

$$J_e = \frac{e\hbar}{m_e}|\psi|^2\nabla\chi - \frac{2e^2}{m_e}|\psi|^2 A. \tag{A7.2.2}$$

Equation (A7.2.1) for the mass current density j_s applies to He II while equation (A7.2.2) for the electric current density J_e applies to superconductors, A being as usual the vector potential. The discussion below follows closely that of Tilley and Tilley (1974).

Equation (A7.2.2) implies the normalisation $|\psi|^2 = n_s/2$, the density of Cooper pairs. The above equations apply provided that the flow velocity is sufficiently small for the superfluid density to remain unchanged. In addition, equation (A7.2.2) is valid only for superconductors of the London type, and not for those of the Pippard (1960) type which have a non-local character represented by

$$J(\mathbf{r}) \propto \int \frac{\boldsymbol{R}(\boldsymbol{R}\cdot\boldsymbol{A}(\boldsymbol{r}'))}{R^4} \exp\left(-\frac{R}{\xi}\right) \mathrm{d}\boldsymbol{r}' \tag{A7.2.3}$$

with $\boldsymbol{R} = \boldsymbol{r} - \boldsymbol{r}'$, $R = |\boldsymbol{R}|$, and ξ an appropriate coherence length.

It is possible with singly connected superconductors to use a (gauge) transformation to eliminate the term in equation (A7.2.2) involving $\nabla\chi$. But for a multiply connected specimen, one with holes in it, such a transformation cannot be made. Below we shall deal with multiply connected regions in He II and (very briefly) in superconductors.

First, consider He II occupying an annular region such as the space between two coaxial cylinders, as shown in figure A7.1. It will be assumed that $T = 0$ so that the He II is pure superfluid. To find the flow pattern, the quantity to examine is

$$\kappa = \oint \boldsymbol{v}_s \cdot \mathrm{d}\boldsymbol{l} \tag{A7.2.4}$$

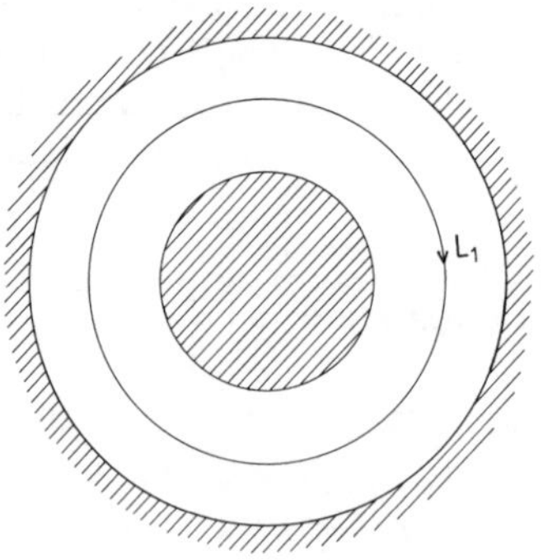

Figure A7.1 Configuration considered in discussion of quantisation of circulation in He II. Shows superfluid occupying an annular (multiply connected) region; contour L_1 is drawn wholly in the superfluid.

the integral being taken round any contour wholly within the liquid.

Now equation (A7.2.1) implies that the superfluid velocity can be written in the form, with M the mass of a ^{4}He atom,

$$v_s = \frac{\hbar}{M}\nabla\chi \tag{A7.2.5}$$

and thus the quantity κ, known as the circulation, defined in equation (A7.2.4), can be expressed in terms of the phase χ of the wavefunction as

$$\kappa = \frac{\hbar}{M} \oint \nabla\chi \cdot d\boldsymbol{l}. \tag{A7.2.6}$$

For the circle L_1 in figure A7.1 the circulation is

$$\kappa = \frac{\hbar}{M} (\Delta\chi)_{L_1}. \tag{A7.2.7}$$

Since the superfluid wave function must be single-valued, a complete circuit round a closed contour must leave it unchanged, and thus the change in the phase χ can only be an integral multiple of 2π or zero. From equation (A7.2.7) we see therefore that the circulation is quantised, taking the values

$$\kappa = nh/M \qquad n = 0, 1, 2, \ldots. \tag{A7.2.8}$$

The quantity h/M is evidently the quantum of circulation and has the value $9.98 \times 10^{-8}\ m^2\ s^{-1}$.

The annulus in figure A7.1 is an example of a multiply connected region, since it contains what can be regarded as a 'hole' in the superfluid. The presence of holes, that is regions which the superfluid cannot penetrate, yet which are completely surrounded by superfluid, ensures that contours such as L_1 in figure A7.1 can be drawn. As we have seen above, this is a sufficient condition for the quantisation of circulation.

In an entirely similar manner, we can use equation (A7.2.2) to discuss the DC magnetic properties of multiply connected superconductors. Just as for He II, it is convenient to define a superfluid velocity v_s by

$$\boldsymbol{J}_e = 2e\,|\psi|^2\, \boldsymbol{v}_s \tag{A7.2.9}$$

and we then rearrange equation (A7.2.2) into the form

$$\hbar\nabla\chi = 2m_e \boldsymbol{v}_s + 2e\boldsymbol{A}. \tag{A7.2.10}$$

Integrating this equation round a contour L_1 encircling a hole in the superconductor leads, essentially, to the same situation as discussed already in Appendix 5.1 and to a flux quantum $h/2e$, having the value 2.07×10^{-5} V s.

A7.2.2 Vortices

In superconductors, as in He II, the holes round which quantisation occurs can be either physical boundaries or the cores of quantised vortex lines. Vortex lines are not found in all superconductors and in particular vortices never appear in type I superconductors. There is a close similarity between vortices in He II and superconductors. But there are important differences; for example vortices in superconductors are coupled to the applied magnetic field whereas a magnetic field has no effect on He II vortices.

Experimentally, the quantisation of circulation in He II was first demonstrated by Vinen (1961).

The final task in this appendix is to give an argument for the form of free energy adopted in equation (5.44) of the main text.

The free energy F of the superconductor is a function of the momentum $\boldsymbol{p}_s$ of a Cooper pair which can be related to the phase as above, namely through

$$\boldsymbol{p}_s = \hbar \nabla \chi. \tag{A7.2.11}$$

Thus in the example leading to equation (5.44) we can write

$$F = F\left(\hbar \frac{\chi_1 - \chi_2}{d}\right). \tag{A7.2.12}$$

In the presence of a vector potential we obtain

$$F = F\left(\hbar \nabla \chi - \frac{2e}{c} \boldsymbol{A}\right). \tag{A7.2.13}$$

But the current carried by a system in the presence of a magnetic field can be calculated quite generally as

$$j = -c\,\delta F / \delta \boldsymbol{A} \tag{A7.2.14}$$

and hence we can write

$$j = \frac{2e}{\hbar} \frac{\partial F}{\partial (\nabla \chi)} \tag{A7.2.15}$$

from which equation (5.44) follows.

Appendix 8.1

ONE-DIMENSIONAL METAL IN THE PRESENCE OF A STRONG ELECTRON–PHONON INTERACTION

Fröhlich (1954b) has shown that if the interaction between electrons and lattice vibrations in a one-dimensional metal is sufficiently strong, a gap in the single-electron energy spectrum results. This is due to very strong excitation of lattice modes with wavenumber $\pm 2k_F$, which we discussed in §4.8.

The system treated by Fröhlich is not an insulator at absolute zero, despite the gap in the energy spectrum, since the periodic variation of electron density is tied to these resonant lattice displacements and not to the lattice itself. In fact, with periodic boundary conditions (i.e. a circular system) states exist in which the electrons and associated lattice displacements move through the lattice in an organised way, carrying an electric current without any resistance.

At finite temperature T, some electrons will be excited into states above the gap, and this will reduce the periodic variation in electron density. This in turn will reduce the amplitude of the resonant lattice modes and thereby result in a narrowing of the energy gap. The gap width is $2E_F|\beta|$, where E_F is the Fermi energy $\hbar^2 k_F^2/2m_e$ and $|\beta|$ is a dimensionless parameter proportional to the amplitudes of the resonant modes at $\pm 2k_F$.

Kuper (1955) has, essentially, applied molecular field theory to the Fröhlich model. The quantity $|\beta|$ plays the role of an order parameter and Kuper shows that a finite temperature T_c exists for which $|\beta|$ vanishes. Above T_c, the level spectrum is that of a normal metal. The specific heat is calculated by Kuper for $T \simeq T_c$ also. He shows that the system exhibits a second-order transition at T_c.

It is true that in order to discuss the transition, an extrapolation is required, beyond the proven range of validity of the model. For as $|\beta| \to 0$, $2E_F|\beta|$ must eventually fall below the energy of the most energetic possible phonon. Such a phonon could then cause a transition across the energy gap. If, however, the transition temperature T_c is very much greater than the Debye temperature Θ_D, then the approximations still hold up to very near T_c, so that only a small extrapolation is needed.

By means of the molecular field approximation, Kuper obtains the result that the specific heat for T near to T_c has the form shown in figure A8.1. The analytic form is in fact

$$\frac{c}{Nk_B} = \frac{c_D}{Nk_B} + \frac{4.45k_B T_c}{E_F} + \frac{15.9k_B(T-T_c)}{E_F} + \dots \tag{A8.1.1}$$

where c_D is the Debye specific heat term. As mentioned in the main text,

Kuper's result appears to have relevance to the thermal properties of one-dimensional conductors.

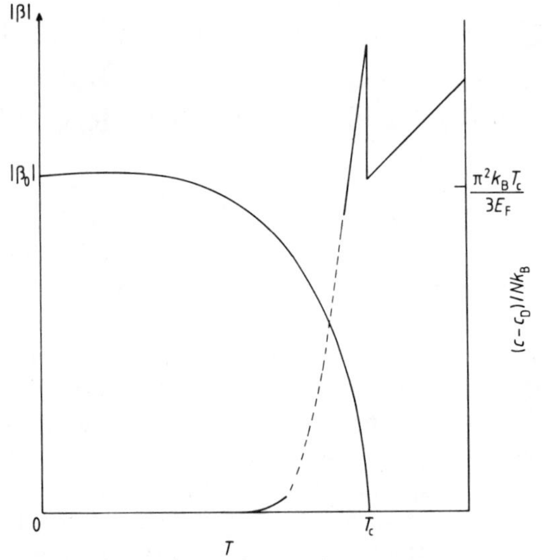

Figure A8.1 Molecular field approximation of specific heat of a one-dimensional metal in presence of strong electron–phonon interaction. β measures the width of the gap, which is given by $2E_F|\beta|$, with E_F the zero-temperature Fermi energy. Above T_c the metal is normal so that

$$\frac{c - c_D}{N_l k_B} = \frac{\pi^2 k_B T}{3E_F}.$$

Finally $k_B T_c = 1.14\, E_F |\beta_0|$.

Appendix 8.2

MODEL OF PHASE TRANSITION IN Nb_3Sn, AND PARTICULARLY ELASTIC PROPERTIES

Noolandi and Sham (1973) have set up a model to describe the Nb_3Sn phase transition. What they obtain is a formula for the change in thermodynamic potential in terms of the deviation of the Nb_3Sn crystal from its cubic structure.

Let us denote the positions of the ions in the cubic structure by $\boldsymbol{x}_{lK}$ where l labels the unit cell and K the ion in the unit cell. We shall confine attention to static and uniform distortions and strains. Thus the displacement of the (lK) ion in the α direction is

$$u_{lK\alpha} = d_{K\alpha} + s_{\alpha\beta} x_{lK\beta} \tag{A8.2.1}$$

where $s_{\alpha\beta}$ denotes the strain, $d_{K\alpha}$ denotes the displacement of the K sublattice and summation over the repeated suffix β is implied.

The contribution to the thermodynamic potential is separated into two parts arising: (a) from the interaction between the ions, including screening from s and p electrons; and (b) from the d electrons in the band (Sham 1972b).

For static displacements, the first part is just the potential energy of the ions screened by the s and p electrons. We expand the potential energy in powers of the displacement of the form (A8.2.1) and neglect anharmonic contributions.

The term in the potential energy linear in the ionic displacement vanishes because of the symmetry properties of the cubic structure. The coefficients of the quadratic terms can be expressed in terms of the coefficients in the long-wavelength expansion of the ionic part of the dynamical matrix (cf Chapter 3; see also Born and Huang 1954, Maradudin *et al* 1971) to yield for these coefficients

$$\begin{aligned} C_{\alpha\alpha'}(\boldsymbol{q}; K, K') = {} & C^{(0)}_{\alpha\alpha'}(K, K') + \mathrm{i} q_\beta C^{(1)}_{\alpha\alpha'\beta}(K, K') \\ & + \tfrac{1}{2} q_\beta q_{\beta'} C^{(2)}_{\alpha\alpha'\beta\beta'}(K, K') + \ldots \end{aligned} \tag{A8.2.2}$$

where $\boldsymbol{q}$ is the wavevector (for metallic crystals, reference should be made to Sham 1972a).

In the harmonic approximation, the lattice potential energy per unit cell is, aside from a constant,

$$\Phi = \tfrac{1}{2}\sum_j \Omega_j^2 d_j^2 + \sum_j d_j P_{\alpha\beta}(j) s_{\alpha\beta} + \tfrac{1}{2} Q_{\alpha\alpha'\beta\beta'} s_{\alpha\beta} s_{\alpha'\beta'}. \tag{A8.2.3}$$

The first sum is from terms quadratic in the relative sublattice displacement

$\boldsymbol{d}_K \cdot \Omega_j$ is the zero-wavevector phonon frequency obtained from diagonalising $C^{(0)}_{\alpha\alpha'}(K,K')$ and

$$d_j = \sum_K M_K^{1/2} \boldsymbol{e}(K|j)\boldsymbol{d}_K \tag{A8.2.4}$$

where M is the ionic mass and $\boldsymbol{e}(K|j)$ denotes the polarisation vector of the phonon with zero wavevector (Sham 1972b).

The second sum in equation (A8.2.3) is bilinear in the sublattice displacements and the strain, with the coefficients given by

$$P_{\alpha\beta}(j) = \sum_{K,K'} e_{\alpha'}(K'|j) C^{(1)}_{\alpha'\alpha\beta}(K',K) M_K^{1/2}. \tag{A8.2.5}$$

The third term is the strain energy with coefficients

$$Q_{\alpha\alpha'\beta\beta'} = \tfrac{1}{2} \sum_{K,K'} (M_K M_{K'})^{1/2} C^{(2)}_{\alpha\alpha'\beta\beta'}(K,K') \tag{A8.2.6}$$

At this point, we must invoke a model for the dynamics of the d electrons, to calculate their contribution to the thermodynamic potential. If we denote by $\epsilon_{\lambda\boldsymbol{k}}$ the d-electron energy in the cubic crystal for the chains in the λ direction, then assuming a simple model in which the lattice distortion only changes the d-electron energy without mixing the different d bands, one can take as the energy

$$E_\lambda(\boldsymbol{k}) = \delta_{\lambda\lambda'}\left[\epsilon_{\lambda\boldsymbol{k}} + \sum_j \sigma_{\lambda\lambda'}(j) d_j + \zeta_{\lambda\lambda'\alpha\beta} s_{\alpha\beta}\right] \tag{A8.2.7}$$

where

$$\delta_{\lambda\lambda'} = \begin{cases} 1 & \lambda' = \lambda \\ 0 & \lambda' \neq \lambda. \end{cases}$$

The second term on the right-hand side is the change in the electronic Hamiltonian due to sublattice displacement, while the third term is due to the strains. Hence the d-electron thermodynamic potential per unit cell is

$$G_e = -k_B T N^{-1} \sum_{\lambda,\boldsymbol{k}} \ln\{1 + \exp[(\mu - E_{\lambda\boldsymbol{k}})/k_B T]\} \tag{A8.2.8}$$

where μ is the chemical potential. Hence the total thermodynamic potential per unit cell is

$$G = \Phi + G_e. \tag{A8.2.9}$$

The most important feature of this is the inclusion of the energy terms due to the sublattice displacement involving the coefficients Ω_j, $P_{\alpha\beta}(j)$ and $\sigma_{\lambda\lambda}(j)$.

A8.2.1 Equilibrium conditions

The spontaneous distortion of the crystal from its cubic configuration is determined by minimising the thermodynamic potential. The chemical potential is evidently given by

$$n_{\mathrm{d}} = -\partial G/\partial \mu = N^{-1} \sum_{\lambda, k} f_{\lambda k} \tag{A8.2.10}$$

where n_{d} is the number of d electrons per unit cell and

$$f_{\lambda k} = \frac{1}{1+\exp[(E_{\lambda k}-\mu)/k_{\mathrm{B}}T]}. \tag{A8.2.11}$$

If one chooses to make the dilatation vanish in the above model for all temperatures it follows that (Noolandi and Sham 1973)

$$\zeta_{1x} + 2\zeta_{1y} = 0. \tag{A8.2.12}$$

There are two possible solutions for $(s_{xx}-s_{yy})$-type shear and the type of sublattice displacement denoted by Γ_{12} in figure 8.14:

$$\begin{array}{lll} \text{(a)} & d(\Gamma_{12}^{(1)}) = 0 & s_{yy} = s_{zz} \\ \text{(b)} & d(\Gamma_{12}^{(2)}) = 0 & s_{yy} = -s_{zz}. \end{array} \tag{A8.2.13}$$

The first type of distortion gives tetragonal symmetry about the x axis; the second type does not. However, a linear combination of the second solution about two different coordinate axes gives the first solution. Since diffraction experiments prove that the low-temperature phase is tetragonal, one writes

$$\begin{array}{ll} s_{xx} = -\epsilon & s_{yy} = s_{zz} = \tfrac{1}{2}\epsilon \\ d(\Gamma_{12}^{(2)}) = d & d(\Gamma_{12}^{(1)}) = 0. \end{array}$$

The thermodynamic potential is then simplified to

$$G = \tfrac{1}{2}\Omega^2 d^2 + 3Pd\epsilon + \tfrac{3}{4}(Q_{11}-Q_{12})\epsilon^2 - k_{\mathrm{B}}TN^{-1} \times \sum_{\lambda,k} \ln\{1+\exp[(\mu - E_\lambda - \epsilon_{\lambda k})/k_{\mathrm{B}}T]\} \tag{A8.2.14}$$

where

$$E_1 = -2\Delta, \qquad E_2 = E_3 = \Delta$$

and

$$\Delta = -\sigma d + \tfrac{1}{2}(\zeta_{1x} - \zeta_{1y})\epsilon.$$

The equilibrium conditions yield the following equations, from which the spontaneous shear strain, the sublattice shift and the deformation potential, can be determined:

$$d = -3\epsilon(P + 8\sigma A/U)/\Omega^2 \tag{A8.2.15}$$

$$\Delta = \tfrac{1}{2}\epsilon U[1 + 48(\sigma^2/\Omega^2)(A/U^2)] \tag{A8.2.16}$$

and

$$\epsilon = (v_1 - v_2)U/12A \tag{A8.2.17}$$

where

$$\nu_\lambda = N^{-1} \sum_{k} f_\lambda(\boldsymbol{k}). \tag{A8.2.18}$$

The combinations of coefficients determining U and A are such that

$$U = \zeta_{1x} - \zeta_{1y} + 6P\sigma\Omega^{-2}$$
$$8A = Q_{11} - Q_{12} - 6P^2\Omega^{-2}.$$

Crystal symmetry has been used to simplify the coefficients of the lattice potential energy (A8.2.3) and the electron deformation energy (A8.2.7). For example (see Noolandi and Sham 1973) $P_{\alpha\beta}(\Gamma_{12}^{(2)})$ is diagonal, with elements $-2P, P, P$. The coefficients $Q_{\alpha\alpha'\beta\beta'}$ have the same transformation properties as the elastic constants and therefore have three independent components. In equation (A8.2.7), $\sigma_{\lambda\lambda'}(j)$ vanishes for all the optical modes except the two Γ_{12} modes. Further, in the one-dimensional model of d electrons which neglects the interchain coupling, $\sigma_{\lambda\lambda'}(j)$ is diagonal, with elements 2σ, $-\sigma$, $-\sigma$. The coefficients $\eta_{\lambda\lambda'\alpha\alpha'}$ with λ, $\lambda' = 1, 2, 3$ and α, $\alpha' = x, y, z$ have the independent coefficients $\eta_{11xx} = \eta_{1x}$, $\eta_{11yy} = \eta_{1y}$, $\eta_{12xx} = \frac{1}{2}\eta_{4x}$, $\eta_{12zz} = \frac{1}{2}\eta_{4z}$ with the rest either zero or obtainable by symmetry from the above. In the model $\eta_{4\alpha} = 0$. The chemical potential contained in equation (A8.2.18) is determined by equation (A8.2.11). The temperature dependence of the tetragonal distortion arises entirely through the temperature dependence of the Fermi–Dirac function in equation (A8.2.18). Thus equation (A8.2.15) shows that the sublattice displacement d has exactly the same temperature dependence as the strain ϵ, since the other factors are independent of temperature. This behaviour is indeed found by comparing (Shirane and Axe 1971) the x-ray measurement of ϵ with the neutron measurement of d. In problem 8.7 you are asked to get the distortion d and the strain ϵ as functions of temperature in the Noolandi–Sham model.

Appendix 8.3

ORNSTEIN–ZERNIKE THEORY OF LIQUID–GAS CRITICAL POINT AND BRIEF REFERENCE TO RENORMALISATION METHODS

The original argument of Ornstein and Zernike (1918) for the form of the pair function $g(r)$ near the liquid–gas critical point started out from the convolution definition (equation (7.86)) of the direct correlation function $c(r)$. They made two assumptions:

(a) That $c(r)$ is short-ranged compared with the total correlation function $h(r)$, which is simply $g(r)-1$. This is only true near to the critical point.
(b) That $h(\boldsymbol{r}')$ in the convolution integral in equation (7.86) can be expanded in a Taylor series about the point $\boldsymbol{r}$.

From (b), the first term $h(r)$ in this expansion gives a contribution $h(r)\rho\int c(\boldsymbol{r})\,\mathrm{d}\boldsymbol{r}$ to the convolution, the term grad h integrates to zero, while the term proportional to $|\boldsymbol{r}'-\boldsymbol{r}|^2$ evidently contributes

$$\text{constant} \times \nabla^2 h \int cr^2\,\mathrm{d}\boldsymbol{r}.$$

There is no reason why $\int cr^2\,\mathrm{d}\boldsymbol{r}$ should vanish and therefore we obtain the following differential equation for $h(r)$:

$$\begin{aligned}\nabla^2 h &= \text{constant} \times \left(1-\rho\int c(\boldsymbol{r})\,\mathrm{d}\boldsymbol{r}\right)h \\ &= \text{constant} \times [1-\tilde{c}(0)]h = \kappa^2 h. \qquad \text{(A8.3.1)}\end{aligned}$$

Since $S(0) \to \infty$ at T_c (from equation (7.14)) and $S(0) = [1-\tilde{c}(0)]^{-1}$, from the Fourier transform of equation (7.86), it follows that κ in equation (A8.3.1) tends to zero as $T \to T_c$. The solution of equation (A8.3.1) which decays to zero at infinity is evidently

$$h = \text{constant} \times \mathrm{e}^{-\kappa r}/r. \qquad \text{(A8.3.2)}$$

This equation leads to an important characterisation of critical behaviour through a correlation length $\xi = \kappa^{-1}$, ξ becoming infinite as $T \to T_c$.

In terms of a conventional critical exponent ν, we write (cf equation (8.39))

$$\xi \propto |T-T_c|^{-\nu}. \qquad \text{(A8.3.3)}$$

Equation (A8.3.2) then leads, at $T = T_c$, to

$$h(r) \sim \text{constant}/r \qquad S(k) \sim \text{constant}/k^2. \qquad \text{(A8.3.4)}$$

While these results are useful first approximations, we shall give arguments below that there is a further exponent η associated with the total correlation function h. This can be defined by generalising equation (A8.3.4) to read

$$h(r) \sim \text{constant}/r^{1+\eta} \qquad S(k) \sim \text{constant}/k^{2-\eta}. \tag{A8.3.5}$$

To see this we can only make some brief remarks about the renormalisation method (Wilson 1972, Wilson and Kogut 1974).

The theory of renormalisation, (with $\kappa = 0$) starts from a result obtained by Gell-Mann and Low (1951). In the present context, this result can be expressed by writing

$$S(k) = k^{-2} s(k/\Lambda) \tag{A8.3.6}$$

and by noting that there is an arbitrariness of scale. This arbitrariness can be transferred to a choice of reference momentum λ, with $\lambda \ll \Lambda$, Λ being some cut-off momentum. The renormalisation approach essentially relates $s(k/\Lambda)$ to $s(\lambda/\Lambda)$. Since s is dimensionless, its k dependence can only involve the remaining length in the problem, the cut-off Λ^{-1}.

The burden of the argument is of course then to determine the form of s. It can be shown (cf March and Tosi 1976, p 242) that

$$S(k) \propto k^{-2}(k/\Lambda)^{\eta} \tag{A8.3.7}$$

with η greater than zero.

A generalisation of this argument can be effected, for $T \neq T_c$, i.e. $\kappa \neq 0$. The interested reader is referred to Pfeuty and Toulouse (1977) for a full account of renormalisation methods in phase transitions. An introductory discussion can be found in the book by Ziman (1979).

NOTES ADDED IN PROOF

Below we record some papers and review articles, related to topics treated in this book, which have appeared since the typescript was completed.

Chapter 2

(i) The review article by K S Singwi and M P Tosi (1981 *Solid St. Phys.* **36** ed H Ehrenreich, F Seitz and D Turnbull (New York: Academic)) deals with correlations in electron and electron–hole liquids.
(ii) A treatment of the x-ray edge problem by D R Penn, S M Girvin and G D Mahan (1981 *Phys. Rev.* **24** 6971) is recommended to supplement the introduction to this problem given in section 2.13.
(iii) An extensive review on excitation of plasmons has now appeared (Raether H 1980 *Springer Tracts in Modern Physics* vol 88 (Berlin: Springer-Verlag)).

Chapter 3

Coherent inelastic neutron scattering in lattice dynamics is fully treated by B Dorner (*Springer Tracts in Modern Physics* vol 93 (Berlin: Springer-Verlag)). Analysis of soft modes is also discussed; this is relevant to structural phase transitions dealt with in section 8.9.

Chapter 6

A further paper by P B Weigmann (1981 *J. Phys. C: Solid St. Phys.* **14** 1463) has appeared on the theory of the Kondo effect (see section 6.12)

Chapter 7

P Schofield has reviewed dynamics and transport in classical liquids in *Theoretical Chemistry* vol 2. (Specialist Periodical Reports, Royal Society of Chemistry (London))

Chapter 8

Progress in the theory of freezing has occurred, both in classical and quantal liquids.

For classical systems, this progress has come via the use of the Ornstein–Zernike direct correlation function (Ramakrishnan T V and Yussouff M S 1977 *Solid St. Commun.* **21** 389; *Phys. Rev.* B **19** 2775: March N H and Tosi M P 1981 *Phys. Chem. Liq.* **11** 79, 89 and 129: Haymet A D J and Oxtoby D W 1981 *J. Chem. Phys.* **74** 2559).

Freezing in the quantal electron liquid has been treated by A Ferraz, N H March and M Suzuki (1979 *Phys. Chem. Liq.* **9** 59) who calculate an approximate melting curve for the Wigner electron crystal. An important quantity in their expression is the mean interelectronic distance r_s at which the ground-state transition from electron liquid to electron crystal occurs. This quantity has been determined in Monte Carlo computations by D Ceperley and B J Adler (1980 *Phys. Rev. Lett.* **45** 568) to lie in the range of r_s from 70–100 a_0.

REFERENCES

Abrikosov A A 1956 *Sov. Phys.–JETP* **9** 1364

—— 1957 *J. Phys. Chem. Solids* **2** 199

—— 1978 *Zh. Eksp. Teor. Fiz. Pis. Mur.* **27** 235

Adler D 1975 *Treatise on Solid State Chemistry* vol 2 ed N B Hannay (New York: Plenum) p 237

Adler J G, Jackson J E and Chandrasekhar B S 1966 *Phys. Rev. Lett.* **16** 53

Allcock G R 1956 *Adv. Phys.* **5** 412

Altarelli M and Bassani F 1971 *J. Phys. C: Solid St. Phys.* **4** L328

Alvesalo T A, Haavasoja T, Manninen M T and Soinne A T 1980 *Phys. Rev. Lett.* **44** 1076

Ambegaokar V and Maradudin A A 1964 *Report 64–929–100*, 2 (Westinghouse Research Laboratory)

Anderson P W 1950 *Phys. Rev.* **79** 705

—— 1951 *Phys. Rev.* **83** 1260

—— 1952 *Phys. Rev.* **86** 694

—— 1960 in *Fizika Dielectrikov* ed G I Skanavi (Moscow: Academy of Sciences)

—— 1963 *Concepts in Solids* (New York: Benjamin)

Andreoni W, Altarelli M and Bassani F 1975 *Phys. Rev.* B **11** 2352

Appel J 1968 *Solid State Physics* vol. 21, eds F Seitz, D Turnbull and H Ehrenreich (New York: Academic) p 193

Argyres P N 1967 *Phys. Rev.* **154** 410

Ashcroft N W and Mermin N D 1976 *Solid State Physics* (New York: Holt, Rinehart and Winston)

Austin I G and Mott N F 1969 *Adv. Phys.* **18** 41

Bardeen J, Cooper L N and Schrieffer J R 1957 *Phys. Rev.* **108** 1175

Bardeen J and Pines D 1955 *Phys. Rev.* **99** 1140

Barton G 1979 *Rep. Prog. Phys.* **42** 963

Baym G 1964 *Phys. Rev.* **135** 1691

Berlinsky A J 1979 *Rep. Prog. Phys.* **42** 1243

Bhatia A B, Leung W B and March N H 1976 *Phys. Lett.* **58A** 205

Bhatia A B and Thornton D E 1971 *Phys. Rev.* B **4** 2325

Bishop A R 1978 *Proc. Int. Conf. on Lattice Dynamics* (Paris: Flammarion Sciences) p 144

Bjelis A, Saub K and Barisic S 1974 *Nuovo Cim.* **23B** 102

Bloch F 1933 *Z. Phys.* **81** 363

—— 1934 *Helv. Phys. Acta* **1** 385

Bohm D and Staver T 1952 *Phys. Rev.* **84** 836

Born M and Huang K 1954 *Dynamical Theory of Crystal Lattices* (Oxford: Clarendon)

Bratby P, Gaskell T and March N H 1970 *Phys. Chem. Liq.* **2** 53

Brinkman W F and Rice T M 1970 *Phys. Rev.* B **2** 4302

Brinkman W F, Rice T M, Anderson P W and Chui S T 1972 *Phys. Rev. Lett.* **28** 961

Brockhouse B N, Arase T, Caglioti G, Rao K R and Woods A D B 1962 *Phys. Rev.* **128** 1099
Brown R C and March N H 1972 *Phys. Earth Planet. Inter. (Netherlands)* **6** 206
—— 1976 *Phys. Rep.* **24C** 77
Bruce A D and Cowley R A 1973 *J. Phys. C: Solid St. Phys.* **6** 2422
Brüesch P 1975 *Proc. German Phys. Soc. Conf. on One-Dimensional Conductors, Saarbrucken* ed H G Schuster
Brüesch P, Strässler S and Zeller H R 1975 *Phys. Rev.* B **12** 219
Brush S G, Sahlin H L and Teller E 1966 *J. Chem. Phys.* **45** 2102
Bullough R K 1974 *Interaction of Radiation with Condensed Matter* vol. 1 (Vienna: IAEA) p 381
Byers N and Yang C N 1961 *Phys. Rev. Lett.* **7** 46
Callaway J 1963 *Phys. Rev.* **132** 2003
—— 1974 *Quantum Theory of the Solid State* Part A (New York: Academic)
Callaway J and Boyd R 1964 *Phys. Rev.* **134** A1655
Carneiro K, Shirane G, Werner S A and Kaiser S 1976 *Phys. Rev.* B **13** 4258
Chambers R G 1961 *Proc. Phys. Soc.* **78** 941
Chang I F and Mitra S S 1971 *Adv. Phys.* **20** 359
Citrin P H 1973 *Phys. Rev.* B **8** 5545
Citrin P H, Wertheim G K and Baer Y 1975 *Phys. Rev. Lett.* **35** 885
Citrin P H, Wertheim G K and Schlüter M 1979 *Phys. Rev.* B **20** 3067
Claesson A, Jones W, Chell G G and March N H 1973 *Int. J. Quant. Chem.* **S7** 629
des Cloiseaux J and Pearson J 1962 *Phys. Rev.* **128** 2131
Cochran W 1960 *Adv. Phys.* **9** 387
—— 1971 *Crit. Rev. Solid St. Sci.* **2** 1
Coles B R 1958 *Adv. Phys.* **7** 40
Combescot M and Nozières P 1972 *J. Phys. C: Solid St. Phys.* **5** 2369
Copley J R D and Dolling G 1978 *J. Phys. C: Solid St. Phys.* **11** 1259
Copley J R D and Rowe J M 1974 *Phys. Rev.* A **9** 1656
Corless G K and March N H 1961 *Phil. Mag.* **6** 1285
Cotterill R M J, Jensen E J, Damgaard Kristensen W, Paetsch R and Esbjørn P O 1975 *J. Physique Colloq.* **36** No. 2 35
Coulson C A 1954 *Valence* (Oxford: Oxford University Press)
Cowley R A 1963 *Adv. Phys.* **12** 421
—— 1978 *Proc. Int. Conf. on Lattice Dynamics* (Paris: Flammarion Sciences) p 625
Craig D P 1974 *Orbital Theories of Molecules and Solids* ed N H March (Oxford: Clarendon) p 344
Cunningham R M, Muhlestein L D, Shaw W M and Tompson C W 1970 *Phys. Rev.* B **2** 4864
Cusack S, March N H, Parrinello M and Tosi M P 1976 *J. Phys. F: Metal Phys.* **6** 749
Daniel E and Vosko S H 1960 *Phys. Rev.* **120** 2041
Davydov A S 1962 *Theory of Molecular Excitons* (New York: McGraw-Hill)
—— 1971 *Theory of Molecular Excitons* (New York: Plenum)
Dean P 1972 *Rev. Mod. Phys.* **44** 127
Deiseroth H J and Schulz H 1974 *Phys. Rev. Lett.* **33** 963
Devonshire A F 1954 *Adv. Phys.* **3** 85
Devreese J T 1972 *Polarons in Ionic Crystals and Polar Semiconductors* (Amsterdam: North-Holland)

Devreese J T, Everard R P and van Doren V E (eds) 1979 *Highly Conducting One-Dimensional Solids* (New York: Plenum)
Dick B G and Overhauser A W 1958 *Phys. Rev.* **112** 90
Dickey J M and Paskin A 1969 *Phys. Rev.* **188** 1407
Dingle R B 1955 *Phil. Mag.* **46** 831
Dirac P A M 1929 *Proc. R. Soc.* A **123** 714
—— 1957 *Quantum Mechanics* (Oxford: Clarendon)
Domb C and Green M S 1972 and 1974 *Critical Phenomena* (London: Academic)
de Dominicis C 1980 *Phys. Rep.* **67** 47
Doniach S and Šunjić M 1970 *J. Phys. C: Solid St. Phys.* **3** 285
Douglass D H 1976 referred to in Sham (1978)
Douglass D H and Falicov L M 1964 *Progress in Low Temperature Physics* vol. 4 (Amsterdam: North-Holland)
Dow J 1973 *Phys. Rev. Lett.* **31** 1132
Dyson F J 1953 *Phys. Rev.* **92** 1331
Economou E N 1969 *Phys. Rev.* **182** 539
Edwards S F and Anderson P W 1975 *J. Phys. F: Metal Phys.* **5** 965
Edwards S F and Jones R C 1971 *J. Phys. C: Solid St. Phys.* **4** 2109
Edwards S F and Warner M 1979 *Phil. Mag.* **40** 257
Egelstaff P A, March N H and McGill N C 1974 *Can. J. Phys.* **52** 1651
Egri I 1979 *J. Phys. C: Solid St. Phys.* **12** 1843
Einstein T L 1978 *Surface Sci.* **75** 161L
Eisenschitz R and Wilford M J 1962 *Proc. Phys. Soc.* **80** 1078
Elliott R J 1957 *Phys. Rev.* **108** 1384
Elliott R J and Gibson A F 1974 *Introduction to Solid State Physics* (London: Macmillan)
Elliott R J, Krumhansl J A and Heath P L 1974 *Rev. Mod. Phys.* **46** 465
Elliott R J and Wedgwood F A 1963 *Proc. Phys. Soc.* **81** 846
—— 1964 *Proc. Phys. Soc.* **84** 63
Englert F 1959 *J. Phys. Chem. Solids* **11** 78
Fermi E 1928 *Z. Phys.* **48** 73
Ferraz A and March N H 1980 *Solid St. Commun.* **36** 977
Feynman R P 1953 *Phys. Rev.* **91** 1291
—— 1954 *Phys. Rev.* **94** 262
—— 1955 *Progress in Low Temperature Physics* vol. 1, ed C J Gorter (Amsterdam: North-Holland)
Feynman R P and Cohen M 1956 *Phys. Rev.* **102** 1189
Fisher M E 1967 *Physics* **3** 255
Flores F, March N H and Moore I D 1977 *Surface Sci.* **69** 133
Flores F, March N H, Ohmura Y and Stoneham A M 1979 *J. Phys. Chem. Solids* **40** 531
Foreman A J E and Lomer W M 1957 *Proc. Phys. Soc.* **B70** 1143
Friedman L R and Tunstall D P (eds) 1978 *The Metal–Insulator Transition in Disordered Systems, Scottish Universities Summer School in Physics, St Andrews*
Fröhlich H 1950 *Phys. Rev.* **79** 845
—— 1952 *Proc. R. Soc.* A **215** 291
—— 1954a *Adv. Phys.* **3** 325
—— 1954b *Proc. R. Soc.* A **223** 296

—— 1970 *Phys. Kondens. Mater.* **9** 350
Fuchs K 1935 *Proc. R. Soc.* A **151** 585
Geballe T and White R 1980 *Solid State Physics, Supplement* 15 eds F Seitz, D Turnbull and H Ehrenreich (New York: Academic)
Gell-Mann M and Low F E 1954 *Phys. Rev.* **95** 1300
—— 1951 *Phys. Rev.* **84** 350
de Gennes P G 1966 *Superconductivity of Metals and Alloys* (New York: Benjamin)
Giaquinta P V, Tosatti E and Tosi M P 1976 *Solid St. Commun.* **19** 123
Gibbons P C, Schnatterly S E, Ritsko J J and Fields J R 1976 *Phys. Rev.* B **13** 2451
Gillan M J 1974 *J. Phys. C: Solid St. Phys.* **7** 11
Ginzburg V L and Landau L D 1950 *Zh. Eksp. Teor. Fiz.* **20** 1064
Girlanda R, Parrinello M and Tosatti E 1976 *Phys. Rev. Lett.* **36** 1386
Goodman B B 1962 *IBM J. Res. Dev.* No. 63
Gorkov L P 1959 *Sov. Phys.–JETP* **9** 1364
—— 1973 *JETP Lett.* **17** 379
Greenfield A J, Wellendorf J and Wiser N 1971 *Phys. Rev.* A **4** 1607
Gross E F 1962 *Sov. Phys.–Usp.* **5** 125
Grout P J and March N H 1976 *Phys. Rev.* B **14** 4027
Gutzwiller M C 1965 *Phys. Rev.* **137** A1726
Haken H and Nikitine S 1975 *Excitons at High Density* (Berlin: Springer-Verlag)
Haken H and Schottky W 1958 *Z. Phys. Chem.* **16** 218
Halperin B I 1973 *Collective Properties of Physical Systems* eds B Lundqvist and S Lundqvist (Stockholm: Nobel Foundation)
Halperin B I and Hohenberg P C 1967 *Phys. Rev. Lett.* **19** 700
Halperin B I, Hohenberg P C and Ma S T 1972 *Phys. Rev. Lett.* **29** 1548
Hansen J P 1976 *Phys. Rev.* A **14** 816
Hardy J R 1961 *Phil. Mag.* **6** 27
—— 1962 *Phil. Mag.* **7** 315
Hedin L and Lundqvist S 1969 *Solid State Physics* vol. 23, eds F Seitz, D Turnbull and H Ehrenreich (New York: Academic) p 1
Heller W R and Marcus A 1951 *Phys. Rev.* **84** 809
Hensel J C, Phillips T G and Thomas G A 1977 *Solid State Physics* vol. 32, eds F Seitz, D Turnbull and H Ehrenreich (New York: Academic) p 88
Henshaw D G and Woods A D B 1961 *Phys. Rev.* **121** 1266
Holstein T 1954 *Phys. Rev.* **96** 535
—— 1959 *Ann. Phys., NY* **8** 325, 343
Hopfield J J 1969 *Comm. Solid St. Phys.* **2** 40
Hulthén L 1938 *Ark. Mat. Astron. Fys.* **26A** No. 11
Inglesfield J E and Wikborg E 1973 *J. Phys. C: Solid St. Phys.* **6** L158
Ivanov V A, Makarenko I N, Nikolaenko A M and Stishov S M 1974 *Phys. Lett.* **47A** 75
Johnson F A 1965 *Prog. Semicond.* **9** 181
Johnson M W, McCoy B, March N H and Page D I 1977 *Phys. Chem. Liquids* **6** 243
Jones R C and Edwards S F 1971 *J. Phys. C: Solid St. Phys.* **4** L194
Jones W and March N H 1970 *Proc. R. Soc.* A **317** 359
—— 1973 *Theoretical Solid State Physics* (New York: Wiley)
Josephson B D 1962 *Phys. Lett.* **1** 251
—— 1964 *Rev. Mod. Phys.* **36** 216
—— 1969 *J. Phys. C: Solid St. Phys.* **2** 200

Kadanoff L P 1966 Physics **2** 263
—— 1970 *Comm. Solids* **2**
Karo A M 1959 *J. Chem. Phys.* **31** 1489
—— 1960 *J. Chem. Phys.* **33** 7
Kasteleyn P 1971 *Theory of Imperfect Crystals* (Vienna: IAEA)
Kawasaki K 1970 *Ann. Phys., NY* **61** 1
Kellermann E W 1940 *Phil. Trans. R. Soc.* A **238** 513
Khurana A and Hertz J A 1980 *J. Phys. C: Solid St. Phys.* **13** 2715
Kittel C 1956 *Introduction to Solid State Physics* (New York: Wiley)
—— 1963 *Quantum Theory of Solids* (New York: Wiley)
Klein M J and Smith R S 1951 *Phys. Rev.* **80** 1111
Kloos T 1973 *Z. Phys.* **265** 225
Knox R S 1963 *Theory of Excitons* (New York: Academic)
Koenig S H 1964 *Phys. Rev.* **135** A1693
Kondo J 1964 *Prog. Theor. Phys.* **41** 1199
—— 1969 *Solid State Physics* vol. 23, eds F Seitz, D Turnbull and H Ehrenreich (New York: Academic) p 183
Krishnan K S and Bhatia A B 1945 *Nature, Lond.* **156** 503
Kuhlmann-Wilsdorf D 1965 *Phys. Rev.* **140** A1599
Kunz C 1966 *Z. Phys.* **196** 311
Kuper C G 1955 *Proc. R. Soc.* A **227** 214
Labbé J and Friedel J 1966a *J. Phys. Radium, Paris* **27** 153
—— 1966b *J. Phys. Radium, Paris* **27** 303
Landau L D 1933 *Phys. Z. Sowj. Un.* **3** 664
—— 1941 *J. Phys. Moscow* **5** 71
—— 1947 *J. Phys. Moscow* **11** 91
Landau L D and Lifshitz E M 1959 *Fluid Mechanics* (Oxford: Pergamon)
—— 1962a *Quantum Mechanics* (Oxford: Pergamon)
—— 1962b *Electrodynamics of Continuous Media* (Oxford: Pergamon)
—— 1969 Statistical Mechanics (Oxford: Pergamon)
Lau K H and Kohn W 1978 *Surface Sci.* **75** 69
Lax M and Burstein E 1955 *Phys. Rev.* **97** 39
Lighthill M J 1958 *Fourier Analysis and Generalized Functions* (Cambridge: Cambridge University Press)
Lindhard J 1954 *K. Danske Mat.-fys. Meddr* **28** 8
London F 1950 *Superfluids* vol. II (New York: Wiley)
—— 1961 *Superfluids* vol. I, reprinted (New York: Dover)
Luttinger J M 1951 *Phys. Rev.* **81** 1015
Lynn J W, Iizumi M, Shirane G, Werner S A and Saillant R B 1975 *Phys. Rev.* B **12** 1154
Lynton E A 1959 *The Many Body Problem* (Paris: Dunod)
Mahan G D 1967 *Phys. Rev.* **163** 612
—— 1974 *Solid State Physics* vol. 29, eds F Seitz, D Turnbull and H Ehrenreich (New York: Academic) p 75
Mannari I 1961 *Prog. Theor. Phys.* **26** 51
Mansfield R 1956 *Proc. Phys. Soc.* **B69** 76
Maradudin A A, Montroll E W, Weiss G H and Ipatova I P 1971 *Theory of Lattice Dynamics in the Harmonic Approximation* (New York: Academic)
March N H 1958 *Phys. Rev.* **110** 604

—— 1968 *Liquid Metals* (Oxford: Pergamon) and *Theory of Condensed Matter* (Vienna: IAFA)
—— 1973 *Band Structure Spectroscopy of Metals and Alloys* eds D J Fabian and L M Watson (New York: Academic) p 297
—— 1975 *Self-consistent Fields in Atoms* (Oxford: Pergamon)
March N H and Murray A M 1960 *Phys. Rev.* **120** 830
—— 1961 *Proc. R. Soc.* A **261** 119
March N H, Suzuki M and Parrinello M 1979 *Phys. Rev.* B **19** 2027
March N H and Tosi M P 1972 *Proc. R. Soc.* A **330** 373
—— 1973 *Phil. Mag.* **28** 91
—— 1976 *Atomic Dynamics in Liquids* (London: Macmillan)
March N H, Young W H and Sampanthar S 1967 *Many-body Problem in Quantum Mechanics* (Cambridge: Cambridge University Press)
Marshall W and Lovesey S W 1971 *Theory of Thermal Neutron Scattering* (Oxford: Oxford University Press)
Mattheiss L F 1961 *Phys. Rev.* **123** 1219
—— 1975 *Phys. Rev.* B **12** 2161
McLean T P 1960 *Progress in Semiconductors* vol. 5, ed A F Gibson (London: Heywood)
Meissner W and Ochsenfeld R 1933 *Naturwiss.* **21** 787
Møller H B and Mackintosh A R 1965 *Phys. Rev. Lett.* **15** 623
Mott N F 1936 *Proc. Camb. Phil. Soc.* **32** 281
—— 1974 *Metal–Insulator Transitions* (London: Taylor and Francis)
Mukherjee K 1965 *Phil. Mag.* **12** 915
Mukhopadhyay G and Lundqvist S 1975 *Nuovo Cim.* B **27** 1
Murray G A 1966a *Proc. Phys. Soc.* **89** 87
—— 1966b *Proc. Phys. Soc.* **89** 111
Muscat J P and Newns D 1979 *Prog. Surf. Sci.* **9** 1
Nabarro F R N 1967 *Theory of Crystal Dislocations* (Oxford: Oxford University Press)
Nagamiya T, Yosida K and Kubo R 1955 *Adv. Phys.* **4** 2
Néel L 1932 *Ann. Phys., Paris* **17** 64
—— 1936 *Ann. Phys., Paris* **5** 256
Nelson D R and Halperin B I 1979 *Phys. Rev.* B **19** 2457
Noolandi J and Sham L J 1973 *Phys. Rev.* B **8** 2468
Nozières P and de Dominicis C T 1969 *Phys. Rev.* **178** 1097
Onnes H K 1913 *Leiden Comm.* **133a**
—— 1914 *Leiden Comm.* **139f**
Ornstein L S and Zernike F 1914 *Proc. Acad. Sci. Amsterdam* **17** 793
—— 1918 *Phys. Z.* **19** 134
Osheroff D D, van Roosbroeck W, Smith H and Brinkman W F 1977 *Phys. Rev. Lett.* **38** 134
Parrinello M and March N H 1976 *J. Phys. C: Solid St. Phys.* **9** L147
Patton B and Sham L J 1976 *Phys. Rev. Lett.* **36** 733
Pauling L and Wilson E B 1935 *Introduction to Quantum Mechanics* (New York: McGraw-Hill)
Peierls R E 1955 *Quantum Theory of Solids* (Oxford: Clarendon)
Peshkov V P 1946 *Sov. Phys.–JETP* **16** 1000

Pfeuty P and Toulouse G 1977 *Introduction to the Renormalization Group and to Critical Phenomena* (New York: Wiley)
Phillips J C 1956 *Phys. Rev.* **104** 1263
Phillips J C 1976 *Solid State Commun.* **18** 831 and references therein
Pippard A B 1953 *Proc. R. Soc.* A **216** 547
—— 1960 *Rep. Prog. Phys.* **23** 176
Pitaevskij L P 1959 *Sov. Phys.–JETP* **9** 830
—— 1966 *Sov. Phys.–Usp.* **9** 197
Raether H 1967 *Springer Tracts in Modern Physics* vol 38 ed G Höhler (Berlin: Springer-Verlag) p 153
—— 1977 *Nuovo Cim.* B **39** 817
Read W T 1953 *Dislocations in Crystals* (New York: McGraw-Hill)
Rehwald W, Rayl M, Cohen R W and Cody G D 1972 *Phys. Rev.* B **6** 363
Reissland J 1973 *The Physics of Phonons* (New York: Wiley)
Renker B and Comes R 1975 *Low Dimensional Cooperative Phenomena* ed H J Keller (New York: Plenum) p 235
Renker B, Rietschel H, Pintschovius L, Gläser W, Brüesch P, Kuse D and Rice M J 1973 *Phys. Rev. Lett.* **30** 1144
Rice T M 1977 *Solid State Physics* vol. 32, eds F Seitz, D Turnbull and H Ehrenreich (New York: Academic) p 1
—— 1979 *Contemp. Phys.* **20** 241
Rickayzen G 1965 *Theory of Superconductivity* (New York: Wiley)
Ritchie R H 1963 *Prog. Theor. Phys.* **29** 607
Ritchie R H and Marusak A L 1966 *Surface Sci.* **4** 234
Ritchie R H and Wilems R E 1969 *Phys. Rev.* **178** 372, Erratum **184** 254
Rowell J M, McMillan W L and Anderson P W 1965 *Phys. Rev. Lett.* **14** 633
Rutledge J E, McMillan W L, Mochel J M and Washburn T E 1978 *Phys. Rev.* B **18** 2155
Saint-James D and de Gennes P G 1963 *Phys. Lett.* **7** 306
Scalapino D J, Sears M and Ferrell R A 1972 *Phys. Rev.* B **6** 3409
—— 1975 *Phys. Rev. Lett.* **34** 200
Schiff L I 1968 *Quantum Mechanics* (New York: McGraw-Hill)
Schmidt J and Lucas A A 1972 *Solid St. Commun.* **11** 415, 419
Schofield P 1969 *Phys. Rev. Lett.* **22** 606
Schrieffer J R 1964 *Theory of Superconductivity* (New York: Benjamin)
Schröder U 1966 *Solid St. Commun.* **4** 347
Sham L J 1971 *Phys. Rev. Lett.* **27** 1725
—— 1972a *Phys. Rev.* B **6** 3581
—— 1972b *Phys. Rev.* B **6** 3584
—— 1974 *Dynamical Properties of Solids* eds G K Horton and A A Maradudin (Amsterdam: North-Holland) chap. 5
—— 1978 *Proc. Int. Conf. on Lattice Dynamics* (Paris: Flammarion Sciences) p 557
Sham L J and Patton B R 1976 *Phys. Rev. Lett.* **36** 733
Shapiro L see reference in Jones and March (1973)
Sherrington D and Kirkpatrick S 1975 *Phys. Rev. Lett.* **35** 1792
Shirane G and Axe J D 1971 *Phys. Rev.* B **4** 2957
Sinclair R N and Brockhouse B N 1960 *Phys. Rev.* **120** 1638

Sinha S K 1969 *Phys. Rev.* **177** 1256
Stanley H E 1971 *Introduction to Phase Transitions and Critical Phenomena* (Oxford: Clarendon)
Stiebling J and Raether H 1978 *Phys. Rev. Lett.* **40** 1293
Stoneham A M 1975 *Theory of Defects in Solids* (Oxford: Clarendon)
Stoner E C 1938 *Proc. R. Soc.* A **165** 372
—— *Proc. R. Soc.* A **169** 339
Sturm K 1968 *Z. Phys.* **209** 329
Su W P, Schrieffer J R and Heeger A J 1980 *Phys. Rev.* B **22** 2099
Svensson E C and Kamitakahara W A 1971 *Can. J. Phys.* **49** 2291
Swift J and Kadanoff L P 1968 *Ann. Phys., NY* **50** 312
Szigeti B 1949 *Trans. Faraday Soc.* **45** 155
—— 1950 *Proc. R. Soc.* A **204** 52
Takeuti Y 1957 *Prog. Theor. Phys.* **18** 421
Tellenbach U and Arend H 1977 *J. Phys. C: Solid St. Phys* **10** 1311
Testardi L R 1973 *Phys. Rev. Lett.* **31** 37
—— 1975 *Rev. Mod. Phys.* **47** 638
Thomas L H 1926 *Proc. Camb. Phil. Soc.* **23** 542
Thouless D J 1978 *J. Phys. C: Solid St. Phys.* **11** L189
Thouless D J, Anderson P W and Palmer R G 1977 *Phil. Mag.* **35** 593
Tilley D R and Tilley J 1974 *Superfluidity and Superconductivity* (London: Van Nostrand Reinhold)
Toulouse G 1977 *J. Physique Lett.* **38** L67
Urner-Wille M and Raether H 1976 *Phys. Lett.* **58A** 265
Van Hove L 1953 *Phys. Rev.* **89** 1189
—— 1954 *Phys. Rev.* **95** 249
—— 1955 *Physica* **21** 517
Vashishta P, Das S G and Singwi K S 1974 *Phys. Rev. Lett.* **33** 911
Vieland L J, Cohen R W and Rehwald W 1971 *Phys. Rev. Lett.* **26** 373
Vinen W F 1961 *Proc. R. Soc.* A **260** 218
Walker C B 1956 *Phys. Rev.* **103** 558
Weger M 1964 *Rev. Mod. Phys.* **36** 175
Weger M and Goldberg 1973 *Phys. Rev.* **74** 1492
Weiss P R 1948 *Phys. Rev.* **74** 1493
Wheatley J C 1975 *Rev. Mod. Phys.* **47** 415
Widom B 1965 *J. Chem. Phys.* **43** 3892
Wiegmann P B 1980 *Phys. Lett.* **80A** 163
Wigner E P 1934 *Phys. Rev.* **46** 1002
—— 1938 *Trans. Faraday Soc.* **34** 678
Williams J M, Petersen J L, Gerdes H M and Peterson S W 1974 *Phys. Rev. Lett.* **33** 1079
Wilson K G 1971 *Phys. Rev.* B **4** 3174, 3184
—— 1972 *Phys. Rev. Lett.* **28** 548
—— 1974 *Collective Properties of Physical Systems* eds B Lundqvist and S Lundqvist (New York: Academic) p. 68
Wilson K G and Kogut J 1974 *Phys. Rep.* **12C** 75
Wohlleben D K and Coles B R 1973 *Magnetism* vol. V, eds G T Rado and H Suhl (New York: Academic) p. 3

Woods A D B, Brockhouse B N, Cowley R A and Cochran W 1963 *Phys. Rev.* **131** 1025

Woods A D B, Brockhouse B N, March R H, Stewart A T and Bowers R 1962 *Phys. Rev.* **128** 1112

Yue J T and Doniach S 1973 *Bull. Am. Phys. Soc.* **18** 465

Zacharias P 1974 *J. Phys. C: Solid St. Phys.* **7** L26

—— 1975 *J. Phys. F: Metal Phys.* **5** 645

Zawadowski A 1978 *Quantum Liquids* eds J Ruvalds and T Regge (Amsterdam: North-Holland) p. 293

Ziman J M 1960 *Electrons and Phonons* (Oxford: Oxford University Press)

—— 1961 *Phil. Mag.* **6** 1013

—— 1976 *Principles of the Theory of Solids* (Cambridge: Cambridge University Press)

—— 1979 *Models of Disorder* (Cambridge: Cambridge University Press)

INDEX

GRADUATE STUDENT SERIES IN PHYSICS

General Editor: DOUGLAS F. BREWER

Professor of Experimental Physics, University of Sussex

This series of books in physics and related subjects is designed to meet the needs of graduate students who are taking postgraduate courses for an MSc or as part of the requirements for a PhD research degree.

Although the books are not primarily research texts, they point out the directions which research is currently taking and where it is expected to lead. They therefore serve as useful introductory surveys for PhD research students, who may otherwise be faced with the formidable task of acquainting themselves with a research topic by reading sophisticated research texts when they have only just finished their first degree.

To take into account the differing background knowledge of students on MSc courses, the books start by introducing the subject at late undergraduate level. They may thus also be useful in some final-year undergraduate courses where, increasingly, optional topics of more specialised natures are taught in the last term.

Collective Effects in Solids and Liquids

This volume offers a comprehensive account of the wide variety of collective phenomena which occur in solids, and to a lesser but still important extent, in liquids. It describes all the major collective effects predicted by condensed matter theory, confronting the theory at each stage with experimental facts. The physical properties and conceptual importance of the quasi-particles associated with many of the phenomena are discussed.

The collective oscillations of itinerant electrons in metals (plasmons) are described and related to observations on energy loss spectroscopy. Some attention is given to plasmons in inhomgenous electron gases, such as exist at metal surfaces, and to bound electron–hole pairs (exitons), and their importance in explaining observed optical properties of solids is treated in full. Quanta associated with misaligned spins propagating through crystal lattices (spin waves) and electrons dressed with their ionic polarisation in ionic materials are also covered. A full explanation of the relation of this latter phenomenon to superconductivity (in which the interaction of mobile electrons with the ionic lattice is crucial) is given. The theory of phase transitions is introduced in the final chapter and its relation to experiment is discussed without the use of detailed mathematics.

Each chapter includes a set of problems and much of the necessary mathematics is relegated to appendices in order to emphasise the physics and chemistry in the main text.

ISBN 0-85274-528-1

ADAM HILGER LTD, BRISTOL